Physicists
Epoch and Personalities

HISTORY OF MODERN PHYSICAL SCIENCES – VOL. 4

Physicists
Epoch and Personalities

E L Feinberg
Formerly of P N Lebedev Physical Institute, Russia

Editor
I M Dremin
P N Lebedev Physical Institute, Russia

Translator
A V Leonidov
P N Lebedev Physical Institute, Russia

NEW JERSEY · LONDON · SINGAPORE · BEIJING · SHANGHAI · HONG KONG · TAIPEI · CHENNAI

Published by
World Scientific Publishing Co. Pte. Ltd.
5 Toh Tuck Link, Singapore 596224
USA office: 27 Warren Street, Suite 401-402, Hackensack, NJ 07601
UK office: 57 Shelton Street, Covent Garden, London WC2H 9HE

British Library Cataloguing-in-Publication Data
A catalogue record for this book is available from the British Library.

PHYSICISTS: EPOCH AND PERSONALITIES

Copyright © 2011 by World Scientific Publishing Co. Pte. Ltd.

All rights reserved. This book, or parts thereof, may not be reproduced in any form or by any means, electronic or mechanical, including photocopying, recording or any information storage and retrieval system now known or to be invented, without written permission from the Publisher.

For photocopying of material in this volume, please pay a copying fee through the Copyright Clearance Center, Inc., 222 Rosewood Drive, Danvers, MA 01923, USA. In this case permission to photocopy is not required from the publisher.

ISBN-13 978-981-283-416-4
ISBN-10 981-283-416-8

Printed in Singapore.

Preface

From the Author

In a phantasmagoric sequence of happy and tragic events, joyful and depressingly grieving feelings which one calls life I have by no means faced only misfortunes but was also astonishingly lucky. Leaving aside even the fact that a nonchalant and soulless 'red wheel' for some reason did not come over me although passed nearby and heavily influenced lives of people close to me, I was lucky in two more most important situations that determined my life. One of them is of a too personal nature to make it appropriate to describe it here. Another one — a lucky ticket that I got in a lottery is a reason for writing this book.

The thing is that at a young unwise age I somehow got attracted to physics which at those times was of little interest to anybody beyond the school walls. After overcoming some difficulties characteristic of those times (and, strangely, to a certain extent due to them) I finally became a student of the physics faculty of Moscow university (at hard times both for the country as a whole and for a student who did not have "working experience"). Luckily from there I went directly to FIAN, the P.N. Lebedev Physical Institute of the Academy of Sciences. For the whole life to come I found myself in an atmosphere of high science, true intelligence, honesty and moraling that was close to me. This atmosphere was determined by people from the generation of my teachers (as well as by some of my coevals). During decades that were tragic for our country and its great culture these people were able to withstand fear and temptation and preserve themselves as persons.

Of course at those times one met many such people: In other sciences, in arts, in literature, among bearers of our culture — 'ordinary' people who were not very noticeable and among the intelligentsia that withstood everything.

It is precisely due to such people that a broad flow of our culture which in the horrible atmosphere of the epoch did at times narrow and become more shallow did not disappear completely under the pressure of ignorant and cruel rulers and imposed 'ideology' befuddling millions. It is precisely thanks to this hunted intelligentsia despised both by authorities and half-educated masses and thanks to an ever living talent of people feeding this culture it survived until nowadays (albeit with some mercilessly torn out sections) and again promises to become an ennobling and refreshing stream.

A poet, Nikolai Glazkov, wrote[1]:

> I watch the world from beneath a table
> The twentieth century, the unusual
> The more it is of interest to historians
> The more grieving it is for contemporaries.

Thanks to the above-mentioned people I did not have to observe the terrible world 'from beneath a table.' Therefore I find it necessary to tell what I know about them, what I was a witness of.

I was close to some of these people, to some — very close. I observed others somewhat sideways for time long enough to understand their salutary influence and at the same time to keep an ability of an individual judgment (they taught me this themselves — of course not by edification and admonition but, without unnecessary words, at an example of their behavior). This does not mean that I did not notice their human weaknesses but all this was of secondary importance with respect to the main feature — their ability to preserve one's personality in a really too 'unusual century.'

Unfortunately totalitarianism was present not only in our country. Therefore it was important for me to compare what we lived through with what happened in Hitler's times. In such a way there appeared an essay on Heisenberg. Here a "memoir content" is naturally more scarce. Nevertheless I gathered this material persistently and for a long time. Therefore it is of more a research character and goes beyond a declared general topic. It describes reasons for Hitler's coming to power, a failure of the German "uranium project" and other questions. It turns out all of them are interrelated.

Almost all the essays and memoirs have already been published in some form. However, a new epoch gave a possibility of their significant enrichment with materials to which previously it had not even been possible to

---

[1] Translated from Russian by A. Leonidov.

allude. In addition new, sometimes astonishing facts and documents were uncovered. Therefore much has been written anew or significantly rewritten.

I have to apologize to the readers for the fact that in some cases the material of one of the essays appears in another one. I tried to diminish a number of such cases but with only partial success. At the same time one should take into account that in the book of this sort the essays are not always read one after another, sometimes even not all of them, and are often chosen according to one's taste.

I allowed myself a little liberty and included into the chapter on A.L. Mintz memoirs of my wife V.D. Konen describing the features of this person that were not touched by myself.

Unfortunately I have not been capable (so far?) to write on all equally remarkable people which one has to write about and on whom I can tell a lot, on L.I. Mandelstam, G.S. Landsberg, A.A. Andronov and others. Let it however at least be what it is now.

E.L. Feinberg, 1999

Foreword to the Second Edition

The second edition differs from the first one in two main points. Firstly, in adding two new essays on Leonid Isaakovich Mandelstam (which was completely absent in the first edition) and the new short one on S.I. Vavilov that adds to what was already written before. Secondly, the essay on Heisenberg has been significantly expanded. In recent years there arose a new boom in a vast literature on him (in which one meets radically different opinions on his behavior during Nazi times for which he was subjected to ostracism by Western physicists) due to an appearance of new previously unreachable sources. The author's point of view differs from these opinions expressed by people who did not have a misfortune to live under an inhuman dictatorship. In the second edition this point of view of the author has been further sharpened.

Besides that I looked through the whole text, corrected the found misprints and stylistic errors and amended the text by small comments reflecting those sent to me by the readers. Using this opportunity I express my gratitude to all of them. Unfortunately I was again not able to write about many one has to write about.

E.L. Feinberg, 2002

Editor's Preface

The history of the twentieth century must yet be written. One of the most important elements in it is the tremendous influence of science compared with previous epochs. In particular, its development in Russia played a crucial role in the transformation of the country from agricultural to a highly industrial one, from the defeat in the First World War to the victory in the Second World War, to the production of the atomic bomb immediately following the US, and then making the first transportable hydrogen bomb, the launching of the first Sputnik (artificial satellite) and launching the first man into space.

How could all this happen and which people did it?

This book is written by a physicist who lived during these times and personally knew many scientists. It answers the above questions by describing the characters and fates of many leading Soviet scientists with wide analysis of that epoch and country rulers.

Among them, most prominent in the Western world is Andrei Sakharov, the father of the Russian hydrogen bomb, exiled to Gorky by the Soviet leaders for his quest for human rights and afforded in 1975 the Nobel Peace Prize.

His tutor, the Nobel laureate in physics (1958) Igor Tamm played an important role in the organization of the hydrogen bomb project and, together with Sakharov, was co-author of a means of producing a controlled thermonuclear reaction. Nowadays it is one of the main efforts in attempts to solve the energy problem of humanity.

Another well known person, especially among physicists, is Lev Landau, the 1962 Nobel laureate in physics, the author of the most popular series of books on theoretical physics.

The portraits of these and some other leading Russian scientists are vividly presented in the book within the background of political and intellectual life in Russia, the role of the revolution of 1917 and subsequent regimes, the immense growth of science and its applications in the twentieth century.

Their close relations with world scientists are described in interesting essays on Niels Bohr and Werner Heisenberg, especially in connection with their attitudes toward the atomic bomb project.

The special intellectual atmosphere in Russia, intermixed with personal tragedies is clearly seen in all essays. Many terms and expressions particular to Russian life have been used. The translation of the book posed severe problems and took a long time because E.L. Feinberg's style of presentation is quite specific to him. He used rather lengthy sentences with additional branches of thoughts, quite uncommon in English. Nevertheless, we tried to preserve his personal style because for us it sounded as we heard it from him. We hope that many Western scientists who were well acquainted with him understand our wish and have a similar impression. It took great efforts and much time of Andrey Leonidov to translate the text in the form which kept its flavor and satisfied us.

This intellectual analysis of life and the role of science in the twentieth century is aimed not only at scientists but at the wide spectrum of people interested in history and personal fates strongly influenced by the ruling regimes of that period of time.

We are sure that every reader will find in this book facts which are unknown to him/her but which are extremely interesting.

We are grateful to Elena Nash and Igor Konstantinov for their help on carrying through this translation.

I.M. Dremin

E. L. Feinberg

Contents

Preface		v
1.	MANDELSTAM, Leonid Isaakovich (1879–1944)	3
	1.1 Progenitor	3
2.	TAMM, Igor Evgen'evich (1895–1971)	49
	2.1 Tamm in Life	49
	2.2 A Life of a Russian Intelligent	79
	2.3 I.E. Tamm and the Formation of Domestic Physics	100
3.	SAKHAROV, Andrei Dmitrievich (1921–1989)	127
	3.1 For a Future Historian[1]	127
	3.2 The Years of 1945–1950	129
	3.3 A Return to "Pure" Science and the Beginning of Social Activity	137
	3.4 A Return to FIAN	146
	3.5 A Hunt	151
	3.6 Arrest and Exile to Gorky	156
	3.7 Trips to Gorky	161
	3.8 Hunger Strikes	164
	3.9 A Release	178
	3.10 Addendum	184
	3.10.1 Letters and telegrams of Gorky period	184
4.	VAVILOV, Sergei Ivanovich (1891–1951)	193
	4.1 Nine Scars on the Heart[1]	193
	4.2 Sergei Ivanovich Vavilov and His Time	209
	4.3 Vavilov and Vavilov's FIAN	218
	4.4 What Gave Birth to the Vavilovs?	242

5. LEONTOVICH, Mikhail Aleksandrovich (1903–1981)   249

    5.1 Mosaic 249
        5.1.1 The first meeting 249
        5.1.2 The war 253
        5.1.3 The 'character' 255
        5.1.4 Second scientific life 258

6. MINTZ, Aleksandr L'vovich (1895–1974)   263

    6.1 Aleksandr L'vovich Tells 263
    6.2 Addendum. Memoirs of V.D. Konen, 'Personality
        of a Scientist' 277

7. BOHR, Niels Hendrick David (1885–1962)   285

    7.1 Bohr. Moscow. 1961 285
    7.2 Addendum: Physicists and Philosophers 300

8. HEISENBERG, Werner Karl (1901–1976)   305

    8.1 Tragedy of Heisenberg 305
    8.2 Bohr and Heisenberg 307
    8.3 Heisenberg and Science During Nazi Times 324
    8.4 Heisenberg and Western Physicists 329
    8.5 Tragedies of the Epoch 340
    8.6 Why Hitler Did Not Get an Atomic Bomb 344

9. What Brought Hitler to Power? And Who?   361

10. LANDAU, Lev Davidovich (1908–1968)   373

    10.1 Landau and Others: '... Verklaerungen
        und Neubegruendungen ...' 373
    10.2 Two Landaus 387
        10.2.1 Landau, Kapitza and Stalin 392
    10.3 Catastrophy 414

Bibliography   419

L.I. Mandelstam

Chapter 1

MANDELSTAM, Leonid Isaakovich (1879–1944)

1.1 Progenitor

A right-hand door under a dust-laden bust of Newton opens in a massive wall of a big physics auditorium in an old building of the Physics Faculty of Moscow University (still standing in Mokhovaya!). A relatively tall, slightly stooping, still dark-haired but already aging man in a dark suit enters the auditorium together with a group of people. Under an unbuttoned jacket which is hanging, due to his round-shouldered figure, one sees a waistcoat. In the waistcoat pocket the man has a watch that he will later check. On the nose he has frameless pince-nez which clip on to the nose bridge. He has soft cheeks and chin with rare but massive deep folds on the cheeks. A flat bag in the hands, he hastily stops behind the end of a ten meter long desk that separates the audience from the lecturer. Behind him are two big blackboards covered by black calico, which one can move, with a slight bow, by rotating by handle a glittering wheel below to the right, and a moving giant white screen between the two blackboards. The entourage (Igor Evgen'evich Tamm, Grigori Samuilovich Landsberg, Mikhail Aleksandrovich Leontovich, Boris Mikhailovich Gessen [philosopher and historian of science, a dean of the faculty at the beginning of the thirties until his, characteristic to these times, death], some more people) rush to their places in the first row. This is a row of chairs specially placed for today in front of the steep amphitheatre which, having altogether about four of five hundred seats, is full and buzzes, but sharply fells silent — a usual accompanying sound of the beginning of a lecture of a respected lecturer. Mandelstam starts at once.

Although he speaks with quite precise phrases, but he begins somewhat awkwardly. Something apologetic in his tone and even pose will also burst open later. However, he gradually warms up and reaches the state in which the only thing that is relevant for him in the world are the words spoken, the thought expressed. His voice is slightly nasal, not loud, and only the wonderful acoustics of the auditorium (subsequently reconstructed and now, unfortunately, non-existent), a clarity of the structure and contents of every phrase make this voice understandable, even for listeners in rear rows. Mandelstam does not make slips while speaking, does not need to correct himself, he pronounces only something he is sure upon and has been thought of. But, until the end of the lecture, he does not leave the saving spot between the end of the desk and the blackboard behind him. On the desk he places his lecture notes which he sometimes bends over or which he, having taken off the pince-nez and holding them with a hand somewhat aside, brings closer to near-sighted eyes. This combination of clarity and firmness in something important and softness of behavior is, as we shall see, characteristic of him. His entire appearance is a variant of that of a Russian-European intelligent[1] of a pre-revolutionary epoch. His entire behavior is that of such an intelligent, unbending in important matters, understanding and yielding in minor. An extraordinary mind power and a high spiritual, moral culture allows him to understand, better and clearer than others, what is truly important, and what is not. Niels Bohr behaves in the same auditorium in the same way a few years later. And, although the facial features of both are very different, and although in comparison to Mandelstam, Bohr is big-headed with bushy eyebrows and looks somewhat like a clodhopper, common generic features are evident.

This was one of the famous Mandelstam 'optional courses' of the thirties. They continued for many years — on theory of relativity, physical optics, theory of oscillations, quantum mechanics. The very word 'optional' always contains a shadow of being not really obligatory, not really useful. It was however sufficient to start thinking on what Mandelstam was talking about to understand its necessity for a physicist striving to 'get to the very essence'.

Mandelstam lectured in a somewhat 'chamber' manner. His formation as a person took place during an epoch in which science in general and physics in particular was the destiny of only a few. An audience of half a thousand

---

[1] The word 'intelligent', widely used in this book, means belonging to intelligentsia, a social layer for which education and moral values are equally important. This term is specifically Russian and is in some way related but not equivalent to 'intellectual'.

people which, together with a lecturer, contemplated on fine details of a basis, historical and logical, of relativity theory or quantum mechanics was something new. The issues discussed were, in addition, an arena of loud fighting among ourselves. Although a whole generation of young scientists had grown up, for whom the 'paradoxes' of these disciplines were as nonexistent as ones of a round form of the Earth or heliocentricity of the world for those that came after Copernicus, an influence of old concepts in scientific community was still strong. And in Russia an "authoritative" support which some (few in number, but loud-voiced) physicists, anti-relativists and anti-quantumists, got from some high-ranked (but too submissive to an official hard ideology) philosophers heated the atmosphere. Mandelstam did not argue, he quietly clarified. He was not afraid of evoking doubt and then having the pleasure of dispelling it by finding exact argumentation. The wide scale of thought, the boldness of following through an argumentation overrode the 'chamberness' and softness of speech, smile and gesture.

The words were pronounced seemingly evenly, but the voice did not sound lullingly murmuring. There were some significant pauses. Some words were stressed. Having formulated some objection to a theory he would say slightly triumphantly, 'But this is not so!', and on 'not' jump softly, apparently vigorously pressed against an invisible spring.

He had his favorite sayings mainly originating from years of studying and working in Strasbourg. "Hier springt der Frosch ins Wasser (here jumps the frog in water)," he said when coming to express an important and sophisticated idea for resolving a difficult paradox. He liked the expression "put a finger on this" that probably replaced the biblical "to put fingers into the wound" (on a doubting Thomas). All this was somewhat old-fashioned for the young people of the thirties but so organic for Mandelstam that did not seem strange. As a witness and participant of great events in science he often did not distinguish between something that was already not worth explaining to the young audience and something that was really difficult. So, giving a detailed explanation of a notion of group velocity of a wave packet he would clarify it by using a complex image: A moving steamboat with "boys and girls" jumping from the water on the stern, running along the ship, and jumping back into the water from the prow. However it was easy to feel sorry for a student who would become so relaxed in the course of this explanation that would miss the subsequent subtle analysis of the "Fleming error" (found by Mandelstam himself in his younger years) or would not be attentive to a discussion of restrictions imposed by causality requirement on a definition of simultaneity in relativity theory. Here something important would be missed. This is understood in the first row of the auditorium: A.A. Andronov, S.E. Khaikin, S.M. Rytov, G.S. Gorelik and everybody else, all professors themselves, are diligently writing down the lecture — to a future benefit.

And the auditorium was filled by students and Ph.D. students, not only from the university, but from other places, and also lecturers and professors from many institutes. Here they all lived in a high spiritual world, here reigned a passion and a joy of scientific understanding, a unity in comprehending the truth, a delight of belonging to a scientific brotherhood. Behind the walls there laid the dreadful world of the Stalin era: a world of lies and hypocrisy, a horror of "big terror", an inhumanity. Inside there was a pure honest world of deep thinking and benevolence. A shelter for a person. A temple.

Only after about ten years since that time did I learn how these lectures had been prepared. When working on a five-volume edition of Mandelstam's scientific work I was offered a honorable task of preparing for print a text of his lectures on theory of measurement in quantum mechanics. The original material was in the form of lecture notes, taken by different listeners, of all five lectures given in 1939, in the first place the especially thorough notes by S.M. Rytov (I found my own notes long after these lectures were published). One lecture (the fourth) was taken down in shorthand. Leonid Isaakovich had never seen neither the notes, nor the shorthand and had not checked them. However his extant working notebooks of the period of 1938–1939 were given to me. These were usual thick (relatively disorderly) school notebooks containing much relating to his work in these years: fragments of calculations without comments, some notes with formulae without clarification and, amongst all this, disjoined pieces of the first three lectures that I needed. Leonid Isaakovich wrote them with full phrases, as if preparing them for print. Each such piece existed in several not too distinct variants. One saw that he was essentially writing with ease, with complete, literary perfect sentences. Very little was crossed out or written between the lines. At the same time a multiple reiteration and variation of the whole excerpts, sometimes with mutually exchanging positions in the text, reflected some sort of indecision, a constant doubt in the readiness, in the finality of the written, a constant care on its improvement. A closeness of these texts to the notes made by listeners allowed to trust the notes of other lectures, for which nothing could be found in the extant notebooks, as well.

There exist two types of thinking and writing people. Some are working out, making things more precise, formulating their very thoughts in the process of writing or discussing or arguing. Others think everything over and make their statements, oral or written, in a quite complete fashion. Mandelstam was probably closer to being of the second type.

The character of Leonid Isaakovich also oozed in these notebooks in the following episode. Each academician had to present a yearly report on what he had been doing during the year (this was done already by Lomonosov). In their majority academicians did not consider this to be something really serious, but rather as something just stealing their time. However also here Mandelstam could not be negligent. Variant after variant of the same report can be found in these notebooks. One gets the impression that this essentially trifling task was tortious and he considered making this report, of one page long, perfect. Let us remember this persevering reiteration. It probably reflects something important in his character.

Let us however temporarily digress from these impressions. Who was Mandelstam in science, why was he infinitely respected, almost worshiped by his students and many colleagues who combined this respect with an admiration of his personality, with a feeling that in many cases could not be defined other than as love?

He was not just a scientist. He was a thinker for whom physics was a way to understand "the nature of things" (in the sense of '*De rerum naturae*' by Lucretius Carus). And, at the same time, physics was an unbounded field of various puzzles arising in the course of studying the concrete physical processes as well as the deep mysteries of nature in general, where Mandelstam was irreversibly taken over by the desire to find the answers and understand their deep meaning. In optics and electrodynamics, in radio- and molecular physics, in quantum mechanics and relativity theory, in general principles of scientific understanding of the world — everywhere.

Besides that, Mandelstam belonged to a rare type of physicist combining a theorist, an experimentalist, an engineer-inventor, a teacher and a philosopher. He was a 'lyric' as well. Mandelstam knew and understood art, adored Pushkin and knew his works at an almost professional literary studies level (a well-known Pushkinist A.B. Derman was his friend). Here are several examples.

As a theorist, based on his experience from studies in radiophysics and optics, he developed a far-reaching general theory of linear and nonlinear oscillations having innumerable concrete applications in various domains of physics and technics. He spoke of an existence of a universal "oscillation thinking". He developed a general theory of an optical image. Based on the newly born quantum mechanics he, together with his Ph.D. student

M.A. Leontovich, discovered a new paradoxical phenomenon, a tunnel effect (see below).

As an experimentalist, he discovered (together with G.S. Landsberg, his junior friend, in some way his student) an important phenomenon of combination scattering of light (the Nobel prize went to C. Raman who simultaneously discovered the effect but did not understand its meaning, see below) as well as other optical phenomena.

As an inventor he owned more than 60 patents, mainly in the field of radiotechnics (half of them together with Nikolai Dmitrievich Papaleksi, his childhood friend who specialized almost exclusively in radiotechnics). As a teaching scientist, not only was he a brilliant lecturer, but (and this could be the most important of what he has done for a development of physics in our country) he created a powerful school of physicists of the highest level, mainly theorists. For example: the, conditionally speaking, first generation (his students and younger scientists) — the outstanding physicists A.A. Andronov, I.Ye. Tamm, G.S. Landsberg, M.A. Leontovich, N.D. Papaleksi, S.M. Rytov; the brilliant S.E. Khaikin, a creator of a principally new system of radio-telescopes (a construction of the first such telescope was completed after his death); an optician V.A. Fabrikant and others. Two among them, talented professors A.A. Vitt and at first his (and then I.E. Tamm's) student S.A. Shubin, had already accomplished much in science in their young years before being killed in the epoch of Stalinist "big terror".

If one takes the next generation of "students of his students", there arises a spectacular geometric progression. In the school of my teacher I.E. Tamm there formed as scientists S.A. Altshuler, V.L. Ginsburg, L.V. Keldysh, D.A. Kirznitz, V.I. Ritus, A.D. Sakharov, E.S. Fradkin, S.P. Shubin and a number of other brilliant theorists.

Speaking on the school of Andronov, one of the founders of a theory of automatic regulation, one has to mention not only the names of his closest students and coworkers M.A. Aiserman, N.N. Bautin, A.G. Meyer and others, but also the fact that, together with Maria Tikhonovna Grekhova he was a "founding father" of an institute that subsequently grew into a powerful constellation of scientific institutes in Gorky (Nizhny Novgorod) having international importance. Here, as students, there listened to his lectures A.V. Gaponov-Grekhov, V.S. Troitsky and many other prominent physicists.

Near M.A. Leontovich there grew E.P. Velikhov, B.B. Kadomtsev, M.L. Levin, R.Z. Sagdeev, V.D. Shafranov and an army of not necessarily highly titled, but very significant scientists.

Near G.S. Landsberg there grew S.L. Mandelstam, I.L. Fabelinski, I.I. Sobelman, M.M. Sushinski, V.I. Malyshev and many, many others comprising his giant school of optical spectroscopy — both purely scientific and, infinite in its applications, extremely valuable applied ones.

With this (and other) schools one can recognize, with varying clarity, an easily distinguishable Mandelstam seal — his scientific style, the norms of human relations. A charm of his personality was combining firmness and clarity of principles and opinions with an unusual delicacy. Softness and tactfulness revealed itself, in particular, in *how* he worked together with a young student or colleague. Often, when somebody had brought him a raw or even wrong result, L.I. began, together with the colleague, to improve the work, but did it in such a fashion that the colleague left convinced that he had done everything himself.

As a matter of fact, our theoretical physics of the middle of the twentieth century was created by two brilliant schools, those of L.D. Landau and L.I. Mandelstam, and by an outstanding world-class theorist V.A. Fock (here I do not mention the third, younger, very strong school of N.N. Bogoliubov, at the beginning of his career more a mathematician, which, with an exception of the founder, "came into play" in physics only after the war, in the second half of the century).

Finally, as a philosopher, not only was Mandelstam convinced that a modern physicist can not escape relating his work with philosophical problems (a conviction expressed in an unpublished manuscript that we shall discuss later), but also formulated (in the same manuscript) his position on the theory of knowledge, on philosophy.

Before elaborating on what was said it is, however, necessary to talk about the real facts from L.I.'s life. From them one can see how drastic were the changes in its conditions due to the events of violent times and how he changed himself, living through several sharply distinct periods before becoming, in the last decade of his life, the person described at the beginning of this chapter still remembered by a few living people that knew him.

Leonid Isaakovich was born in 1879, in the same year as Einstein, but also as Stalin (and Trotsky!). More than two decades had passed before he enthusiastically experienced a powerful influence of the greatest scientific genius, and more than four decades had passed before a bloody hand of his contemporary fell on him and the whole country.

He grew up in Odessa, one of the few Russian towns of European type, with characteristic features of a port that had grown on exporting Ukranian bread, with merchants that had become rich on this, with the ragged dockers and fishermen, bandits, but also with a lot of intellectuals, a magnificent opera building, in which there performed the best singers of Europe. In Odessa that would soon give to the country so many virtuoso musicians who conquered the world, and also writers, the cream of the Russian literature of the period, who are still extremely important at present, and also many scientists.

The father of L.I. was a well-known and successful gynaecologist. Women in labour from the whole Ukraine came to him and in the tram one could hear people ask a conductor for a ticket "to doctor Mandelstam". The childhood and youth of L.I. were serene.

After finishing gymnasium Leonid Isaakovich went to the University of Odessa (at that time the Novorossijskij) but soon was expelled from it because of participation in student political upheaval[2] and went to continue his studies abroad, as did almost all young Russians that took a great interest in science. As did many, he went to a main scientific country of the time, Germany. L.I. had already selected physics and chose the excellent University of Strasbourg, where physics was headed by Carl Brown, one of the founders of radio engineering. Together with Markoni, Brown received a Nobel Prize in 1909 (for "Brown tube", with corresponding improvements this led to an oscillograph, a television set and a computer monitor). Many Russian physicists — P.N. Lebedev, B.B. Golytsin, A.A. Eichenwald and others studied or worked with Brown. In this university there also worked outstanding mathematicians. L.I. became very close to one of them, Richard Mises, a like-minded person in understanding the foundations of statistical physics and philosophy. Another attraction to this university was that Aleksandr Gavrilovich Gurvich (later a well-known biologist), an uncle of L.I., although he was only five years older than his nephew, also worked in Strasbourg. He became a lifelong friend of Mandelstam. Another lifelong friend was Nicolai Dmitrievich Papaleksi, who came to Strasbourg three years later, also to continue his education.

Here L.I. spent 14 years and went orderly through all levels of the scientific hierarchy. He returned to Russia as a full professor (titular professor) at the day when the First World War was declared.

---

[2]It is interesting that in the autobiography written at the end of 1917 [Frenkel (1997), p. 68], he simply writes "In 1900 I left from the IV semester and went to Strasbourg". Leaving aside an uncertainty in timing (N.D. Papaleksi writes that this was in 1989 [Prokhorov (1979)]), it is worth noting that L.I. does not mention anything about the political background of his "leaving". Why? It is possible that he did not want, out of self-esteem, to create for himself a "good" political past at times of already established Soviet power or, possibly, already wanted to distance himself from it.

It is natural that his first scientific works were done on Brown's theme, radiophysics (which was at that time inseparable from radio engineering). This refers not only to the diploma work containing a result important for radio communication that at once made him known among the experts. Very soon he became "the first assistant" of Brown. This meant that he began supervising scientific work, giving the research topics to young scientists from all over Europe that were coming to Strasbourg. Papaleksi recalls 12 names [Prokhorov (1979), p. 21].

Mandelstam's scientific authority grew. When he obtained a title of a docent with an accompanying right of delivering courses of lectures, Brown himself often attended them and made notes in his notebook. Together with Brown he participated in his works (related to the firm Siemens and Galske) in practical tests of their system of radio-telegraphy. He made an acquaintance with A.S. Popov and other Russian pioneers of radio engineering.

Therefore, at the most important age for formation of a scientist, in particular a theoretical physicist, L.I. studied and worked in one of the best universities of prospering Europe which only at the end of this period began to feel the approach of stormy times. He made acquaintances with many physicists, visited many countries and became in the full sense a European and a mature scientist well known in the scientific world. An illustration is a postcard from Einstein received in July 1913 in which he writes, "Dear Mr. Mandelstam! I have just reported at colloquium your beautiful work on surface oscillations about which Ehrenfest[3] told me earlier. It is a pity that you are not here. With best regards, A. Einstein." On the postcard one sees 16 other signatures of the participants of this colloquium (see a figure in [Prokhorov (1979), p. 59]).

Already from this one sees that in 1913, at the age of 35, Mandelstam went far beyond the borders of radiophysics and radio engineering that remained his first love. On this basis there grew his interest in a theory of oscillations in general, first of all in optics. At that time optics played a fundamental role in understanding the foundations of physics. An appearance of quantum physics was tightly related to it. It was optics where M. Planck introduced in 1900 a notion of quanta, i.e. of a portion — like emission

---

[3] A friend of Einstein, one of the most prominent physicists of his time, Ehrenfest had not got a job in his native country after graduating from the University of Vienna and at that time worked in Petersburg where he became close to Russian physicists and did much good for developing modern physics in this city. He also kept these ties later, after the revolution. Ehrenfest was very seriously interested in a discussion between L.I. and Planck (see below) and expressed a wish of coming to Strasbourg to work with Mandelstam which at some point came true.

of light and Einstein in 1905 introduced a notion of quanta of light. But then L.I. was interested not in quantum problems, but in those in classical optics, mostly a problem of light scattering in homogeneous medium, e.g. atmosphere.

One should imagine a "childish" situation in this field at those times, when this problem was to a large extent unclear. An illustration for readers who are physicists can be that L.I., following Stark (later a Nobel Laureate) erroneously thought that molecules can scatter light only if being electrically charged [Rytov (1948), p. 121]. At the same time the scattering in atmosphere, assumed to be quite homogeneous, was studied by the Englishman Rayleigh, the physicist highly esteemed by Mandelstam. Planck himself published a paper on this topic in 1902. Mandelstam, however, argued in 1907–1908 that a dense medium can scatter light only if its density is inhomogeneous. This was a correct, deep idea, but its validity was finally proven much later by Smolukhovski and Einstein. And the work by L.I., as is now clear, contained an error[4].

I would ask a non-physicist reader to forgive me such an extensive deviation towards professional issues. It will immediately become clear, why this is essential for understanding the personality of L.I. of that period. Those knowing him only as much more mature, in Moscow after 1925, would be surprised.

The issue is that a young, having had rapidly established a name, L.I. refuted both Rayleigh and Planck by launching (in four papers) a discussion in a tone impossible in later times. At the beginning L.I. was still reserved: 'Therefore it seems to me that to reduce, as done by Planck, an absorption in an optically homogeneous body to scattering by particles constitutes a delusion' [Rytov (1948), p. 118]. In the same paper he objects to Rayleigh (p. 116).

Two and a half months later he publishes another note exclusively devoted to Rayleigh: it turns out 'inadmissible to reduce the blue color of sky to the scattering of light by air molecules themselves' (i.e. individual molecules — E.F.) [Rytov (1948), p. 190], as done by Rayleigh.

Soon (1908) Planck refutes the conclusion by L.I. in a new paper, but L.I. responds and at this time does not restrict himself to a single restrained sentence, but hurls at Planck a flood of attacking statements.

"Mr. Planck in his theory of dispersion answers this question affirmatively. On the contrary, by following two different lines of reasoning I come

---

[4] I am grateful to I.I. Sobelman for a clarifying discussion on this issue.

to a conclusion that in the optically homogeneous medium such scattering does not take place. In other words, I have come to a conclusion that the Planck model can not provide any understanding of the attenuation of propagating wave in general.

"In particular, in my second paper it was, in my understanding, shown that an opposite conclusion of Mr. Planck should be explained by an imperfect calculation." [Rytov (1948)]

"Mr. Planck considers my calculation to be wrong. From his motivation I conclude that is was not correctly understood" [Rytov (1948), p. 162].

"As shown at the beginning, a principal question of absorption is treated oppositely to Mr. Planck," [Rytov (1948), p. 163], etc.

We have to recognize an important fact: the great Planck objects to Mandelstam on a question that should have been settled after his work. In a few months L.I. publishes a fourth paper filled by equally harsh (but formally polite) phrases. For example, "The settlement of the issue would be much easier if Mr. Planck would want to find a mistake in my calculations as I have done with respect to his" [Rytov (1948), p. 172]. Planck did not respond to this paper.

We see that a young cockerel L.I. just flew up! I have described this episode in such detail on purpose. To all of us who knew Mandelstam in the thirties it is difficult to believe that he could be so opinionated, could behave in such an aggressive way. In all recollections of his students and colleagues, simply those having known him, the first quality mentioned is his softness, pliability, nice smile, immediately-acting charm, modesty. And all this is literally true. One knows an episode at the Fourth Congress of Russian physicists in Leningrad (1924) narrated by I.E. Tamm to V.Ya. Frenkel [Feinberg (1992), p. 366]. There was a discussion on the talk by Ehrenfest concerning one difficult question from optics. Tamm sat next to Mandelstam somewhere higher than the tenth row of the auditorium ascending as an amphitheatre. At some moment Ehrenfest said, "Let the most prominent expert on optics Professor Mandelstam express his opinion on this issue" and started to search for Leonid Isaakovich with his eyes. Mandelstam, strongly embarrassed, to Tamm's astonishment literally slipped down the bench so that he could not be seen from below where Ehrenfest was standing.

But then, in his younger years, the above-described episode of L.I.'s opinionated aggressiveness was not the only one. In the same period the English radiophysicist and radio engineer Fleming published two papers on a question that L.I. also worked upon. L.I. immediately pounced upon

Fleming, found errors and concluded, "... although Fleming considers in both papers the same case, he arrives at different (in both cases erroneous) results"; "Fleming uses the formula (3) keeping the terms of the second order which is clearly inadmissible and makes the above-mentioned error in sign" [Rytov (1948), p. 141]; "Summarizing the above discussion: the calculations by Fleming are wrong" [Rytov (1948), p. 149].

We see that he did not spare accusing phrases, did not soften them when talking again and again about the author's mistake. This is so unlike the "Moscow" Mandelstam!

Whereas the critique of Fleming was essentially grounded, in the case with Rayleigh and Planck this was not quite so.

I dare make a (perhaps too bold) suggestion that *a realization of his error and of an inadequacy of his behavior in younger years exerted a big influence on L.I.'s later life, both scientific and everyday.*

It is well known how insistently did he demand from himself, his students and coworkers to repeat and make more precise experiments, calculations, understandings. On more than one occasion this costed him dearly. In such a way, essentially because of their usual style of behavior after obtaining a new result and its publication, he and Landsberg did not get a Nobel Prize for a discovery of combinational scattering of light (Raman effect).

A student and a collaborator of G.S. Landsberg of many years, I.L. Fabelinski witnesses.

"Investigations of G.S. Landsberg and L.I. Mandelstam were always characterized by particular thoroughness, carefulness, deep understanding of a studied subject and deliberation in publishing the results obtained. Moreover, when the work was completed and even written for publication, it was put for some time into a desk drawer. For the case if a different idea would come into head or it would be necessary to make the text more precise or change nuances in argumentation. And in general, it was necessary that everything had settled, after that one could send the paper for publication".

"At the same time (continues I.L. Fabelinski. — E.F.), ... after working with G.S. Landsberg for 20 years and having discussed the questions related to the origin of the new phenomena (discussed in this chapter. — E.F.), I never heard that G.S. Landsberg regretted their line of behavior" [Fabelinsky (1978), p. 6].

Precisely this style showed itself in the story of the discovery of combinational scattering and of not getting the Nobel Prize. Having had made reliable observation already on February 23–24, they first sent a report for publication only on May 6, after having had fully understood the essence of the phenomenon, developed its theory and been convinced that their observations agreed with theoretical predictions. Because of this Raman, who sent a report into a journal right after getting the first result, although having an incorrect understanding of the essence of the phenomenon at that time and in the two subsequent publications, was ahead of them *in publication*.

In fact they had observed the effect in the experiment even *earlier* than Raman.

This is clear from a letter by L.I. sent to O.D. Khvolson. Answering his direct question L.I. wrote: "We first noticed an appearance of new lines on February 21, 1928. On negatives dated February 23-24 the new lines were already clearly seen." (see below, Fabelinsky (1978), (1998), (2000), (1982)).

Raman, in his turn, wrote (in the *Indian J. Phys.*, p. 287, 1928): "First time we noticed an appearance of new lines on February 28, 1928. The observation was made public". Thus the first actual observation of the new effect by Landsberg and Mandelstam (*not* the date of sending a report for publication and *not* the publication date) *was ahead* of the observation by Raman at a week[5].

It was already mentioned how, in the course of preparing his lectures in the thirties, L.I. many times rewrote in his notebook almost the same fragments. In such a way he sort of sealed and established his confidence in the correctness of what was written, in the perfection of the form. He always strived for completeness, overcame his uncertainty. In his mature years everybody knew him as a quiet expert, always definite and assured in his statements, and only the members of his family (they told me about this) knew how he suppressed a neurasthenic in himself. On lecture day his condition was awful. At the lecture itself, however, no traces of this could be spotted[6]. When he had to go to Leningrad in the evening then, as known only by the people closest to him, the clocks in the house were put one hour ahead so he would not be late.

---

[5] I am grateful to I.L. Fabelinski for full clarification of this issue.

[6] However, it is worth paying attention to some sentences at the beginning of this essay: "He hastily stops behind the end, of a ten meter long desk" (p. 5) and "..until an end of a lecture, he does not leave the saving spot between the end of the desk and the blackboard behind him" (p. 6), and also "Something apologetic in his tone and even pose will also burst open later" (pp. 5-6). They are taken from my very brief recollections of 1979 ([Feinberg (1992)], see p. 235, line 13 from below and p. 236, lines 5-6 and 13-14 from above) and were written 20 years before I learned about his neurasthenia. It is possible that I felt its signs already then. It is like written by Fazil Iskander about a boy Chick: "He knew it, but he did not know he knew."

I was told about only one episode in which he exploded. In his office he had a discussion with his favorite young student A.A. Andronov. Suddenly Andronov, with a crimson face, rushed out of the office and through the common office, where there were other people, and ran away. The breach with L.I. lasted for several months but then, of course, everything was settled. One assumes that a young, then of pro-communistic views, Andronov tried to turn L.I. to his beliefs, provoking his fury.

But was Mandelstam always unerring in science? Alas, those writing about Mandelstam, justly enraptured with him as a physicist of an exceptionally high class, with what he accomplished and his scientific work, with creation of such an amazing phenomenon as the Mandelstam school with its successors, with his charming personality, often create an impression on some never erring deity. We have already described a story with the scattering of light in opaque media. A mistake was that at that time, before Smolukhowski, neither he, nor other physicists knew about an existence of (density) fluctuations in continuous media (Rayleigh obtained the correct formula in some sense by chance). Characteristically, in the 5-volume collection of the Mandelstam scientific works the corresponding papers are not commented upon and the partial fallacy of his position in a dispute with Rayleigh and Planck is not mentioned (although L.I. himself understood it and a later paper on other subject contained a piece on Rayleigh's rightfulness [Rytov (1948), p. 146]).

There was another case when following an idea by Max Laue (who received Nobel Prize for this) his younger students, studying in 1913 a scattering of X-rays in crystals arrived at a result which (as especially clearly shown by L. and G. Braggs who received a 1915 Nobel Prize for this) proved that rays are oscillations, waves, and not a flux of particles which could have been the case[7]. Mandelstam, not waiting for Bragg's result, put forward an assumption that the observed scattering of X-rays was caused

---

[7] A curious episode (completely unrelated to L.I.) was told to me by Yu.B. Rumer, a well-known physicist who lived in the twenties in Germany and thus knew much that remained unpublished. "Then, in 1913, it still remained a long time until quantum mechanics and physicists did not know that a flux of particles has wave properties as well. At the nearest meeting of German physicists one of the participants told Laue that to fully prove the wave nature of X-rays one should repeat the same experiment with a flux of electrons and convince oneself that in the latter case the scattering does not give a Bragg picture. However, as agreed by both participants of the conversation, this was surely unnecessary". In reality the wave properties of electrons could lead (at appropriate electron energies) to the same picture as with X-rays. Would not the wave properties of electrons had been discovered 10–11 years before it actually happened?

simply by microscopic cracks on the crystal surface and even tried to prove this experimentally, mentioning in the same paper that *a described experiment is still not completed*! Can one imagine that in a paper by "mature", "Moscow" Mandelstam there appeared argumentation based on an unfinished experiment [Rytov (1948), p. 242].

Could one, should one accuse Leonid Isaakovich for such mistakes? This is a principal question for a scientific researcher which is fully answered by a brilliant aphorism by V.I. Vernadski (in his diary before the Second World War): "A freedom of creative work is a right to err". And this is undoubtedly true. There erred a great Newton who considered light to be a flux of classical particles. There erred a great Maxwell who believed in the existence of an all-penetrating mechanical ether and thought that his ingeniously guessed "displacement current" introduced into his theory of electromagnetism reflects a real displacement of ether particles (that was not only wrong but reflected a retrograde understanding). It is absurd to reproach Mandelstam for the mistakes of his younger years.

A.A. Andronov in his unusually capacious and meaningful characteristics of L.I. [Frenkel (1997), p. 190], writes, in particular, that L.I. "... did not like making mistakes and almost never erred. When having made a mistake, and this happened very seldom, then, when he realized that he had made a mistake, he became very anxious about it, started calling you or asked, through other people, to visit him to improve a little mistake". Andronov, however, knew Mandelstam only after 1925. A great Turkmen poet of the eighteen century Makhtumkuli characterizes, in his verse "Humanity II", distinctive features of each decade in human life in their turn [Makhmutluli (1948)]. In particular, he writes[8]:

> At twenty life is not the same,
> Bursts away a youthful flame,
> Every day he's drunk with dream,
> Every moment captures him.
> At thirty is his life diverse,
> With temptation's idle growth,
> He is thinking clear though
> He has understood himself.

(In the period of disputes with Planck and Fleming, L.I. was 25–30 years old.)

---

[8] Translated into English by A. Leonidov.

> Youth experience is cheap,
> No medicine when eye's not sharp,
> Thoughts cohere only at forty,
> With life experience in veins.

In 1914 Mandelstam was 35 and by the time of his return to Russia, almost at a doorstep to an age of 40, the Greek *akme* (flowering, flowering force), an experience of life had penetrated into the veins of this remarkable man. He had already "comprehended himself".

And here the epoch delivered him a heavy blow. There came an 11-year period in which all this "flowering" could not be realized in the fruitful scientific work for which he was destined and so well prepared.

A young but already experienced, "having comprehended himself", talented scientist known and cherished by a scientific world even embodied by its highest representatives such as Einstein and Ehrenfest, Sommerfeld, Brown and Mises, a thoroughly educated European at the peak of his creative power returns to the motherland, to Odessa, after 14 (15?, see the footnote on page 18) years of absence. First of all absurd delights of tsarist Russia pounced upon him. To have the right of lecturing in a Russian university one had to have a Russian master's degree. To get this he had to defend once again a thesis, and for this to get a diploma (L.I. did not have such a one). Thus formally L.I. was a nobody. However the university charter contained certain reservations and the Novorussian university elected him as a private docent in physics. This was a non-permanent staff position of an instructor who was allowed to conduct seminars and even deliver lectures but usually on optional, facultative courses. Besides that an approval of the minister of education was required.

It is clear that in such conditions L.I. could not launch real scientific research so at the end of the year he undertook a decisive step by accepting an invitation from Siemens and Galske firm in Petrograd (with which, as already mentioned, he and Brown had already collaborated) to become a consultant on their radiotelegraph plant[9].

Two war years in Petrograd were filled with consultations on research development, sometimes down to purely engineering, to working out a

---

[9] Let us notice a curious detail: this was a German firm (perhaps under temporary administration by the Russian government or temporarily confiscated) and this was a time of war with Germany. After 30 years already after the death of L.I. this provided, in the disgusting after-war soviet atmosphere of the attacks on "low bowing towards abroad" and state antisemitism, a ground for some careerist physicists to accuse L.I. in spying for Germany (!).

technology of wire oxidation, construction of rheostates and launching their production. A future coworker of Mandelstam and Papaleksi, E.Ya. Shegolev, who also worked there, recalls [Prokhorov (1979)] that this work was also done by L.I. with such a brilliance that not only did he get patents for his inventions, but the sagacity of his approach evoked an admiration of radio-engineers. At the same time he generously shared fruitful ideas. Here his "third" above-described talent, that of an engineer-inventor, showed itself. This was, however, not the kind of science his soul was longing for.

At the end of 1916 L.I. approached one of his friends, T.P. Kravets, with a letter in which he recognized that without having a Russian scientific degree, scientific work would probably be impossible. He asked for help in getting the degree without passing an exam and defending a thesis, replacing it by some of his published papers (why had not he done this before? Perhaps he had considered this humiliating? Or, as it had always been in the past, new exams would have become a tortuous experience for his weak nervous system?). Kravets, a cordial person, answered with a warm brief telling [Prokhorov (1979), p. 6], that he had already discussed this issue with a radiophysicist from Kharkov, Professor Rozhansky[10], and had found out that such a possibility is granted by the University Charter: a scientific council had the right of approaching, with a corresponding application, a Ministry of Education. "For some reason the physicists used this option most often and it was in this way that P.N. Lebedev, N.P. Kasterin, A.G. Kolli and A.A. Eichenwald (who had also studied abroad) got a scientific degree", writes Kravets.

I am describing this in such detail to show a corporate solidarity of Russian intelligents-physicists, although we will soon meet with an opposite phenomenon during Stalin times.

There came, however, a year of 1917. After the February revolution the disturbances, fueled by a diarchy of an Interim Government and Soviets, grew, the food in Petrograd was becoming scarcer and scarcer and a winter of cold and starvation was looming ahead. Many who could, including scientists, tried to leave for the southern provinces, at that time still quiet, warm and having plenty of food. L.I. was simultaneously invited by the Council of Tiflis University and the council of the private Polytechnic Institute in

---

[10]This person is also worth mentioning. Later, in 1931, when working in a Leningrad institute of A.I. Ioffe he dared *not to vote* at a staff meeting for a demand of capital punishment for another group of "people's enemies" (from "Prompartiya"; even a tribunal did not sentence them to death by shooting). Rozhansky was sentenced but, thanks to the efforts of influential Ioffe, his imprisonment was not long and he survived.

Ekaterinoslav and was elected as an acting ordinary professor of physics. He chose Tiflis and was approved for this position by the Minister of Education in July. In the autumn of 1917, L.I. moved, together with his family, to the Menshevist Georgia which, in May 1918, declared itself an independent democratic republic. However, here there was also no peace. Already in the first half of 1918 there were Bolshevist uprisings around Tiflis and in June, due to a treaty with the Menshevist government, German troops entered Tiflis. However, in November the First World War ended and in December these were replaced by English troops.

It is easy to understand that already in the autumn of the same 1918, a year before the bolshevization of Georgia (as in many other cases it was simply a capturing of Tiflis by the Red Army waiting at the Georgian border and "coming to help" a bolshevist committee organized in the city that declared itself a government with the menshevist one still being in power) L.I. preferred to be in a native although, as it turned out, even more starving and, in addition, freezing and having been torn into parts by changing authorities. Odessa, first Soviet, it was after some time occupied by Germans and a Ukranian government of Skoropadski supported by them, then again taken by reds, then by a White army of Denikin. Then it followed a French Entente intervention. Around the city raged Makhno, "Greens" (not to be mixed with modern ones), Petluyra.

Finally, Odessa became Soviet. After the expulsion of Vrangel from the Crimea, the Civil War in the south came to an end.

In these conditions one could not even dream of a quiet atmosphere and scientific work. Still, an appearance of many professors and students from the north allowed, even in these circumstances, the organization of a new Odessa polytechnical institute. L.I., together with friends who also came there, actively participated in it. N.D. Papaleksi became head of the physics chair. Astonishingly it turned out possible to assemble relatively good laboratories and deliver lectures in the freezing auditoriums. L.I. was a brilliant lecturer ('talent number 4'). It is sufficient just to look through his introductory lecture on a topic "Why engineers need physics" [Prokhorov (1979)] to see his talent.

When the soviet power was already established there arrived a yet unacquainted, but recommended by A.G. Gurvich, young (16 years younger than L.I.) Igor Evgen'evich Tamm who became a lifelong closest student, friend, colleague. Together with his wife, Tamm settled in the house of the Mandelstam family where they occupied an empty room of A.S. Isakovich, a brother of L.I.'s wife. The room was empty due to a then usual reason.

Its inhabitant was imprisoned by Cheka, probably because of being a "burzhui". He was a lawyer, had contacts with an international "Society of Indo-European Telegraph" and was relatively wealthy. In a few months or a year he was released. At the same time, saving themselves from starvation and in order to get additional payment and food rations, Leonid Isaakovich and N.D. Papaleksi organized, at a radio-telegraph plant, a "vacuum artel" producing radio valves. Its staff included radio technicians and the new friend I.E. Tamm.

It is easy to understand that a prolonged absence of regular scientific work tormented L.I. It is however equally clear that it was impossible to fully stop his head from working. As we saw, already in Strasbourg he expanded his scientific interests in the fundamental issues of optics (a remarkable mathematical theory of optical images, etc.). It was in these starving years (N.D. Papaleksi writes, in 1918 [Prokhorov (1979)]) that he understood that light scattering in the medium can occur on elastic waves creating a necessary inhomogeneity. Then it should be accompanied by a small shift in the frequency of the scattered light. In other words, a spectrum of this light should contain two lines, not one. This work was however published only in 1926. Clearly, such pause could not be explained only by the Civil War conditions. Probably, the L.I.'s hesitation played a role here. He was sure that calculations were correct but could also doubt a necessity of publishing an article predicting a very small effect (the frequency change should have constituted 0.003 percent) before it could be verified in an experiment. This turned out possible much later, in a completely new period of L.I.'s life. We shall return to this later. Here we just note that independently this effect was predicted in 1922 in France by a French physicist L. Brillouin. Therefore in physics it is known as a Mandelstam-Brillouin effect. Its studies continue even nowadays. Using lasers it turned out possible to explore it in more detail and use it in other studies, also for applied purposes.

This theoretical work of 1918 was for L.I. the only one in the first eight years of living in Russia. This did not alter a general situation of isolation from scientific research. Therefore when, in the summer of 1922, a transition to NEP[11] had already begun making its beneficial influence on a real life of the country, L.I. accepted an offer from a board of an electrotechnic trust of weak current plants (a NEP child!) to supervize (as a scientific consultant) together with Papleksi the scientific and scientific-technical research in the

---

[11] New economic policy. A.L.

Moscow radio-laboratory of the trust. In October, L.I. Mandelstam and his family and N.D. Papaleksi moved to Moscow. Soon (in March 1922) L.I. made a scientific trip to Germany where he met Einstein and many other scientists and returned to Moscow with a baggage of scientific news from the nine years of his absence and the newest scientific literature.

The next year the laboratory moved to Leningrad where L.I. had much better conditions for work but ... only on radiotechnics. Certainly this was also important work for him as in this field he had made, with Papaleksi, many useful things (new ways of radio-telegraph and radio-telephone communication, frequency stabilization, etc.). However, there remained a question: Why had the best universities been silent? Is it possible that when L.I. was in Moscow (1922–1924), the Moscow University did not realize what a promising possibility was at hand?

At that time a physics faculty of MSU (just separated from that of physics and mathematics) was in a state of despondency. When in 1911 P.N. Lebedev (together with the best professors from other faculties) had resigned, the level of teaching and research work had sharply declined. Many rooms of the wonderful newly-built building of the physics faculty were simply empty (this was in the twenties). It is true that also in the NEP years there remained several true physicists (a corresponding member of the Academy, Lebedev student V.K. Arkadjev and his wife A.A. Glagoleva-Arkadjeva; a young N.N. Andreev who had just graduated from Basel University; equally young G.S. Landsberg, S.I. Vavilov, S.T. Konobeevski). However, not them, but quite different people determined a general level and, most importantly, its general inadequacy with respect to the comprehensive scientific knowledge. For example, the only course on a special theory of relativity, created in 1905, was given in 1918 by N.N. Andreev (at that time the only case in Russia!). The majority, even when including some gifted people, fought with relativity theory and later with quantum mechanics with all means, tried to "simplify" these disciplines, already reigning in the world, down to the level of notions of classical physics. Some of them declared the new physics to be a bourgeois distortion of science.

At the same time not only the best young professors (S.I. Vavilov, G.S. Landsberg, N.N. Andreev and others), but also other young people, students and Ph.D. students that were highly sensitive to new tendencies felt a clear dissatisfaction with this situation. At the faculty there began a furious struggle for making an offer to Mandelstam. Among the argument "against" were also anti-Semitic ones (these were at those times not common at all). S.I. Vavilov, visiting his friend G.S. Landsberg whose apartment was at that time in the yard of the university building, told him, with hands perplexedly stretched out: 'P. told me, "How don't you understand, Sergei Ivanovich, that if Mandelstam comes, he will bring others like him, of his kind". I do not understand such people'.

At that time an important role in decisions on all issues was played by the young. Students, Ph.D. students (A.A. Andronov, M.A. Leontovich and others) essentially influenced the positive decision on the matter. In 1925 L.I. finally became a professor, a head of theoretical physics chair of the physics faculty and a staff member of a then existing at the faculty scientific research institute, NIIF. Soon the responsibilities of the head of the chair were transferred to I.E. Tamm. There began a totally different life.

Leonid Isaakovich, together with his wife Lidia Solomonovna and his son Sergei Leonidovich, who later also became a physicist, obtained an apartment with three large rooms in the building of the physics faculty. Its important distinctive feature was that along with a "parade" exit to the stairs common for all the professor's apartments and separate from the auditoriums and laboratories it had another exit directly to a corridor at the first floor of the "working" part of the building. Right behind this door there followed, one after another, those to the rooms of the laboratory of G.S. Landsberg and others. Straight ahead was a X-ray laboratory of S.T. Konobeevsky, etc. For sure at those times there were no guards at this door and a communication of the apartment with the rest of the building was completely free. L.I. used to work at home and when it was necessary to participate in an experimental work in the laboratory, or conduct seminars and lectures in auditoriums. He went there without leaving the building. At the same time his colleagues had direct access to him.

Gradually there formed a following habit. At the end of a working day, at 5 p.m., his closest friends gathered at his place around a tea table. Here they discussed a wide range of topics: scientific, political issues, discussion of students that were of interest to somebody, of mutual acquaintances. Here everybody trusted each other and was open-hearted. The new life was, scientifically, extraordinarily fruitful. If one ignores a lack of equipment[12] and a depressing social and political atmosphere in the country (up to an end of NEP, still not completely horrible) it could be considered as an almost happy one for L.I. At last he had lived up to the time when the scope

---

[12] Even in 1930 when I came to the university I saw many empty rooms and empty shelves for instruments, but very soon these started to fill with instruments made in the country.

of his scientific work was becoming broader and broader, an entourage of talented colleagues and young scientists was expanding fast and he was deeply and sincerely respected and esteemed by them.

In 1925 Leonid Isaakovich was 46. Wars and revolutions had, without sense, practically taken away eleven most precious years from the life of an outstanding scientist at the peak of his talent. These were also taken away from a large part of the young university population in Moscow who had been deprived of modern science. Now they were passionately striving after the Mandelstam school and his colleagues. In spite of sixty patents and a theoretical foresight and comprehension of the Mandelstam-Brillouin effect, these years had not been those appropriate for the physicist who had accomplished what he had done in fourteen Strasbourg years. This was evident.

Things became even better when in 1930 B.M. Gessen, a close friend of I.E. Tamm from gymnasium years but, unlike him, a convinced bolshevik and, at the same time, a highly intelligent person, became a dean of the faculty and a director of NIIF. He graduated from an institute of red professorship and for some time even was its director. In this entourage he was certainly distinguished by a general education, general culture. He studied natural philosophy. It is suffices to mention that his talk at the 2-nd International congress on the history of science in 1931 in London in which his Marxist position was presented drew serious attention and was broadly quoted afterwards in western literature (for more details on him see [Gorelik (1995)]).

Gessen deeply respected Mandelstam, admired him. I recall a scene in the cloak room when he helped Mandelstam to put on his coat, literally blowing off specks of dust from it. It was not surprising that with such a director the most favorable conditions for L.I. that were possible at that time were created. This was, however, the time of the beginning of horrible processes followed by the "big terror". In 1936 Gessen was (with no reason) arrested and executed. This brought new troubles to everybody close to him. We have, however, run far ahead. Let us return to the scientific work of L.I.

The university and the presence of many coworkers provided L.I. with the possibility of expanding it simultaneously in several directions. Besides giving a course on a theory of electromagnetism in a completely new style for the university, he began theoretical and experimental studies in various directions. With Andronov, Papaleksi, Vitt, Khaikin, Gorelik, Leontovich and Rytov, work on theory of oscillations in general, of nonlinear ones in particular, was carried out. Together with G.S. Landsberg, L.I. began expanding experimental studies in optics with a focus on finding the Mandelstam-Brillouin effect. With Tamm he finished work on a relativity theory for anisotropic media.

During the Strasbourg period, after beginning with radiophysics and radiotechnics, L.I. had already significantly expanded an area of his research by examining, as already mentioned, the most serious problems in optics. Now he went even further.

Right before this time, in 1925–1926, a new quantum mechanics of Heisenberg-Schroedinger (in two apparently different but equivalent forms) had appeared and had shaken the foundations of physics. And, as has been already mentioned, L.I. published with M.A. Leontovich an important paper in which remarkable properties of the main equation of quantum mechanics, the Schroedinger equation, were studied in detail. Discovered, in particular, was an astonishing paradox: A quantum particle can go through a "potential barrier", through the domain, in which its kinetic energy is smaller than its potential one! This phenomenon is completely impossible in classical physics, makes no sense in it, violates the law of energy conservation. However, due to the wave properties of a particle, it takes place. Called later a "tunnel effect" (in the paper of our authors this term is still absent) it plays nowadays a great role in physics and technics[13].

G.A. Gamov (at that time a Soviet, later an American physicist) was the first who noticed the discovery of Leontovich and Mandelstam. He applied it, very elegantly, to explain a radiative decay of atoms and nuclei, the phenomenon discovered already at the end of the nineteenth century which was totally incomprehensible in classical physics. Although, as was well known in the physics community, Gamov knew about the work of Leontovich and Mandelstam before publication and relied on the properties of tunneling effect established in it but, unfortunately, did not cite it, so up to now a dominating opinion of physicists around the world is that the author of "tunneling effect" is Gamov. Mandelstam never argued about authorship. Some have an opinion that he lacked ambition that a scientist nevertheless needs. I believe, however, that he considered such disputes humiliating

---

[13]This discovery was for L.I. not accidental at all. In Strasbourg, when studying optics, he had already demonstrated both theoretically and experimentally that optical waves that should have experienced a so-called complete internal reflection from a border of a solid body, e.g. glass, in which they propagated, with air, did in fact partially jump through the gap with air if the same glass was placed nearby. Such integral understanding of classical and quantum physics was characteristic of Mandelstam.

(people like him think that if you are really worth something as a scientist then all of your results will not be stolen anyway so, instead of arguing, it is better to write another good paper).

This was by no means a unique episode. When an extraordinary difficult experiment on finding the Mandelstam-Brillouin effect had finally been started L.I. and Landsberg were not satisfied with how things proceeded. A spectrometer they had used was not good enough and, although an observed broadening of a spectral line constituted a clear indication of the sought phenomenon, they wanted something better. State Optical Institute in Leningrad possessed a better instrument. They turned to its director, an outstanding scientist D.S. Rozhdestvensky, with whom they had perfect relations based on mutual respect, to ask him to charge one of his young researches to repeat the measurements at this better device. This was done in 1930–1932 by E.F. Gross who studied the effect in detail. During a parallel work in Moscow and Leningrad there was an extensive exchange of letters. L.I. visited Leningrad. One thought this was considered a common work that should have resulted in two publications, one by Mandelstam and Landsberg, another by Gross. However, when Mandelstam and Landsberg sent him a manuscript of their paper, Gross surprised them by telling that his own paper, covering all relevant questions, had been already in press. Thus the publication of the Moscovites turned out to be unnecessary.

It was perhaps one of similar episodes when L.I., replying to someone convincing him to protest, told a phrase quoted in the memoirs of S.M. Raisky [Prokhorov (1979), p. 16]: "One does not educate an adult person. One either deals with him or not. One should not deal with N".

In fact L.I. could have "good" ambitions. But above them was his self-respect that did not allow him to mix a humiliating "fight for priority" with questions related to science, to comprehending the truth, even when he himself internally suffered from a complete injustice. I.E. Tamm, and almost all the physicists of Mandelstam school (including those, unfortunately much more rarely, belonging to the next generations) also behaved in the same way.

These cases of the unjust oblivion of Mandelstam's name seem to constitute some kind of a pattern. An even more important episode took place in connection with a very big discovery of Mandelstam and Landsberg, the combination light scattering that has already been mentioned. As mentioned, they did not receive a Nobel Prize for it. It went to an Indian physicist Raman who gave a wrong interpretation but was faster in publishing a paper while our physicists were polishing their experiments and theoretical understanding to achieve a perfect sheen.

The thing is that in the process of experimental studies of light scattering and Mandelstam-Brillouin effect, over a long period of time many versions of these experiments were performed. In the course of this work it was discovered that apart from the Mandelstam-Brillouin scattering happening due to an interaction of light with elastic acoustic waves in the crystal as a whole, there existed a scattering of light with higher frequencies in which a structure of individual molecules of a body is of importance. Here there also appeared new spectral lines with a frequency depending on intramolecular vibrations and not only on that of the scattered light. Therefore it was called a combination light scattering. The authors had a clear understanding of the importance of this discovery that led to many practical applications because, by studying new lines, one could learn a lot on the nature and structure of scattering molecules.

To catch these lines at that time was, however, difficult; their intensity was too weak. With the technique available one had to photograph the spectrum with an exposition of tens of hours. Nowadays, due to photoelectric registration and lasers, this is much more easy.

Finally, at the beginning of 1928, L.I. and Landsberg obtained photos of a quality that was satisfactory to them[14]. This was achieved in difficult conditions when at times the most necessary materials were lacking. For example, a required quartz tube and many other supplementary materials were brought by Landsberg from a scientific trip abroad with his personal saved money. High quality quartz crystals in which the scattering was studied were found in second-hand shops by buying the quartz seals used before the revolution to seal letters by the sealing wax.

Having had already surely discovered the new phenomenon in February our physicists, following their usual above-described line of behavior, did not, as already written, hasten to publish. So the first public talk was made by Landsberg in the Institute of Physics of Ministry of Health (that had a big optical department) on April 27 and sent the paper, as

---

[14] A history of these experiments and discovery is diligently studied and explained in the papers of I.L. Fabelinsky [Fabelinsky (1978, 1998, 2000)]. In particular he reproduces a photo-plate with a spectrum in which the combination scattering lines are clearly seen. On the plate it is written, by the hand of Landsberg, the date: February 23–24, 1928. This is before an oral presentation of Raman in the Indian Physical Society (which was of course published much later) and long before publication of Raman's paper sent to the journal on March 8, 1928. Our physicists sent their first paper later, on May 6. One could think that they had other successful observations earlier than on February 23–24.

already mentioned, on May 6 after constructing a theory of the phenomenon that agreed with their experimental observations. By the time their paper came out in print (July 9) many physicists had already understood an importance of this discovery as described in Raman's talk. There appeared 16 papers by different authors, these were already calling the discovered phenomenon a 'Raman effect'. Our authors themselves, with their characteristic decency wrote in their first paper that Raman (and his student and coworker Krishnan) had published in *Nature* the work "describing an observation of the same effect" and in the second paper "the authors refer to the work by Raman and Krishnan which was known to them before sending the paper for publication" [Singi and Riess (2001)]. A public opinion on the precedence of the Indian scientist was already formed. Only special study could restore the story in its details. Who would have done this? It was too late....

There was, in fact, another reason for the delay that has to be mentioned.

The Mandelstam clan included, besides a wife and a son, also L.I.'s sister, a nephew (a physicist M.A. Isakovich), two nieces, a husband of one of them (a student of G.S. Landsberg) S.M. Rajsky and a Gurvich family. Having no close relations with L.I. himself I had, after his death, friendly relations with this circle. I do not remember who it was (most probably it was S.M. Rajsky) who told me about a characteristic episode that took place at the end of the study.

Once L.I. returned home from Landsberg laboratory holding in his hands a still wet, after development, photographic plate and told his wife with a squirming smile: "Imagine, Mizya (an intrafamily name of Lidiya Solomonovna), one gets Nobel Prizes for such things". L.S. replied indignantly: "How could you think about such things when Uncle Lyova is in jail and they have already stopped taking parcels!"

L.I. squirmed even more and there began a discussion on what could be done for "Uncle Lyova"-L.I. Gurveich, a close relative of A.G. Gurvich, who was already sentenced to death. This was 1928 and at that time one could still petition for mitigating a person's fate. Very soon it became totally forbidden, but in that situation L.I. and A.G. Gurvich dared to talk with a then rector of MSU, A.Ya. Vyshinski. He was still not a prosecutor general at the bloody trials of the thirties (although probably already in contact with 'organy'[15]). Improbable as it seems, after knowing about him what we

---

[15] The word 'organy' is a Russian colloquial for inland security bodies that were, in particular, instrumental in Stalin's repressions.

now know, Vyshinski said he would try to help. As a result the execution was replaced for "Uncle Lyova" by exile to Vyatka (now Kirov)[16]! This gives a clear idea on how already at those still "mild" NEP times did the "organy" "administrate justice". It would seem that an inveterate outlaw and an enemy of Soviet power should have been sentenced to a capital punishment. It turned out however that without harming the state interests it could be replaced by the peaceful exile.

So, what did happen with the Nobel Prize? Having made their discovery Landsberg and Mandelstam surely did not make a secret of it. In summer of 1928 there was a sixth congress of Soviet physicists at which many scientists from abroad, including the most famous (Dirac, Darwin, Born, Pauli, Debye, Peierls and others), were invited. The sessions first took place in Moscow, then on a ship sailing down Volga and in several towns down the way. At this congress Landsberg gave a talk on this work that made a strong impression (in their correspondences Darwin [in the English journal *Nature*] and Born [in the German *Naturwissenschaften*]) put a special emphasis on this talk using for its characterization the highest epithets [Fabelinsky (1978, 1998, 2000)].

Therefore the discovery of Moscow scientists became known to the world scientific community very soon (although after the publications by Raman).

In 1930 Raman received the Nobel Prize. Why him alone? This injustice caused a painful reaction in our country. There appeared a lot of versions: an anti-Soviet sentiment of the Nobel Committee, etc. The question was clarified 50 years later when, in accordance with a Nobel Committee Charter, all documentation related to this matter that previously had been kept secret was published[17]. It turned out that in 1928, our physicists were not nominated by anybody while Raman was suggested by Bohr and by another physicist. The 1929 prize went (quite justly) to Louis de Broglie, an author of an idea of wave properties of an electron on which a Schroedinger picture of quantum mechanics is based. In 1929 Raman was nominated not only once again by Bohr but also by Rutherford and other, in total 10 (!) authoritative physicists. At the same time Landsberg and Mandelstam (Raman as well!) were nominated by O.D. Hvolson, our senior remarkable physicist, a honorary academician, an author of a five-volume course on physics issued

---

[16]These details were communicated to me by the daughter of A.G. Gurvich, Natalia Aleksandrovna Belousova-Gurvich. I am very grateful to her for them.

[17]We should clarify a mechanism behind a decision on Nobel Prizes. A year before the decision (in our case for the nearest possible decision in 1929 this was a year of 1928) the Committee invites a group of prominent scientists of its choice to present their candidates ("nominees") and, in the end of the next year, announces a final decision.

in several languages who, at his age of 76, could understand the new physics (later he wrote a remarkable book on it) and appreciated the new discovery under discussion, and N.D. Papaleksi who due to some reason nominated only Mandelstam. Some other of our prominent physicists did not nominate any of them but gave other names [Singi and Riess (2001)]. The Committee, apparently guided by the difference in the dates of sending the papers to the journals (March 8 for Raman and May 6 for Mandelstam and Landsberg), made a simple decision which was certainly influenced by the difference in number and international prominence of those who nominated.

What could have been expected with such a difference? But the difference itself is not easy to understand. Everybody who nominated Raman did already know about the "brilliant" work from the Landsberg talk in the summer of 1928. Why were our other physicists silent and modest in nominating the candidates (everyone had a right of nominating several works)?

The first question is easy to answer: an "organizational" work among the scientists the committee chose for nomination was required. Alas, we know that also nowadays some nominees are intensively working in this direction as well. I was told about this by my Western friends as well. An extraordinarily intelligent, late remarkable Italian physicist, Jiuseppe Occialini, "fell off" two Nobel Prizes given for works in which he co-authored. He was strongly reproached with this by another similarly unlucky participant. Occialini told me how some scientists "struggle" for the prize. He himself was completely incapable of doing this because of his character and did not worry. It would have been crazy to expect such activity from our two Moscow physicists with their decency, intelligence, feeling of self-respect (I am by no means generalizing this on all Nobel Prize winners but 10 nominations for Raman and 1–2 for Landsberg and Mandelstam is too an absurd, blatant difference).

The reason lied primarily with the fact that Raman did not hesitate to publish his papers, even the first three in which he gave a totally erroneous interpretation of the phenomenon considering it to be analogous to Compton effect. He could not wait until these papers came out in press. Having informed about his discovery at a meeting of the Indian Physical Society on March 16, on the next day he prints thousands of copies of this talk and sends them around the world. When an Indian journal containing this talk comes out he gets 2000 reprints and mails them again to all prominent physicists in different countries [Singi and Riess (2001)]. He had been in correspondence with Borh, Rutherford and other influential people before that. On December 6, 1929 he writes a letter to Bohr in which he directly asks Bohr to nominate him for the Nobel Prize [Singi and Riess (2001)]. His contacts with a member of the Nobel Committee Siegbahn are also established [Singi and Riess (2001)].

In general, while Fabelinski relied upon what he witnessed, the authors of the important paper [Singi and Riess (2001)] made a detailed study of Western papers and materials from numerous archive funds that were opened 50–70 years later. They clearly see an injustice of giving the prize to Raman alone. "He knew how to fight for precedence", they conclude.

It is more difficult to explain the silence of our physicists many of whom could have nominated Landsberg and Mandelstam for the prize but did not do it. Perhaps in spite of existing achievements (and they did exist) in our then young physics a feeling of lagging behind, a kind of an inferiority complex led to an underestimation of an importance of this discovery.

But above all there was a mere delay in publication and this was decisive. Let us repeat what was already said. In his ripe years, L.I.'s striving not just to "get to the very essence" but to reach an unbreakable confidence in being right was particularly strong. Thus the delay of publication until full clarity was reached. Thus a multiple rewriting of fragments of the forthcoming lecture as described at the very beginning of this article. As I have already written I dare to think that this could be considered as a distant consequence of the mistakes of his younger years and a strange undue opinionated familarity in a tone of his polemics with Rayleigh, Planck and Bragg's. For a person with such a sensitive and vulnerable nervous system, which was a characteristic of L.I. (although only the most close to him new about it, for others in his Moscow years he always seemed to be calm and self-assured), an evaluation of this behavior in his young years should have meant a lot.

To all this there added external, societal conditions of existence at those times — from the lack of instruments to concerns, to worries of "Uncle Lyova". Raman did not have such problems, hence his unrestrained strive for a Nobel Prize. A completely different person. *Not* a Russian intelligent.

After a discovery of the combination scattering for Mandelstam and his school which was rapidly expanding by attracting, as it turned out, quite broad circles of talented young people, there came at first sight happy times. In all the branches of physics that interested him the work was progressing and each of them could be committed to some of his brilliant disciples. In addition, as already mentioned, in 1930 B.M. Gessen, who took care of

Leonid Isaakovich, became a director of NIIF and a dean of the physics faculty. Accepting his support was not shameful. I listened to his lectures on philosophy of natural history when studying at MSU in the beginning of the thirties. They were distinguished by their high level and a definiteness of thought. Lectures of stupid staff "diamatchiks"[18] were of no comparison to them.

This same period was characterized by a constantly growing autocracy of Stalin, intensification of terror and ideological pressure. It was from 1930 when Stalin was called nothing else but a great leader of all working people, ingenious, sage, etc. An atmosphere was building and the murder of Kirov on December 1, 1934 became a signal for a beginning of "big terror" in the face of which there faded everything, terrible enough by itself, that had happened before. Almost on the same day came changes in a criminal code and a code of practice which are impossible either to forget or to underestimate. In cases of terrorist organizations to which almost all arrests were tied to it was now prescribed: investigation should be completed in the shortest time possible, within several days; cases could be heard in the absence of the accused (this had been a case in Russia before the reforms of Alexander the Second); appeals or pardons were not allowed. Death sentences should be carriesd out immediately. The newspapers published lists of hundreds of sentenced and executed. And how many were executed without an announcement!

As was common in those times when in 1936 Gessen was arrested in the University there began meetings at which staff members, especially those close to him, had to confess a lack of watchfulness (they had failed to recognize an enemy!) and absurd "facts" of his sabotage activity. Few could keep their human dignity in this atmosphere of horror (as could, for example, G.S. Landsberg, see below p. 238). Mandelstam, who was about the only one who did not attend these meetings (at least I just can not remember him present; he did not like meetings and sessions but these were, of course, singular cases), was probably pardoned for doing so, so big was a respect for a man who breathed new life into university science which had been falling into decay for years before him.

Then came a horrible epoch. Other arrests followed. There disappeared young and very talented disciples of L.I., S.P. Shubin, (who was also a disciple of Tamm) and A.A. Vitt who, in co-authorship with A.A. Andronov and S.E. Khaikin had just finished a fundamental work summarizing the results

---

[18]Lecturers in dialectic materialism.

of joint research with L.I. on theory of oscillations, especially nonlinear ones, for which new methods of studying numerous problems of practical importance were developed. In particular Andronov introduced a notion of "auto-oscillations". This was a new breakthrough in one of the most important domains of physics and gave birth to a school of Andronov that formed in Moscow and Gorky. The book, however, could not be printed with the name of "the enemy of people" Vitt on the cover, but not to publish it was a crime against science. So a heavy moral sacrifice, to leave only the names of Andronov and Khaikin on the cover, was necessary. The fact that these highly moral people and L.I. decided to do so (it was without any doubt a sacrifice for them) shows how necessary this book was! After the war it was translated in the USA (it seems to me, without the author's permission) and after Stalin's death was reprinted with Vitt's name restored (more than 20 years after the first edition which shows by itself that this was a classical work that remained important for a long time). This is also a characteristic episode from the history of both our epoch and the Mandelstam school.

In spite of all, even with "a noose around a neck", the school of Mandelstam developed and worked.

Nobody could say, however, for how long they could have endured all this if an unexpected even had not taken place. In Moscow there appeared a new scientific center in physics that turned out to be an oasis, a salutary pillar for Moscow physics.

The thing is that in 1934, according to government decision, the Academy of Sciences and many of its scientific institutions that since the times of Peter the First resided in Petersburg-Petrograd-Leningrad were transferred to the capital, to Moscow. Among those was a small (about 20 researchers and Ph.D. students) Physics Institute of the Academy of Sciences, FIAN, that had separated from an Institute of Physics and Mathematics. The director of this institute was a young academician Sergei Ivanovich Vavilov. He immediately invited to the institute the best Moscow physicists, first of all Mandelstam, his main collaborators, his school (but certainly not only them). They kept their university positions and also became heads of the main laboratories of the new institute or researches in them (see an essay Section 4.3 in this book). There arose an institute (the number of staff grew about 10 times) where there dominated an atmosphere of devotion to science, decency, intelligence, good will, mutual help and cooperation.

My usage of the word "dominated" is not occasional. In the reality of our country in those days it was impossible to be completely free from the pressure of evil characteristic for its life. But the most clever organizer Vavilov did not only take a blow by himself when it was necessary. A younger "helping" and "directing" party section that was at those times mandatory in such institutes was selected to include, in particular, several (three or four) furious supporters of the regime. A distinguishing feature of these people was however that they really loved science, were in some cases talented physicists and already because of this could not but respect Mandelstam, Landsberg, Tamm and others as scientists. Therefore their fury could not completely spoil a general style even in most difficult situations when, in the epoch of "big terror", they (as it was in the university and basically everywhere) attacked their scientific mentors with accusations that were at those times dangerous. This left Vavilov a possibility of softening all the blows and to not allow an establishment in the institute of an atmosphere of baiting which these scientists experienced in the MSU where the same people that had previously fought against inviting Mandelstam and their disciples were now in power.

Very soon all the "Mandelstamovists" transferred their main scientific activity to FIAN (for example, Tamm moved his weekly seminar there; it was FIAN where, together with N.D. Papaleksi, Mandelstam launched large-scale research on a new subject of radio-geodesy, propagation of radio waves around the Earth, new methods of radio-location, etc.). In the university they basically limited themselves to lecturing.

Here we should pause to discuss how Mandelstam was lecturing.

Perhaps only the first course on field theory given at the very beginning of his work at the university (1926–1927) could basically be described as a usual university lecture course (although these lectures bore a seal of an individuality of L.I.). The rest of his lecturing and seminar activity during 20 years (1925–1944) was completely unusual. These were not courses that ended by exam and they did not conform to a usual set of university courses. Mandelstam simply chose some problems, domains of physics, that were in his opinion relevant, contained unclear points or were not covered in the literature with a sufficient depth and were of great importance for understanding of physics in general. Often it was a "horizontal" scientific slice. For example the biggest course "Lectures on oscillations" (1930–1932) covered theoretical consideration of general and specific properties of oscillations in many different domains of physics, in hydrodynamics and electrodynamics, in mechanics and optics, even in quantum mechanics. They constitute the thickest, fourth volume of his collected works. The most important thing in the character of his lectures that often included research results of L.I.

himself (without mentioning it) was that in these lectures pedagogics was inseparably tied with scientific research in general. A wonderful characterization of these lectures is given in [Prokhorov (1979)], see the Introduction. We will just simply quote from it.

Lectures by L.I. were a bright and non-disguised demonstration of a process of physical thinking as such. There one saw how a physicist stumbles against an obstacle, how paradoxes and contradictions accumulate on the way and how he manages, sometimes through a mental feat, through pushing aside the habits most rooted in human consciousness to free oneself from contradictions and reach a new previously unreachable height from which there open new horizons. Not a single detail in the lectures of L.I. was vapid, lifeless; in each problem he was able to find and bring to the audience some special sharpness and beauty. Not only did he force by the perfect logic to agree with his statements but tried hard, with success, to find a common language with his listeners, convince them "from inside", remove those difficulties formed due to psychological protests that so often hinder understanding in physics. Taken together all this created some unusual emotional richness and ensured that everything heard from L.I. reached the deepest layers of consciousness.

When Mandelstam analyzes at a lecture a scientific question he first casts doubt on everything, shakes loose usual understanding. Nevertheless this generates firm knowledge. Here is, for example, a problem of a Heisenberg microscope used to justify an uncertainty relation (vol. V, p. 396): "I would first criticize what is usually told about the microscope and what always shocked me '... each word in this reasoning is, in my opinion, incorrect (although in essence it is the case)', writes Mandelstam and then completely destroys the usual line of reasoning. Mandelstam mentions at least three flagrant absurdities. When they are pointed at one feels awkward for not noticing them before. Having seen any of them an ignoramus not possessing an ingenious intuition of Heisenberg has a right of declaring that the uncertainty relation is not proved or, at least, can not be proved *like this*. Then, however, Mandelstam gives a correct proof for the same microscope and there opens something more essential and it becomes clear why he did confidently tell that "in essence it is the case". Mandelstam's confidence was always based on his own, well thought, judgement.

As a physicist who had been growing in parallel with relativity theory and quantum concepts he, it would seem, had more reasons to loose his head from the new miracle and could have praised the science of twentieth century as something fundamentally different from all the previous physics than the scientists of forthcoming generations. Even now, half a century later, one often encounters an opinion that this physics is something totally special. One refers to a supposedly new rejection of a principle of intuitive obviousness, etc. How weighty there sound the sober and exact words of Mandelstam:

... *principles* of constructing quantum theory or, if one might put it so, a structure of a frame with which the quantum theory is framed by, is the same as in any other physical theory. However one can not deny that a structure of the picture itself is quite different from the classic one and a statement that here we are dealing with a new physical paradigm could hardly be considered as an overstatement (vol. V, p. 402).

This statement is then substantiated in details. Mandelstam explains what "intuitive obviousness" means and shows that its rejection and a change of paradigm took place at each turning point in physics, in particular when a Maxwell electromagnetic theory established itself.

This somewhat old-fashioned and, in the eyes of my generation, soft person spoke for many decades to come.

To these lectures and also to seminars (at which L.I. always gave an introductory lecture) there came, as was already mentioned, different listeners, from students to academicians, sometimes from other cities. They were catching every thought, many made notes. The texts of lectures at some seminars are, however, lost irretrievably.

Seldom there were also other lectures — talks. So on April 28, 1930 (let us parenthetically mention an unrelated significant fact: Landau had been arrested at previous night) L.I. gave a talk at a general assembly of the Academy of Sciences on a seemingly very special subject, a radio interferometry, i.e. on measuring distances on the Earth with the help of radiowaves (as was told above large-scale research on this and related topics was launched by L.I. together with N.D. Papaleksi and a large group of co-workers at FIAN). It is clear to everyone that this can be of practical importance, but how could this talk be of interest to the general assembly of academicians including those specializing in liberal arts, chemistry, biology[19]? Nevertheless after the talk a mineralogist academician A.E. Fersman summarized his impression in one word "Poem!" (and then sent Mandelstam a short exalted letter [Prokhorov (1979)]).

---

[19] I.E. Tamm told me that after one of his talks at a general assembly an academician in liberal arts told him: "From your numerous referrals to some beta-rays I conclude that there also exist alpha-rays and, possibly, gamma-rays as well."

On September 26, 1943, also at a general assembly of the Academy devoted to the eightieth anniversary of academician Aleksei Nikolaevich Krylov, a mathematician, mechanician, ship-builder, translator of Newton from Latin, engineer, an amazing person (L.I. had become very friendly with him during the wartime two-year evacuation of a large group of old and ill academicians to a resort, Borovoe, in Kasakhstan). L.I. gave a talk "On scientific works of A.N. Krylov" on a subject seemingly very distant from his domains of physics. But, as told about Mandelstam by A.A. Andronov [Prokhorov (1979)]: "For him there were no locked doors in the huge building of physics".

Is it then surprising that, when congratulating him on some occasion in 1944, a philologist-sinologist academician V.M. Alekseev adds:

On many occasions I had a pleasure of sharing with you my limited opinions on my unlimited admiration of everything in your talks and speeches that was understandable for me. You are probably belonging to the rarest kind of scientists that profess and preach science as a clear, not cumbersome, thought and treat its difficulties as circumstantial and not essential [Prokhorov (1979)].

This was told not without ground. Back in Borovoe, Alekseev had listened to the talks by L.I. "Optical works by Newton" and that on the works of academician A.N. Krylov.

Academician P.L. Kapitsa was answering questions on different scientists and when it came to L.I. he exclaimed: "Oh! This is an aesthete!" [Prokhorov (1979)].

Andronov, Rytov and others underlined that an emphasis on a logical structure of a theory was an important feature of his lectures. Andronov thinks that a "watchful and consistent attention to questions of epistemology" belongs to most characteristic features of L.I. He was interested in "how do physical notions arise, develop and transform, how are they related to objective reality ... From his lectures and statements it was clear that he studied a logical structure of physical theories in depth" [Prokhorov (1979)]. Therefore his purely physical lectures possessed a philosophical flair.

At that time physicists did not completely realize the importance of a simple fact that experience is necessarily limited. Conclusions therefore can not be considered as universally applicable. Indeed, after many repetitions and variations of an experiment producing results conforming to a conclusion there always comes a moment when a researcher should say: "Enough, now *I am convinced* that these results reflect a true property of nature". This "I am convinced" is, however, an act outside logic and therefore its unbounded unconditional applicability is not guaranteed. Same is true for conclusions of a "collective researcher" when, based on experimental data, science recognizes a validity of a law of nature, of mathematical axioms, etc. Each natural mathematical science arises through consecutive logical construction based on a *out-of-logic intuitive statement* adopted in an above-described way. It is synthetic by its very nature because of being made based on taking into account knowledge of different sorts, half-knowledge, estimates, guesses, etc.

Physicists either did not realize or did insufficiently realize that scientific knowledge is necessarily built as a combination of logical and out-of-logics elements (this is meant when one says that a criterion of practice is always relative, does not have an absolute validity). Only due to this, only because earlier out-of-logic established principles (axioms, laws of nature) are not unconditionally valid for all times and can be changed when new facts appear, can a development of science consisting in finding out (based on new experimental knowledge) more general laws, for which previous knowledge turns out to be a particular case valid in specific circumstances, take place. This more general understanding formed (although by no means becoming universal) only in the twentieth century (see, e.g. [Feinberg (1992)]).

At the beginning of the twentieth century an overwhelming majority of physicists and mathematicians thought that a presence of out-of-logic elements in their sciences was an evil from which they ought to and could be freed. In mathematics this conviction was shared by such people as a great mathematician David Gilbert and a philosopher and mathematician Bertrand Russell. In physics this direction of thought became dominant basically after Mach. Such tendencies backed by a powerful development of mathematical logics were very useful in stimulating a detailed analysis of notions that science operates with, of definitions introduced. Therefore, in particular in the lectures by Mandelstam (especially in those on relativity theory and quantum mechanics) such great an attention is paid to a question of defining the notions and restrictions thereby imposed on them.

On the other hand this movement led to a domination of positivist views of various kinds. Einstein, who had at first also been under the influence of Mach, soon parted from him. It is most clearly seen from his discussion (in 1931) with Rabindranat Tagore [Tagore (1931)]. Tagore's perseverance finally causes an openly strong irritation of Einstein. Not going into detail, he just insistently repeats phrases like "... this table will stay on its place even when nobody is at home" [Tagore (1931), p. 132]. This thought of his is an example of an *out-of-logics intuitive statement*. One can neither prove nor disprove it. Nevertheless Einstein just accepts it as a reasonable conclusion from experiment accepting thereby that knowledge necessarily includes out-of-logics intuitive statements. What about Mandelstam? In his published works there is nothing on this. However, about 20 years ago (I do not remember it more exactly), his son Sergei Leonidovich handed me three school copybooks clumsily sewn with white thread. In them, written by the handwriting so familiar to me, one finds a very simple exposition of the philosophical views of L.I. on the above-discussed issues. It was written in one of the last years of his life in the wartime evacuation in Borovoe. It was thus like settling the accounts. Several typewritten copies of these copybooks had been made and they had been kept in absolute secrecy up until new times came.

It is in these copybooks where L.I. says that a physicist can not avoid considering philosophical questions. He continues with saying that an understanding of objective reality should have its origin from those elements that constitute facts one can not doubt. To his opinion such facts are in our experience and our feelings (more often L.I. uses the word experience). Here one should not ask what these are. This is a basic element with the meaning clear to any normal person. We do not have a ground to speak on "taking place outside us" as of some material reality, it is not given to us, only a combination of feeling and experience is. *Only this set of feelings can be considered as objective reality*. Correlation of these feelings is being studied by us and helps to establish what we term 'laws of nature'.

After all, Einstein, in a phrase preceding the above-quoted one says: 'Even in our everyday life we "have to" ascribe to things we use the reality that does not depend on a person. We do it in order to establishe in a reasonable way an *interrelation between the data provided by our senses*. (NB. Author's italics). This is, of course, another intuitive statement. One has an impression that a difference with L.I. is in that L.I. does not think that "we have" to do it and so does not do it. Therefore Einstein is a materialist, although on the preceding page he says: "I can not prove that my concept is true but this is my religion". Here the word religion does not, of course, have anything in common with a standard notion, Einstein uses it purely metaphorically; on the principal distinction between the religious faith and the trust into an intuitive statement in science see [Feinberg (1992), Ch. 6]. One can not say the same thing about Mandelstam. In his words when we say "a tree" this makes sense *only* as a short "shorthand" notation for the complex of corresponding feelings.

Nevertheless, in spite of an aspiration of cutting away all the out-of-logic statements, L.I. could not avoid them. He says, for example, that complexes of feelings of different people coincide. This can be judged from the coincidence of external manifestations of these impressions. A number of external manifestations is however always limited and we still have to make an intuitive statement on the sufficiency of a set of reactions in question and this introduces an out-of-logic element.

This is of course not an appropriate place to consider the manuscript by L.I. containing many sophisticated, instructive and very interesting arguments in more detail. For example, the following phrase is worth quoting: "Not only do I not deny the existence of the external world and its reality but... give it the above-mentioned *definition*" (in terms of complexes of feelings). Yet one should say that a feeling of dissatisfaction does nevertheless remain.

Similar to Einstein discussing a table that remains in the room in his absence, L.I. discusses, among possible objections, the following. One says that from defining "a tree" as a complex of feelings there follows that "objects of the outside world discontinue their existence as soon as we turn away from them". This is a "misunderstanding". "The complex of feelings that I call a tree contains a feeling-confidence that if I turn away and after that do not hear, say, the sound of a falling tree, then, when I turn my head back, I will see this tree again. *This* feeling constitutes an *important* part of a notion of a real tree". Here everything seems to be consistent. Still there arises a bewilderment. The feeling-confidence is of a principally different nature than the feeling. It is a product of a functioning consciousness that joins together some limited number of indirect indications of the non-disappearance of the tree, each of them individually not constituting a proof. The confidence arises as a out-of-logic *synthetic intuitive statement*. One can not avoid an out-of-logic *statement*. Is it not simpler to use it from the very beginning, as Einstein does, without introducing a feeling complex. At the beginning of the twentieth century all this was not sufficiently clear to physicists and mathematicians. Yet a rapid development of mathematical logic led, already at the beginning of the thirties, to the most important result: A mathematician Goedel *proved*, using the apparatus of this discipline, that it is impossible to banish out-of-logic elements from mathematics. In the course of the development of mathematics there necessarily come moments when it is necessary to choose, in the out-of-logics way, one of the possible directions of its further development. It is *proved* that the development of mathematics will be characterized by an *infinite* number of such moments. A further study of foundations of mathematics showed, by the end of the century, that it is rich with arbitrary ("intuitive") definitions of notions that can not be logically proved. Some authors considered this to be a catastrophe (see [Klein (1984)], others came to a different understanding of mathematics as the the same science as physics [Klein (1984)]), even as a part of theoretical physics based on an out-of-logics generalization of experimental data [Arnold (1999)].

L.I. formed as a thinker at times when all this was alien to the spirit dominating in science, in the hazy atmosphere of competing ideas. There is nothing surprising in his exposure to this atmosphere, basically inclining to one of versions of positivism. Nevertheless in his lectures he never speaks about this position. At that time it would have been a suicide. Soviet philosophers were violently chasing any "deviation", of positivist type, from the official ideology. This fury can be characterized by paraphrasing the well-known command of a guard in a Stalin camp: "Step to the right, step to the left is considered to be an idealism, shoot without warning". One philosopher dared, after Stalin's death, to tell: "There practically existed a cause-effect relation: ah, idealistic deviation, therefore "popovshina", therefore an "enemy of the people" — arrest — camp — an end". These philosophers, like hounds, had good scent. They were seeking in remarkable lectures by L.I. suspicious places and pounced on them with ferocious energy. In 1950–1953 there were several populous meetings in FIAN specially devoted to "ideological mistakes of Mandelstam and his disciples". In spite of the statements, sometimes harsh, of these disciples (L.I. himself was already dead) in defense of "suspicious points", menacing and at those times dangerous resolutions were adopted. This was the time when many branches of science — genetics, cybernetics, physiology, — were smashed (in addition in 1951–1953, anti-Semitism reached its peak).

Fortunately the "scent" alone of the persecutors was not sufficient. All more or less qualified Marxist philosophers had been killed a long time ago and the existing ones had a poor knowledge of both physics and philosophy. One could argue with them.

For example, after the war there began the preparation of the five-volume edition of works of L.I. A careful preparatory work (a titanic labor!) was carried out by the editor-in-chief S.M. Rytov. The main difficulty was in publication of lectures. Only a minuscule number of lectures were stenographed, but these were neither read nor proofread by L.I. himself. Everything was based upon the notes (very detailed) of Rytov himself and also of Andronov and many others. One had to compare different notes.

Lectures and seminars constitute the last two volumes. After publication of the first three volumes the printing was stopped by ideological bosses. Nevertheless it turned out possible to resume and complete the publication. On the one hand this was due to the fact that a deep admirer of Mandelstam, S.I. Vavilov, was a President of the Academy and used his influence. On the other hand this was due to some tricks: Firstly, Rytov slightly edited some places causing a special fury of philosophers (this was possible because, as was already mentioned, shorthand notes were effectively absent and, with a knowledge of a true opinion of L.I., it was possible to slightly change the text but preserve the meaning put in by Mandelstam). Secondly, it was written that the editor-in-chief of the last two volumes was not Rytov but M.A. Leontovich (he was more suitable, perhaps because of being an academician and not being a Jew — unlike Rytov).

So, here is a depressing fact. An outstanding scientist that made so much for science itself and for creating a huge school of leading scientists of our country, that thought so much on fundamental philosophical questions and developed certain views in this field did not even dare to openly hint at this, had to keep them in absolute secret, although this was about science, not terroristic preparations. Terrible evidence of a horrible time. A "loop on the neck" did not loosen even if a scientist managed to survive.

Now, what was the attitude of Mandelstam himself to what was going on in the country; what was his social and political position? How did he behave himself in this horrible and complicated epoch?

We saw that in the life of L.I. there were several well-separated periods. First, a safe youthhood in Odessa, ending in participation in student political disorder in the university from which, for this, he was expelled. At the same time everybody who knew him in his mature Moscow period stress that he distanced himself from any type of social activities outside science. Others, almost everybody, had to yield to the humiliating norms of social behavior. Not all of them, especially in the twenties, did this out of fear. Yet almost everyone in the best case were "fellow travellers" of Soviet power, sometimes even its supporter, appreciating the positive development (general education, intensive development of science, rapid reconstruction of the economy that had been totally destroyed during the civil war, etc.). Nevertheless, already during the next decade, in the epoch of the "big terror" of the thirties and even earlier, more and more intelligents came to a sharp denial of the Stalin regime but did not show it, the fear "holding in leash", while the number of adaptive careerists did not decline.

L.I. himself was in political questions "completely buttoned" with respect to everybody save for the people closest to him and did not reveal his position.

Of course he could not escape the influence of his 14-year stay in Germany where, up until Hitler times, the scientific world traditionally stayed away from any kind of political activity and even from being interested in political questions. Academic life did not have anything in common with them (except for the problem of anti-Semitism which bothered, for example, Einstein). In Germany a mass protest of scientists against government actions like the one that in 1911 forced 150 progressive professors, indignant at the actions of minister Kasso, to leave Moscow University was impossible. This was however not sufficient.

In fact the political position of L.I. was extremely resolute and clear: he fully and sharply denied the Soviet regime and all the ideology and practice of social life enforced by the party.

Many Russian intelligents, in particular scientists, that already in tsarist times were "infected" by liberal and even socialist ideas found in the revolution and the societal structure that followed, with all its horrible features, positive sides. Some, like I.E. Tamm, participated before the revolution in political life, in revolutionary movement and, even when this participation had been over, remained faithful to some of the socialist ideas of their youth. Others, like S.I. Vavilov, had at first deeply accepted and highly esteemed the positive features of the new life and changed their attitude later. Third, like the young L.D. Landau, were charmed by high communist ideals and in the course of almost 20 years were ardent supporters of the Soviet power. One had to experience (often at oneself) the horrors of the "big terror" of the thirties in order to understand the essence of Stalinism and go into "internal emigration".

For Mandelstam everything was clear from the very beginning. In spite of participating in student disturbances in his youth, after experiencing life in civilized Europe and living through the Bolshevist revolt and horrors of the civil war in the Ukraine, he once and for all completely rejected the Soviet system. He had kept his disgust to it inside and only after its breakdown did we learn the strength of this disgust from those close to him. It turned out, for example, that in December 1922, I.E. Tamm had written about him from Moscow (where at that time, as mentioned before, L.I. worked in the trust of weak current plants):

Lidiya Solom[onovna] tells that his nerves are in extremely bad shape and finds it very worrying... By the way the disgust to everything bolshevistic, in spite of things going well for him, became for Leonid Isaakovich quite extreme, up to the fact that a necessity of sitting at dinner (in different corners and not talking) with a communist, with this communist behaving, in L.I.'s own words, very decently, caused a terrible migraine for the whole following night [Andreev (1998)].

Of course such an extreme attitude towards Communists, such a sharp reaction, were not characteristic of all years in Moscow. From Tamm's letter it is clear that his nervous system was at that time in particularly bad shape. Besides that, a party member and a participant of the civil war, S.E. Khaikin (later falling victim to the merciless "ideological" witch-hunting), belonged to his closest disciples and his attitude to him was very warm. He was in very good relationship with B.M. Gessen. Among his last Ph.D. students was a feverish Bolshevik Maksim Andreevich Divilkovski. (I do not think, however, that L.I. forgave his active, one can say, leadership in the attack of 1936–1938 on those close to the "enemy of the people" Gessen. Probably he, with difficulty, tolerated him because of professional capabilities and a serious attitude to work.)

In the everyday behavior of L.I. in all situations where political issues showed up, one could observe nothing but his complete absenteeism. He could say, as some Germans in Hitler times: *"Ohne uns!"* ("Without us!").

This total clarity of position, together with unshakable moral values of a Russian intelligent, in addition a European, helped L.I. to gain a psychological stability that overcame both the ever present elements of nervous sensitivity of his finely organized soul and the difficulties of existence of that time.

The last "Moscow" twenty years of his life can be considered as almost happy ones. An unusual intensity of his scientific and of the, inseparable from this, pedagogical activities (what were his above-described brilliant lectures, a scientific or pedagogical creative activity?), a fabulous growth of a number of disciples surrounding him, from Ph.D. students to academicians, their respect and love — all this protected his "temple" from the horrors of the surrounding world, softened its blows. The power of his thought did not weaken with the years.

Here is characteristic evidence by I.E. Tamm [Prokhorov (1979), p. 134]. As is well known, Einstein, who had introduced a notion of light quantum in 1905 and can thus be considered to be one of creators of the fundamental principles of quantum theory, considered the quantum mechanics worked out in 1924–1926 (still developing and fruitful up until now) to be incomplete in its foundations. To prove this he, in the course of several years, invented various experiments in which a quantum-mechanical treatment led to senseless results. There arose a paradox (see also a chapter on Bohr, p. 293). Discussion, both oral and written, occurred mainly between him and Bohr. A deep analysis, in particular for two last very difficult paradoxes, always led to their full clarification and after a reply by Bohr appeared in the journal the question was settled. 'Because of the features of his character L.I.', says Tamm, "did not publish anything on Einstein paradoxes but to us, his disciples, he explained their complete resolution a day or two after appearance of the next paper by Einstein". One has to mention that when telling this Tamm added that he and other people urged L.I. to publish his reasoning but L.I. only smiled and said that Einstein and Bohr are very clever people and this reasoning is surely known to them. (It is interesting to compare this behavior of L.I. with his own actions of his youth in the above-described arguing with Planck. It can be that precisely this unpleasant experience of his youth explained his behavior with respect to Einstein — Bohr discussion.)

Yes, almost happy years, with a correction "only" on them being filled with Stalinism having reached its climax, Hitlerism and war that probably took away only a little fewer lives than terror.

Moscow Zoo has a terrarium. In niches in the wall, separated from visitors by thick glass, one sees snakes. Here a gigantic, coiled-up python sleeps peacefully in the light of a low electric lamp hanging from the ceiling. And under the lamp there gather, to warm themselves in its light, mice for the python's future meal. They are really happy, because they do not understand the situation. For people of the twenties, thirties and later periods this lack of understanding was not that easy.

What about L.I.'s everyday life in these "almost happy" years?

In August of 1938 my wife and I did in a "savage" fashion spend our vacation in Teberda in the Caucasus. At that time there existed a, very popular among the scientists and indeed a very good, sanatorium of CAS, Commission for Assistance for Scientists (a governmental organization helping the scientists in both everyday and professional matters). In this sanatorium there spent their vacation L.I., Papaleksi and Tamm. Once, obeying to a burdensome tourist ritual, we went on a hike to some miserable sources. This turned out to be an uneasy enterprise. I was already quite exhausted when I saw an unusual picture. In the opposite direction, returning, rode on horses Leonid Isaakovich and Nikolai Dmitrievich. At that time they were both in their sixties and rode well. We were already sort of acquainted and the riders stopped smiling — it seemed to me, somewhat awkwardly.

It can be that recalling this encounter now, after many years, I am imagining something. It can be that Leonid Isaakovich did not wear leggings, did not have a riding-crop in his hand and a flat kepi on his head but this is an image I have before my eyes even now, that of a rider from some old pre-revolutionary photograph. On a natural question on how far do we still have to walk we got a reassuring answer: no, do not worry, not so far. We parted. At first we were heartened but we walked and walked and, in the heat, finally dragged ourselves to these sources only after about an hour and a half.

The next day, when visiting Igor Evgen'evich in the sanatorium, I met Leonid Isaakovich and wondered why had he cheated us. He answered with a disarmingly kind smile and an explanation: "If somebody is so tired, how could one tell him that he still has a long way to go?" One can also see another aspect of this: he probably could not imagine that one can have his intentions and give up reaching the set goal. We could have turned back though.

There passed only a little more than five years after this encounter but these were hard, wartime years. The health of Leonid Isaakovich declined, he did not leave his home, was sad. There remained less than a year until the end (November 27, 1944). Sergei Leonidovich told me once that his father's spirit could be improved by good music, for example, Beethoven quartets that he liked. At that time there tape recorders were not available and good vinyl discs were rare. My wife and I were very lucky: At that time she worked at the Conservatory and in her classes with students made wide use of a rich collection of vinyl discs of a gramophone recordings fund. She had no doubt that a supervisor of the fund would give her, for some time, any disc. It was less than half-kilometer between the Conservatory and Mandelstam's apartment. It remained to understand more precisely what Leonid Isaakovich would like to listen to. The answer, corresponded by Sergei Leonidovich, was unexpected: bringing discs was completely out of question. "It is inadmissible to use state property for private purposes. It can be used only by those for whom it is destined". It turned out this was an "iron", unbreakable rule.

As already told the Mandelstams lived in the apartment where one exit led directly into the corridor of the University Physics Institute. This door was used, all day long, by friends, colleagues, disciples of Leonid Isaakovich, by himself and his family. The apartment was perceived as a part of the institute. In fact an invisible border existed in this case as well. "Do you really think", Sergei Leonidovich explained to me, "that if our radio set does not function and one has to check voltage on one of the valves it is possible to bring, for a minute, a laboratory voltmeter? This would cause a real anger by the father. Principle is a principle — state property is untouchable.

A very similar case is recalled by a close acquaintance of the Mandelstams, I.O. Vilner [Prokhorov (1979), p. 2]:

Once when I visited the Mandelstams, L.I. told me that although he did not feel well he would like to go and see a tennis match. Because it was difficult for L.I. to go by public transport I tried to get a taxi but did not succeed and, after returning empty-handed, suggested to call the garage of the Academy and ask for a car. L.I. gave me such a look that I could have rid from embarrassment: "What are you talking about? To go to see a tennis match in an academic car? How could such a thing occur to you?".

Many decades passed. Much has changed around us. It turned out easy to replace an old-fashioned watch in the waistcoat watch-pocket by modern electric watch with a LCD-screen. However, the "old-fashioned" principles and high moral characteristic for L.I. remain irreplaceable and keeping them intact turns out to be an uneasy business.

The contrast, so striking for me, between the softness of his manners and the complete clarity, sureness of his statements showed up in everything — in life situations, in general moral problems, in science.

Moral norms that he undeviatingly followed were those of the Russian intelligentsia, Chekhov and zemsky doctors. They were shaken within intelligentsia in Soviet times. Even these norms were sometimes betrayed due to the dominating general fear or due to careerist intentions. Nevertheless, of course not only due to such master exponents as L.I., much was carried through to new times, including the whole great culture of the country. What will be their fate in the new century, in new Russia?

I. E. Tamm

Chapter 2

TAMM, Igor Evgen'evich (1895–1971)

2.1 Tamm in Life[1]

Igor Evgen'evich Tamm was born and grew up in the Russia of the last emperors whose reign seemed unshakable to almost everyone. Cinema and automobile were still unknown. A slow replacement of rustic torch by kerosene lamp presented a scientific and technical revolution. The radio had just been invented. Such novelty as a telephone set was installed in a flat of a minister of foreign affairs to provide a direct link to the Gatchina residence of the Tsar. The minister, however, could not use it and called for his nephew's help when necessary. The Russian army was in the process of getting the newest equipment — a trilinear gun. In the minds of intellectuals there reigned Tolstoy and Chekhov; artists — peredvizhniks and Chaikovsky; Darwin and Marx; Rhoentgen and Curie. Russia tried to make up for possibilities missed during the centuries of serfdom retardation. Civil consciousness was forming and revolutionary movement was developing. The great Siberian railroad was being built. The picture galleries were opened.

A stratum of intellectuals experienced a rapid expansion. On one hand these were the people of highest technical and scientific qualification often playing a prominent role in the state structures (I.A. Vyshnegradsky, one of the inventors of a theory of automatic regulation, was known not through this activity but rather for becoming an outstanding minister of finance; A.N. Krylov, a mathematician and theorist of ship-building, a translator of Newton from Latin, played a big role in building the modern navy). On

---

[1] Published in Feinberg, E.L. (1981). *Memoirs on I.E. Tamm*. Moscow: Nauka, reprinted in 1985 and 1995. The text is expanded with additional material.

the other hand, these were generations of devotees, zemsky doctors, rural teachers.

Tamm was born when a scientist seemed to be an oddity and the word "physicist" did not mean much. He died when a radio control at a distance of millions kilometers was not surprising to anyone, when a man walked on the moon and Igor Evgen'evich could hear from a little girl (my grand-niece) posing, like a grown-up, a justified question: "So what is special about it? As far as I remember myself people always were on the moon." He was a witness of two world wars, a grandiose revolution, totalitarianism with its horrible concentration camps, Hiroshima. He died in a country with a social structure as different from that of the Russia of his youth as a "trehlineika" gun differs from not even uranium, but the hydrogen bomb.

By the end of his life many millions of people on Earth had lost their belief in old moral values. At the same time Igor Evgen'evich himself already belonged to a category of scientists — physicists who were deified and cursed, respected but also rejected. Often it was considered as an embodiment of a culture antagonistic to an old humanism but romanticized by young people.

Can one sustain such transformation of the surrounding world and protect oneself as a person? What should happen to one's character, beliefs and views on life when in his childhood he rode horses and lived in a dusty provincial town and in his senior years flew, in a few hours, to another hemisphere to discuss possibilities of preventing a thermonuclear war?

It seems that most characteristic of Igor Evgen'evich was precisely the fact that his attitude towards life, people, science and himself had already formed in his young years. It stayed unperturbed through all transformations happening in the world, through all changes in his personal life, in joy and sorrow, in an atmosphere of negligence and in that of worship. His firm position in life served as a basis for developing spiritual and practical activity, not stubborn stillness.

I was happy to know Igor Evgenjevich for almost forty years, until his death. I heard a lot on the preceding years from him and those close to him. But only recently, when I had the possibility of reading his letters from 1913–1914 to his future wife Natalia Vasiljevna Shuiskaya did I understand how conscious, purposeful and sometimes painful was the formation of his personality. Already at an age of 18 did this young man understand what one should, in his opinion, be. Already at that time he was essentially the forty-year-old man that I met.

The main features of this person were the best of those characterizing Russian intellectuals of the beginning of the twentieth century.

This remarkable layer of society was far from being homogeneous. I have already mentioned such seemingly distant groups as highest technical intellectuals and zemsky doctors. There were also those belonging to an "underground" for whom revolutionary struggle justified life. Should we discuss something that is everybody's knowledge? There were rich, fashionable and often really talented engineers, doctors and lawyers. There were hectic enemies of the "repleted" turning to religion and antiquity, symbolism and futurism, terror or martyrdom. It is impossible to list them all. There appeared poets, determined revolutionaries and practical engineers convinced that the most essential thing was to build, create, make something useful for people.

With all this diversity there existed however the basic, most important and reliable — the middle-class working intelligentsia with its firm foundations of spiritual world. The most expressive representative of this very intelligentsia was Chekhov, not the "weeper" seen in him by some short-sighted contemporaries, but an active and perfectly truthful, sensitive but not tearful and sentimental, sophisticated, reserved, even reticent in expressing personal feelings, but also cheerful. The Chekhov seen in his writings and letters so well described by K. Chukovski.[2]

Igor Evgen'evich as a person comes precisely from this intelligentsia — with all its merits and drawbacks. The most important feature of his character was perhaps an internal spiritual independence, in large and small, in life and science. It was by no means accompanied by pugnacity, dissidence, protest for the sake of a protest or scoffing which often replace a thought-of firmness of a position. Already in the letters from Edinburgh Tamm was writing a lot, albeit in a youth-like manner, on his conviction that independence is the most important thing for a person.

> We all care too much about opinions and thoughts of others... To me people differ from these bushes between a door and a gate... only in that bushes need a gardener to cut them while a man, equipped with reason, learned himself to cut off... features of the character that do not match a standard. Everybody possesses such features, not only "remarkable" people.

---

[2]Chukovski, K.I. (1962). *Contemporaries, Molodaya Gvardia*, Moscow, p. 704 (in Russian).

As soon as you have these branches cut off, your happiness flies away from you", wrote Tamm, feeling himself somewhat shy of his "depth of thought".

Igor Evgen'evich dreamt of devoting himself to revolutionary activities. On a way to this goal he faced, however, the resistance of his parents. His father, a "city engineer" in Elisavetgrad (he was building a water-supply system, an electric power station, etc.) was overburdened with work and Tamm's mother's argument was that his sick heart would not stand if something would happen. Igor Evgen'evich rushed about in fury being unable to take on himself such a responsibility. A temporary compromise was achieved when in 1913 he went to study in Edinburgh to, in his terms, a "voluntary exile" (to be far from revolution). However there he, keeping it a secret from his parents, established contacts with socialistic groups of different kinds, with Russian political emigrant organizations, and studied life conditions of the poor. Igor Evgen'evich did not want to be an engineer because, as he once wrote in a letter, "to be a factory engineer definitely means to be against the workers. Perhaps I will at some point of time distance myself from politics — this is, unfortunately, possible, — but will by no means change sides and will not fight against my present comrades".

The pause did not change his intentions and when in February 1917 there burst a revolution he dived into it with all his passion. He spoke, with success, at mass meetings against continuing the imperialistic war, he was a deputy of Elisavetgrad Soviet and a delegate of the First Congress of Soviets in Petrograd. He always followed his anti-war line. He was a Menshevik-internationalist.

This was most organic for Igor Evgen'evich. With all the spiritual independence characteristic of him, he was, nevertheless, a son of his time and belonged to a vast layer of Russian intelligentsia having a characteristic "guilt feeling" with respect to the humiliated, half-hungry working class. This feeling essentially underlies the Russian literature of the second half of the nineteenth and the beginning of the twentieth centuries, from Tolstoy and Dostoevsky to Alexander Block, who wrote in 1911[3]:

> At life's impenetrable horror
> It's time to open your eyes.

A Russian intelligent could not neglect such "life's horror". The question was only in *how* to overcome it. He could not forget about this question

---
[3] Translation into English by A. Leonidov.

and live peacefully knowing that a bottle of champagne consumed with friends was worth a horse or a cow that could save a rural family from poverty.

Different solutions were possible — from that of Tolstoy to that of Block ("Let the righteous rage ripe, prepare your hands for work..."), an actively revolutionary-terroristic, for narodovoltsy and esery; propagandistic, preparing for revolution, for social democrats. Or evolutionary, a patient active participation in improving life conditions as chosen by Chekhov, zemsky activists and many others.

The character of Igor Evgen'evich was such that his choice could have been foreseen. He found a rational basis for his choice in Marxism and was true to its principles until the end of his life. He remained on basically Marxist positions even when, on the one hand, noble ideas were used by "daemons" in building, before his eyes, an awful totalitarianism, and, from the other, capitalism itself did, somewhat later, essentially changed by adopting some most important elements of the socialistic idea such as a responsibility of state and society for the personal well-being of citizens (in old age, in sickness, etc.), an essential role of the state regulation of economy, a state responsibility for ensuring personal rights and freedoms, etc. In other words, the capitalism of the nineteenth century was in some sense transformed into the capitalism "with a human face".

Igor Evgen'evich clearly saw all the horror of the Stalin and after-Stalin periods. He suffered in these times himself, but even a year before his death, when we discussed all this with sorrow, he said: "Yes, but it is impossible to deny that an economy is built on socialist principles".

However, as a witness of October fights in Moscow, he was horrified by what the Bolsheviks did and wrote about this in a letter to Natalia Vasil'evna.

After the October revolution Tamm distanced himself from political activity. The years of civil war were not helpful for scientific work. He graduated from Moscow University, taught for some time together with many outstanding scientists at the Crimean University in Simferopol. Only later, after leaving for Odessa, where he met L.I. Mandelstam, could he, with all his passion, devote himself to science. It is not surprising that his first scientific paper was published in 1924, when he was 29. Nowadays his chances of becoming a theorist would be estimated as negligible. The formation of Tamm as a scientist in these years was clearly influenced by his older friend, worshiped by Tamm until the end of his days, the outstanding physicist and deep thinker, Leonid Isaakovich Mandelstam.

In the middle of the thirties Igor Evgen'evich once told me: "I think that if Pushkin lived now, he would become a physicist". And, having read by heart the verse "Motion",[4] added with admiration: "What an understanding of a relativity of motion, of an uncertainty of evident!" However, it was he who, twenty years later, said that there comes an epoch in which a major role will belong to biophysics and biochemistry and that if he were young he would become a biophysicist.

A period between the end of the twenties and the end of the thirties was marked by a powerful ascent of his scientific activity. Works on quantum theory of optic phenomena in solids, quantum theory of metals, relativistic particle theory, nuclear forces, Vavilov-Cherenkov radiation, cosmic rays followed one another, all being important. They gave him a feeling of personal satisfaction and wide recognition in the scientific community.

This very time was, however, darkened by a growth of tragic events in the country and in the world. For Igor Evgen'evich, a social optimist, a believer in steady humanistic progress, this was a source of heavy emotional feeling. The tragic events also directly intruded into his personal life. A merciless totalitarian system kept him under unremitting control. Here is tragic evidence for this. During the period of collectivization and famine, perhaps most awful in Ukraine, Igor Evgen'evich took into his family a nephew, a school boy from Odessa. In 1934 the boy was called to "organy" and was demanded to report on everything happening in Tamm's family After returning home, the boy told everything to a grandfather (who was at this time alone in the apartment) and took his life by throwing himself from the window.

In two or three years the beloved younger brother of Igor Evgen'evich, a well-known engineer, was arrested and forced (one can only guess by what methods) to present false evidence on a public court process in 1937 (of Bukharin, Pyatakov, etc.) and was afterwards executed. Some other people

---

[4]Of course, it is known to everyone, but let us still give it here to make clearer what did Igor Evgen'evich have in mind:

> There is no motion, said a bearded sage.
> Another one stood up and began silently walking before him.
> Everybody praised this elaborate answer,
> But, gentlemen, this curious episode
> Does bring another situation to my mind:
> There walks the Sun before us every day,
> But still was right the stubborn Galilei.

(Translated into English by A. Leonidov.)

very close to him were also killed: a childhood friend B.M. Gessen and an extraordinary talented student and friend S.P. Shubin.

How was it possible to bear all this and even continue a creative work? Probably because it provided a saving shelter. The emotions caused by all this could be guessed from seeing how more and more wrinkles appeared on his face of only forty-years old, how seldom did he laugh and how deep was his concentration, together with the same energy of motion. Thus began to shape his appearance for years to come.

Life was not such that Igor Evgen'evich could allow himself to avoid from expressing, on a number of burning issues, his civil position. In arising philosophical discussions on problems of new physics Tamm was incessantly fighting for its correct understanding. He did not fear heavy, even dangerous at those times, but unjustified accusations in idealism. Tamm's position was on the whole easily vulnerable with respect to attacks. However, he did not do anything to contradict his own understanding of decency, but potentially allowed to make his life easier and "improve his reputation". He continued caring about friends and relatives in peril and, most importantly, remained the same Tamm for whom, with all attention and even respect towards the opinions and authority of others, his own opinion remained most important for him.

His personal independence also showed itself in his organic atheism. In his youth years this atheism showed itself in almost childish quarrels with a teacher of religious law, a priest, but when Igor Evgen'evich grew up he developed a full spiritual clarity with respect to this problem. Of course, he never showed even a slightest disrespect to believers, but was never able to understand how could a person entrust to somebody (even to a "higher creature" or its representatives on Earth) to establish the norms of his own behavior. Moral foundations of life should be worked out by everybody individually and be well preserved. If believers could offer him some ideas, he was ready to listen, but a decision on their appropriateness and value he left to himself. His personal conclusion was already made at a young age.

Now on another field: when at the end of the fifties and the sixties there appeared a Paguosh movement, Tamm, understanding how little could be expected from it, how much naivety and hypocrisy was brought into it, considered nevertheless an active participation in it as his duty. He thought that if a minuscule usefulness was possible, one should not turn away, even if many would laugh at the attempt.

Tamm's independence also revealed itself in other very different issues.

For example, when his seventieth birthday was approaching, there appeared an idea of presenting him a sculptural bas-relief bust of Einstein. To understand whether he would like this present his two younger friends invented a "clever trick". Not saying a word on the present they invited Tamm to visit together a workshop of the maker of the bas-relief. This bas-relief was made in quite modern a manner, and this appealed to many, including these two friends. The taste of Igor Evgen'evich in art had however been formed 15–20 years earlier and although, when once listening to the music of Shostakovich, he took it seriously and with interest,[5] and during the visit to the workshop he was attentive and concentrated, he did not like the bas-relief at all. In the workshop he was silent and serious and only after leaving it he shortly said, "No, I do not like anything there at all". He did not pretend to share the tastes of the young, as some wanting to be "up to date" would do. This independence of thinking and behavior was perhaps playing a decisive role in his scientific achievements. It happened so that on more than one occasion his works met a sharp critique from colleagues. Here are just two examples.

After the neutron had been discovered and it became clear that an atomic nucleus is built from neutrons and protons there arose a problem of making this model consistent with the, shortly afterwards, measured magnetic moments of nuclei. Already an experimentalist Baucher noticed that one could understand the nuclear magnetic moments if one ascribes a magnetic moment, the negative one, to the neutron itself. Igor Evgen'evich, together with his Ph.D. student S.A. Altshuler, analyzed the available experimental data and came to the same conclusion.[6]

Nowadays, when we are so accustomed to a picture of spatially extended hadrons with a complicated distribution of electric charges and currents it is quite difficult to understand why this was considered as an absurd heresy, still forgivable for an experimentalist, but shameful when expressed by an educated theorist. At that time it was considered indubitable that the only possibility to be consistent with the relativity theory is that elementary particles are pointlike and that neutron, having no electric charge, can not have a magnetic moment.

At an international meeting in Kharkov in 1934 where this work was reported, many well-known physicists, some of them famous, both foreign (e.g. Niels Bohr) and Soviet, attended. Tamm told me how softly, and even

---

[5] See recollections of V.D. Konen (*Three Episodes*) in the book mentioned on page 49.
[6] See recollections of S.A. Altshuler in the book mentioned on page 49.

with some compassion, these people who esteemed and loved him and, in turn, deeply esteemed by him tried explaining to him, in several languages, the absurdity of his conclusion. He listened to them with attention, passionately argued and could not give up his point of view: he did not see a convincing disproof. Later, it became clear that he had been right.

Twenty years afterwards when pion physics on accelerators in the energy range of one GeV had started its development, Fermi found that a scattering of pion on nucleon is of a resonant character. Igor Evgen'evich understood Fermi's result as an evidence of an existence of short-lived unstable particles. He was passionately carried by this idea himself, infected with his enthusiasm a group of young theorists and launched at the Physics Institute a wide-rangeing investigation of pion processes. Based on an idea of an existence of the resonance states in the pion-nucleon system pion-nucleon scattering, photogeneration of pions on nucleons and proton-neutron interaction were successfully considered. He called them "isobars". In Tamm's Theoretical Department of the Physics Institute an intense work was underway, isobars were a "hit of the day". Semen Zakharovich Belen'ky (who successfully applied "isobars" in studies of multi-hadron production at high energies in the framework of Fermi statistical theory) wrote a comic verse in which words were rhymed with the word "isobars" in every second line: 'Students, students, range in pairs, you will study isobars! Tary-bary, tary-bary, study, study "isobars"', etc.

It immediately became clear that a satisfactory quantitative description of all processes could be given only if one assumes that this state has a large resonance width of more than a hundred MeV, i.e. only a little below the magnitude of the level itself. This caused a sharp scepticism of some of our best theorists outside the Physics Institute (i.e. of L.D. Landau and I.Ya. Pomeranchuk whose personal attitude towards Tamm was very friendly).

Indeed, could one speak of such resonances as of something real? Calculations are done in approximate way (multipion states are not taken into account), "one can not believe numbers". A calculational work undertaken was however enormous, and Igor Evgen'evich had a good feeling from estimates of a stability of conclusions with respect to the assumptions made. And, with the skepticism and authority of the critics being as strong as it was, he did not yield to it and continued to insist on the reality of these objects. Some time passed, and the resonances (including the one he had begun working with) became fully acknowledged (and enjoying the full rights) members of a family of elementary particles.

It would be, however, completely erroneous to think that Igor Evgen'evich never agreed with his critics. His strength was precisely in thinking over, with full seriousness, an opinion of an opponent and immediately recognizing his mistake when facing a convincing argument. How many times did he disprove himself, how many times, having presented at an internal "Friday" seminar a result obtained, did he passionately and recklessly repent a week after. Walking hastily along a blackboard, eager to express himself, he cursed himself for "having told nonsense previously", that he was ashamed, etc. Sometimes this happened after one of the younger colleagues had privately pointed him out an error. Published was, however, only something he had re-checked many times and on which he was sure. I can not recall any published work that turned out to be wrong. There were one or two papers with very particular hypotheses that he himself considered to be shaky (he published them only to wait for a response of experimentalists) which turned out to be erroneous.

He remembered his errors better than some of his old works that were no longer of interest to him, remembered and did not conceal them. Sometime at the beginning of the sixties he told me about his political dispute with Bohr in 1934. They were returning by train from Kharkov to Moscow. These were alarming times with Hitlerism impending internationally. Bohr said that it could be withstood only by uniting all the anti-fascist forces, communists, socialists, liberals. He was convincing Tamm: "You do not understand, Tamm, this is necessary."

In essence, Bohr had described what soon came true in the National Front in France, in the Spanish Civil War, in the Resistance movement. Tamm, however, was of, at that time, widespread opinion, inspired "from above", that such a union would only weaken the anti-Hitler struggle. They argued for almost the whole night standing by the window in the corridor. With what a bitterness did he recall his (if it would have been only his!) blindness!

Igor Evgen'evich did not need to "squeeze out a slave from himself by drops" as did Chekhov who grew up in a terrible environment of obtuse shopkeepers and philistines. He could err, he could overly trust the habitual words that changed their meaning with time, but even subduing to the insurmountable, he was not a slave.

Igor Evgen'evich was a courageous person. He was brave also in a literary sense. He behaved himself with calmness and dignity at the battlefront, under bombing, during the First World War. Describing one such bombardment in the letter to his future wife he wrote with satisfaction:

"It is terrifying to hear, when standing in the open spot, an ominous hissing. However it is easy to control oneself" (23 May, 1915). During the civil war, when travelling between Crimea, Odessa and Elizavetgrad, he often faced authorities of all kinds (including such as "batka" Makhno). Tamm recalled the episodes when he found himself in mortal danger and only his self-control saved him. He was one of our most experienced mountain climbers, faced danger many times, but returned to the mountains again and again.

One episode showing his own understanding of courage is worth recalling. His son also became a mountain climber, of the higher rank (a distinguished master of sport), and frequently headed unique extremely dangerous climbs. I never heard Igor Evgen'evich praising his son, whom he loved very much, on any occasion. He never "boasted" about him.[7] He could only tell, in his usual rapid speech: "Zhenya[8] has returned, there was a winter traverse, difficulty grade 5b", the highest possible. I remember only one case when he could not restrain himself. A unit ("svyazka") from the son's expedition conquered the top of Khan Tengri (7000 m) and he himself, climbing in the next unit, waited for two days of bad weather several hundred meters from the goal and, after calmly estimating the situation, gave an order to give up and go down. Having heard about this Tamm was delighted. "What a man," he told me, "what courage, what a spirit does one need to assume such a responsibility and deprive himself and his comrades of the joy of a possible victory being just two steps away from it!"

Special courage of a person with high intellect was particularly characteristic of Igor Evgen'evich, especially during his last years, years of grave incurable illness. For all his life Tamm was uncommonly healthy, he was never seriously ill. He was already over 60 when he joyfully told me after one weekend, "Yesterday I learned how easy is it to gain one second on

---

[7] In general it is worth noting that any kind of boasting on Tamm's part was completely impossible (if only as a joke — when playing volleyball, etc.). I never heard him saying when listening to some scientific talk: "I have said this before", or the like. This was an organic feature of him and his family. Once his grandson, after returning from the kindergarten, answered to grandfather's question on what had they been doing there today: "We braided a basket", and carelessly added: "My basket was the best!". And everybody's laughter fixed this phrase forever. The grandson grew up, he is himself a father and a grandfather and Igor Evgen'evich is no longer with us, but the phrase: "My basket was the best!" lives, as a characteristic of the boaster, in the family up until now.

[8] E.I. Tamm (1926–2008), a physicist-experimentalist, doctor of physical and mathematical sciences, worked in FIAN from 1949. In 1982 he headed the Soviet expedition climbing Everest.

one hundred meters. One just has to once run the same hundred meters in advance."

And then this mobile man whose gait was such as if he rushed to overtake himself underwent, because of the degeneration of a nerve that controls the diaphragm, an urgent surgery and was switched to artificial respiration. A metallic tube that was inserted from outside into his trachea was connected to a respiratory machine which, in the rhythm of natural respiration, uniformly blew air into his lungs.

I was waiting for this moment with horror. I was almost sure that this very courage of Igor Evgen'evich would impel him to tear out the tube and put an end to this half-life. My understanding of the nature of his courage was, however, oversimplified. After several days I could visit him in the clinic. This was a day when he, for the first time, sat for an hour in the chair and still had not learned to speak — he had to pronounce words only when inhaling. I barely spent two minutes with him when the head of respiratory department, professor Lyubov Mikhailovna Popova,[9] who had been in charge of the surgery and supervised the treatment until the end invited me to her office and asked with anxiety: "Have you noticed that Igor Evgen'evich was writing today? When we helped him to the chair he asked us with signs to take a drawer out of the table, put it upside down on his knees, put paper on it and began writing some mathematical signs. Have you seen them? Is it adequate?". This meant: "Has he possibly gone crazy?"

Most probably I did not manage to completely convince her that he simply continued calculations, related to the work that interested him very much, which he had terminated in the hospital before the surgery. Evidently when lying motionless for all the days after the surgery he had realized something and wanted to check, as soon as possible, whether he was right. "But this is impossible! Everybody else who underwent this surgery suffered a psychological shock, "fell into pieces" and needed a long time to recover!"[10] Igor Evgen'evich did not "fall into pieces", he found his way out in work.

During several years one respiratory machine stood at his bed and another on his desk. He stood up from the bed, went to the table, switched to the other machine and worked for several hours. A metallic connection

---

[9]L.M. Popova, a physician, doctor of medical sciences, at that time a head of department in the Institute of Neurology of the USSR Academy of Medical Sciences.

[10]Clearly this case of psychological stability was exceptional but not unique. Usually, however, if a certain balance does come, it happens much later.

of a man with a rhythmically puffing apparatus made a grave impression on everybody seeing this for the first time. Nevertheless he was not broken. Possessed by some idea related to particle theory which, as usual for him, was cardinal and pretending to resolve fundamental difficulties of the theory (he cursed himself for not being able to leave it before being convinced that it is good or bad), he calculated and calculated, numbering only the pages he kept with four-digit numbers.

In 1968 the Academy of Sciences awarded Tamm with its highest award, the Lomonosov Gold Medal. According to the charter a laureate has to give a talk on his work at a general assembly of the academy. Igor Evgen'evich, though bedridden, decided to follow this rule within the limits of possibility. He wrote a report some 20 pages long that was particularly remarkable in not focusing on the past work (as was common) but to what he was currently working on, what he expected and what in his opinion were the perspectives of a theory of particles. This talk was read to the assembly by his disciple and friend Andrei Dmitrievich Sakharov.

When the participants of the assembly were taking their seats in the auditorium Dmitri Vladimirovich Skobeltsyn (who had earlier asked me with astonishment whether the report would really be presented) told me with regret: "This will of course be formal, the talk was prepared for him". However when the report was presented he told me, when passing by: "This is surely a talk by Igor Evgen'evich. It is clear that it was done by him, all of it".

Igor Evgen'evich was an active man. The banal words "one lives only once" that, for many, justify a consumer attitude towards life were never pronounced by him. It always seemed, however, that these constituted for him a basis determining his self-demands: To accomplish the maximum of worthy things you can do; to leave something to science; to help people around you; to do everything you can, irrespective of how small is this at the scale of mankind. His usual complaint in the letters to Natalia Vasil'evna, already in his young years, was on time lost without purpose, on his forced, due to some reason, inactivity.

A feeling of connection with the fate of mankind was characteristic of him. It determined his involvement in politics in his younger years, a struggle against all forms of fake science, a participation in taming the atomic energy in order to establish an equilibrium of forces in the world that is, as he and many others thought, necessary in order to prevent a nuclear catastrophe. It often happens, however, that when taking care of mankind one does not think about men.

Here is, however, a fact. In 1953, after a successful test of a hydrogen bomb, in the construction of which he played an active and very important role, a flood of honors and awards showered down upon Tamm. He had never experienced anything even remotely similar to this before. Once he took me to his office and said: "I received a very big premium. I do not need this money at all. Do you know some young people who need help to be able to do science?" Recently I learned that this question was posed not only to me and that the practical implementation was in all cases very swift.[11]

This was in general a common sight: after a scientific discussion was over Tamm suddenly took out a cigarette box (or an envelope of a letter he had received) covered with notes understandable only to him and (a cigarette in the corner of the mouth pointing upwards, the cigarette smoke irritating the squinted eye) recalled: "Aha, I should call this one, I should solicit for this one, ask about that", etc.

In science his active mind forced him to work all the time. He liked to work at night, an immense portion of what was done was not reflected in publications: He published only real results and the number of his published papers is, by modern standards, improbably small (if one excludes popular papers, reviews and reprints in foreign languages, one ends up with only 55 papers).

Sometimes, usually after a failed attempt to solve some big problem that had taken away much energy, there came disenchantment and there was no new idea. Then he felt devastated and unhappy. He came to the institute and asked the younger scientists: "Toss me some little problem". He called this "... freshening a nip after a hangover". So there appeared unexpected papers on concrete problems: on elasticity theory (together with L.M. Brekhovskih[12] on a localized impact on an elastic plate) and the one together with V.L. Ginzburg on electrodynamical theory of a stratified core. Tamm's range as a theorist was immense. He had a firm professional grip and could easily work in different domains of physics.

The work on stratified cores refers to war years though and here the situation was even more serious. Being in the evacuation with the institute in Kazan from August 1941 till September 1943 Igor Evgen'evich was deeply unhappy. In this time so extremely difficult for the country he was not

---

[11] See the memoirs of S.M. Raisky in the book mentioned on page 49.
[12] L.M. Brekhovskih (b. 1917) — a theoretical physicist, academician of RAS. At that time (1946) a candidate of physical and mathematical scientists, a staff member at FIAN.

involved in things that were most important at that moment[13] Helping A.P. Aleksandrov and I.V. Kurchatov in their work on protecting ships from magnetic mines he participated in calculating magnetic fields with complex configuration and was glad to be involved. Then he calculated an optical system for spectral devices that were very important for defence industry[14] and helped in studying the properties of explosives.[15]

This work was however too easy for him. He was not satisfied with it. He felt that his talent and qualification were not properly used. I never saw him so, almost permanently, irritated and worried. It seemed that Tamm, always so simple-tasted and almost ascetic in his everyday needs, felt a necessity, in conditions of a hungry evacuation life, of taking care of elementary needs for him and his family humiliating. With a mortal danger to the country in the background this was torturious for him.

When the First World War broke out his attitude to it was, as was told at those times, "anti-patriotic". He understood that this was an "alien" war and did not want, like some other students, to go to the front as a volunteer (as a student of the Moscow University he was, during the first years, free from the draft). His wife Natalia Vasil'evna recalls his furious arguments with the "defense"-minded members of their family.

However, at the front there flew blood and, having finished his classes somewhat earlier, in the spring of 1915 he went as a volunteer hospital attendant to a field hospital. He saw floods of blood and during fighting hundreds, thousands of wounded went through his hands. He was a witness of sufferings that were impossible to forget and the underlying reasons could not be left without analysis. At this time, however, the situation was different and he deeply suffered from being pushed aside from the common cause. Of course, he was working intensively all the time — Igor Evgen'evich could not exist without scientific work. This was however a theory of elementary particles, a theory of nuclear forces and other similar questions that in the wartime seemed to be incredibly distant from practical applications. At the time few could foresee that in just several years these "abstract", "irrelevant" questions would be among the most burning and vitally important ones.

---

[13] With bitter humor did he recall the order given by Napoleon in the Egyptian campaign under Mameluke attacks: "Range in square, donkeys and scientists to the middle" (donkeys were precious transport, scientists accompanied the expedition).
[14] See the memoirs of S.M. Raisky in the collection cited on page 49.
[15] See the memoirs of V.A. Krainin in the book mentioned on page 49.

I must use another phrase, worn out from being too easily used, by saying that Igor Evgen'evich was true to principles in his behavior. If one removes from these words a habitual touch of the empty jubilee solemnity one would see how precisely do they define what will be told below. Of course much of what is written above can prove this statement but it is worth emphasizing *how* did Igor Evgen'evich argue, how did he fight fake science.

Unfortunately one usually understands a polemic talent as an ability of startling an opponent with blazing formulations, eloquence, pointed attacks, sometimes even an ability of humiliating an opponent, of "unmasking" him. All this was totally alien to Igor Evgen'evich. He, who with such excitement enthralled the listeners with colorful stories of his and other's travels, adventures, comic, tragic and tragicomic episodes (he always had plenty of these at his disposal), was strict and even reserved in public speeches or discussions. His aim was to reveal, to find the truth and convince his opponents, introduce them to this truth, only with thought, argumentation, knowledge of facts. All private moments were completely excluded. Being honesty and truthful himself he assumed the same honesty and truthfulness in his opponent in advance. Of course this was in most cases naive. Here are three examples.

Improbable as it seems (this phrase is used here so often but there is nothing to be done about this: During Tamm's life there happened much of what is now a distant past and is nowadays difficult to believe) even in the thirties we had titled scientists who considered the electromagnetic field as a manifestation of the mechanical motion of ether. Most active propagandists of this view, rejected by science already at the beginning of the century, were perhaps the physics professors A.K. Timiryazev and N.P. Kasterin as well as an academician V.F. Mitkevich.[16] A particular difficulty of the situation was that they, in claiming that all this follows from the dialectic materialism and, as was mentioned before, they were trusted by people who did not know physics but self-assuredly ruled over science and decided the fate of scientists.

Igor Evgen'evich could not stand aside — by virtue of his temperament, as a creator of a course on the theory of electromagnetic field that he gave many times at Moscow University, as somebody who, already in his younger years, had studied Marxism and Marxist philosophy in particular. However

---

[16] V.F. Mitkevich, and electrical engineer, and academician of AS USSR.

he confronted the biting demagogic formulations of these persons only with the seriousness of argumentation.

To demonstrate a conditional character of the concept of force lines and the fictitious nature of a notion of their number, Tamm invented and calculated a beautiful example. In a system of two electric currents, the linear one surrounded by the circular one, the resulting magnetic field has a toroidal shape: The magnetic force line coils around a "bagel", a torus surrounding the linear current. If the forces of two currents are in the rational relation with each other the force line, after making a certain number of rotations, will close on itself. However if one makes an arbitrary small variation of the strength of one of the currents so that this ratio becomes irrational the force line will never close on itself. Then an infinite number of force lines will go through an arbitrary cross-section of the torus. There forms a continuous toroidal magnetic surface.

It would seem that each physicist should understand that a density of the number of force lines can therefore be only a conditional measure of the strength of the field and a separate line can not be a real mechanical motion of ether. It is worth adding that the above-described physical example placed already into the first edition of the course by Igor Evgen'evich took on, within the recent decades, a practical significance. The magnetic surfaces of this type play an important role in stellarators (one of the intensively investigated variants of controlled thermonuclear fusion).

The second example refers to the year of 1936. In Moscow, in a giant auditorium of the Communist academy in Volkhonka, there was a meeting of the General Assembly of the USSR Academy of Science at which academician A.F. Ioffe, a director of the Leningrad physico-technical institute, gave a report on its work. Long debates transformed into a discussion of generic organizational and scientific problems of physics in our country. Sharply critical speeches accusing A.F. Ioffe of being overly optimistic were given, in particular, by young L.D. Landau and A.I. Leipunski. At some moment (I do not remember exactly, when) Igor Evgen'evich confronted V.F. Mitkevich who once again defended a mechanical theory of electromagnetism. As usual Tamm's speech was motivated, clear and concentrated. The audience was heated from the previous debates, the amphitheatre was full, many (including myself) sat on the stairs in the passages. Clarifying an inapplicability of certain mechanical notions to electromagnetism and answering, in particular, to the question repeatedly asked by his opponents on what substance does move in space between two electric charges when one of them moves[17] Igor Evgen'evich said, "There exist questions that do not have a reasonable answer, for example, whether the color of the meridian in Pulkovo is red or green."

Here academician Mitkevhich says loudly: "Professor Tamm does not know the color of the meridian he is standing at but I do know: I am at the red meridian".

Igor Evgen'evich, surprised, glanced at the speaker, shrugged his shoulders and did not pursue the argument.[18]

Finally, the third episode. In the mid-fifties Tamm, together with a number of biologists, physicists and mathematicians, fought an intense struggle against "lysenkovshina" and for development of scientific genetics, once having had leading positions in the world, in our country. In October 1956 a meeting of the General Assembly of the Academy of Sciences for reelecting for a new term its president A.N. Nesmeyanov was convened.[19]

It seemed that the issue was not complicated. Tamm, like many other members of the Academy, highly esteemed A.N. Nesmeyanov as a scientist. Their relations were perfect, their families were friendly. In spite of this Tamm took the floor and gave a big and firm speech. He expressed his general and very positive opinion of the president and his confidence in his progressive scientific news but voiced disagreement on several issues, especially on, in his opinion, insufficient activity on developing biological science. Tamm suggested to postpone the reelection and ask A.N. Nesmeyanov to present, in advance, a clear list of actions that he planned to undertake in the General Assembly. Emphasizing his respect to A.N. Nesmeyanov again and again he told that such a decision would help him because a decision of the General Assembly would give more authority and effectiveness to a difficult job of the president.

Igor Evgen'evich had made this suggestion already at the preceding session of the section of physical and mathematical sciences where it had been approved. But all other sections were for simple reelection without any conditions (everybody understood that a candidature of A.N. Nesmeyanov was proposed by the party Central Committee). Tamm's speech at the

---

[17] These opponents asserted that a mechanical displacement of substance is a necessary primary element of any materialistically understood physical process. In reality variation of electromagnetic field is accompanied by displacement of energy stored in vacuum even in the absence of bodies in this space.

[18] This episode is described by V.F. Mitkevich in his 1936 book *Basic Physics Views* (2nd edition, p. 161). See also *Izvestia AN SSSR*, ser. fiz. N1/2, p. 118.

[19] A.N. Nesmeyanov (1899–1980), a chemist, an academician of AS USSR, President of the Academy (1951–1961).

General Assembly caused a storm. It made such a strong impression that a compromise decision to reelect A.N. Nesmeyanov but to hold, in the nearest future, a special session of the General Assembly to discuss his proposals was taken. This meeting took place in December and about 30 members of the Academy participated in the discussion. It was a long time since the Academy had had such a broad, frank and often sharp discussion.

It is well known that Igor Evgen'evich was in general uncompromising with respect to fake science. At the end of the forties, during Zhdanov years of stalinshina that were particularly gloomy for science, some authors resumed their attacks on the theory of relativity and quantum mechanics as "bourgeois-idealistic" ones. In this atmosphere some professors tried to adapt these theories, to make them acceptable to the critics even at a price of vulgarizing the science. Of course, modern science was at that time rooted deeply enough and was intensively used in research important for the country. An authoritative defending of science by I.V. Kurchatov, V.A. Fock, S.I. Vavilov and many others also had its effect. There were also quite a few ordinary scientists who were not frightened by the attacks but the situation was nevertheless very difficult. Tamm did not forgive betrayal of science born from careerism or timidity, even for those previously close to him, and cut his personal relations with them.

The behavior of Igor Evgen'evich surprisingly combined cheerfulness, liveliness, openness, impetuosity, even shortness of temper and irascibility (sometimes ungrounded) with restraint, almost insularity, tactfulness, correctness. Cheerfulness, liveliness, impetuosity were for small talk, for relaxation, for lectures, especially popular ones. Shortness of temper and irascibility were only related to insignificant and of secondary importance, everyday, domestic, unworthy, hindering life and work. In matters related to something significant, serious, of real importance this was another person — only thought out words, only full-weight reasoning, only justice in relations and statements — no hurry, nothing unrelated, nothing empty.

In society or, as it is less ceremoniously said now, in company Igor Evgen'evich was an inexhaustible storyteller enjoying his stories himself.[20] He easily became a focus of everyone's attention, was ready to passionately participate in any kind of ideas, charades, games, half-serious sporting competitions. He was happy when winning and violently cursed himself when

---

[20] I confess it seems to me that in these stories he often exaggerated. He could have known a phrase by Bohr (unfortunately known to me only through his disciple): "If a story is really good, does it really matter that everything in it should be true?"

losing. But even here, in noise and fun, his born and cultivated tactfulness remained unchanged. He never shadowed another person, was ready to listen to stories told by others without interruption, inserting his remarks in such a way and at such moments that they did not disturb but rather help the storyteller and the other listeners.

Alas, this culture of behavior is by no means characteristic of people of later generations. Once A.I. Solzhenitsyn, to whom Igor Evgen'evich also had a big interest, came to make acquaintance. Tamm invited his two younger friends (V.L. Ginzburg and the author of these notes). Very soon, impulsively interrupting each other, they got a discussion with the guest under their control and Igor Evgen'evich could only ahem, gleam with eyes, smile and occasionally put in a couple of remarks. In essence he did not talk to his guest himself. Of course he did not reproach his friends. Moreover, as we learned later, he did not even notice that he was, without ceremony, pushed aside.

With all his sociability Igor Evgen'evich was very sparing in expressing his deep feelings. It seems to me that one can judge upon hidden features of a person by, for example, his behavior when playing chess (like a couple of hundred years ago when a character showed itself when playing cards). Emotional and intellectual concentration possessing a theoretical physicist when he works calculating at the desk (and, perhaps, all scientists) is, in my opinion, close to what one feels at the chess board. One has to overcome the resistance of an "adversary" (a problem at hand), foresee the events for "many moves ahead" — possibly without completely calculating some variant, to understand which difficulties, which tricks one could possibly meet, to estimate the most dangerous, "weak" points of the position at hand in different possible situations.

When playing chess one has to constantly keep in mind the chosen general plan and the goal. One has to avoid simple calculational mistakes, keep everything in the head and act, move forward, continuously processing information on the changing situation. A separate issue is a visible reaction of a player on a successful or bad move, the behavior of the opponent, etc.

Chess was among the passionate hobbies of Igor Evgen'evich. He was quite a middling player, not better than that of the second or third grade. The play, however, showed much of him. First of all there was a remarkable transformation from the liveliness and fun of small talk to the maximum concentration and seriousness as soon as the first move was made. Further during the play one saw a full mobilization. If somebody, an opponent or a spectator, made a joke, Tamm did not notice it or, in rare cases, was distracted for a second and artificially smiled only with his lips.

He made the best moves in difficult positions. Sometimes it seemed that he did not have a way out but a long tense think and a passionate wish to stand or win produced completely unexpected results. Having made a good move in such a difficult position he moved a cigarette into the other corner of his mouth, clenched his wrists between his knees and, moving eyes back and forth from the board to the opponent, he waited with the same tension for the response or began nervously searching for a cigarette box and matches which always happened to be not in place. He perceived a loss as a big trouble. However, like in situations in real life, he was very sparing in revealing his feelings although his passionate nature did not make this easy. Here suffered his striving for self-affirmation which played a big role in his life in general. One can think that a chess situation is a good model of his behavior in the process of scientific work.

It has already been mentioned that he never splashed out his problems onto others. Their sufferings caused his deep compassion but this was also expressed in restrained words and in a restrained way. Only in his last years, the years of illness, there appeared some traces of softness and sensitivity.

Years passed by, there multiplied troubles intertwined with periods of satisfaction; there came true joy and big disaster but Igor Evgen'evich himself remained in the same. His character, a liveliness of his reactions, an interest to the world, a devotion to science, benevolence and implacability — his inner essence, — remained unchanged. Heavy emotional experience, however, left permanent traces on this same clever and dynamic face.

A characteristic feature of Igor Evgen'evich was his profound dignity. I would even dare to say he was a proud person. Using this word one has, however, to explain a lot. This was not pride that some people often associate with arrogance. The Russian intelligentsia which Igor Evgen'evich belonged to worked out its own special gauge. When seeing how Igor Evgen'evich rushed towards a person sympathetic to him, how he bustled when "coaxing" him, scattered words — pausing in speech, agitated, — be it Niels Bohr or a non-famous comrade mountain climbing, some elementary feeling and non-clever observers watched, even with some compassion, at this "absence of the feeling of self-esteem".

Some were tricked by this spontaneity of behavior. A rough and rude administrator that allowed himself to talk, lazily lounging on the sofa, to Tamm, who was standing in front of him and at first so polite (perhaps the ill-bred administrator took this for acceptance) suddenly jumped up when this polite person burst in anger. Close associates can remember other similar scenes of Igor Evgen'evich's temper even on less clear occasions.

Nevertheless one can count only about three or four such episodes when Tamm lost control of himself when facing disrespect or direct rudeness in several decades. We around him felt awkward for him and Tamm himself considered such behavior unworthy and was ashamed of it. Here I am convinced (and have grounds for this) that at such moments he was usually in a state of nervous tension because of other really serious reasons which he, as usual, did not reveal to those around him (according to the proverb "Shouts at the cat and thinks of the daughter-in-law").

If Igor Evgen'evich himself would have heard that the word "pride" was associated with him he would probably have laughed or would have been surprised or would have expressed indignation. He mocked such big words. But how can one characterize, for instance, this independence and firmness of his position, as described before? How should one describe his quiet reaction on both the official neglect he faced for many years, and the official awards?

Igor Evgen'evich was elected a corresponding member of the Academy in 1933. By the mid-thirties he had made almost all his major works: A theory of light scattering in crystals including the combinational scattering (Raman effect) in which the lattice vibrations had been consistently quantized for the first time and a notion of "quasiparticle", phonon, had been introduced; a consistent second-quantized theory of light scattering on electrons that had proved, in particular, that negative energy levels in Dirac theory can not be neglected — this had been of a deep principal meaning; calculations of positron lifetime in the medium; theoretical prediction of surface levels of electrons in crystals — "Tamm levels"; a fundamental work on photoeffect in metals and, finally, a theory of beta decay. In 1937 there came an explanation and development (together with I.M. Frank) of the complete theory of Vavilov-Cherenkov radiation (that would afterwards bring him a Nobel Prize).

The period of 1930–1937 was the period of an incredible creative ascent. Tamm's might showed itself with impressive productiveness. All physicists considered him as one of the most prominent theorists. Ehrenfest, who intended to leave his chair in Leiden (that he took after Lorentz), named Tamm as the most desirable successor. After Igor Evgen'evich's work on beta forces (1934) Fermi was of a very high opinion on it and on Tamm himself as a big theorist (evidenced by B.M. Pontecorvo,[21] at that time a collaborator of Fermi). Nevertheless the Academy of Sciences did not elect Igor Evgen'evich to full membership. This was by no means an underestimation of his scientific achievements.

At that time Academy elections were strictly controlled by the party Central Committee. It is known that before the elections in the mid-forties Zhdanov personally erased Tamm's name from the list of those allowed to be elected. Tamm was "listed as an idealist" up until Stalin's death. He was elected only in 1953. However, nobody ever saw him expressing bitterness on this, being disturbed or offended. He was only surprised when observing such a reaction from others. Problems in developing a full theory of nuclear forces worried him incomparably more; these really distressed him.

Here is an "inverse" situation. In 1958 he (together with I.M. Frank and P.M. Cherenkov) was awarded the above-mentioned Nobel Prize. Since then our scientists have received several Nobel Prizes but at that time this highest international recognition of scientific achievement was still sensational. Among Soviet scientists only N.N. Semenov[22] had been awarded the prize. Having learned about the decision of the Nobel Committee I rushed to Igor Evgen'evich's office and, excited, began congratulating him. Calmly and even more slowly than usual, traversing the office with hands behind the back he seriously replied:

"Yes, this is of course very pleasing, I am glad ... Very glad ... But, you know, this joy is mixed with some dissatisfaction."

It was easy to guess what. "Because this prize was awarded not for the work you yourself consider to be the best, not for beta forces."

The ultimate manifestation of his dignity or pride (one can name it as one wishes) was however one particular feature of his scientific activity; he always chose the, in his opinion, most important at that time, directions of research, although usually these were also the most difficult.

I do not know whether for him this was a consciously formulated principle or if this was an inevitable consequence of his tough character, of his striving to accomplish something almost impossible, to "jump above one's head". If he would have decided to step away from it then, with his qualification and erudition, with his brilliant professionalism, his capacity to work, ability to make error-free calculations, a wonderful power of a master, he would have easily produced good works in incomparably larger

---

[21] B.M. Pontecorvo (1913–1994), a physicist-experimentalist, academician of AS USSR.
[22] N.N. Semenov, a physicist and chemist, academician of AS USSR.

quantitites.[23] This is seen for example from such works as an investigation of the width of the shock wave front, a study of magnetic confinement of plasma in controlled thermonuclear fission, etc. Apparently, however, these did not satisfy him. One should not wonder that he perceived the age-related decline of scientific potency as a tragedy.

Only at the end of the sixties did he come across a new idea of giant scale — the thought of basing elementary particle theory on a concept of nonlocality with non-commuting operators of coordinates of four-space (and four-momentum) and with an elementary length. The new element was constructing the theory in momentum space with variable curvature. First general thoughts and first attempts were reported at the international conference in Dubna in 1964 and at the conference dedicated to the jubilee of Yukawa meson theory in Japan in 1965 (it is worth reminding that at this time he was 70).

The accomplishment of this idea turned out to be extremely difficult both mathematically and from the point of view of physics ideology. Tamm had a perfect command over the mathematical toolbox that a theorist needs and worked, as he told himself, as a 'hard drinker'. He continued this work in the hospital and at home until the last months of his life. He was surrounded by the skepticism of many theorists but working in this atmosphere was not new to him. The work remained unfinished. It is still unknown whether this "crazy idea", one of the many studied by theorists all over the world, could lead to something useful.

The same feeling of self-esteem defined Tamm's attitude towards such delicate problems as that of precedence in science. A scientist probably needs some ambition. The question is only how it shows itself and how does it influence relations with others. It seems to me that for Tamm ambition was completely confined to self-assertion and, most importantly, to self-assertion in his own eyes. Not to elevate so that others see it and express delight but rather convince oneself: "I made it".

An inner recognition for achieving a difficult goal was something that gave him satisfaction while external signs of success were only a pleasant addition to this. It is therefore impossible to recall a single case when he complained, even quite moderately, on somebody having used his idea or did not appropriately refer to his work. At the same time such complaints

---

[23] This was written before I learned about a similar phrase by Landau cited by S.A. Altshuler (in a collection of memoirs, see reference on page 61).

and offences are, unfortunately, quite a widespread phenomena. Some are infected by these as a heavy illness.

Tamm's attitude to the precedence problem is illustrated, for example, by one episode worth describing. At the beginning of the thirties an idea came to his head. He realized it and made a wonderful work that exerted a big influence on the subsequent development of the theory of this problem. He completed the research, the most difficult and extensive calculations, during one conference working, as almost always, at night. When everything was done it turned out that the final formula did not justify the initial hope on providing a quantitative description of the phenomenon. Nevertheless, as mentioned, the work turned out to be important and Tamm prepared a short note for a journal publication.

At that moment one young theorist who was coming every morning to the hotel to enquire how successful was the night work asked Tamm: "Would you mind if I also send a letter to the journal? We have so often discussed this problem together." Tamm was surprised but could not answer with refusal. So it happened that there simultaneously appeared a note by Igor Evgen'evich containing, besides a clear physical formulation of the problem, the final formula and a negative conclusion from it and, nearby, a letter to the editorial board of this young theorist containing only general considerations, "an idea", giving him nevertheless a dubious right to claim that his name as a co-author of the whole theory should always be mentioned along with that of Tamm.

Tamm told me this story a quarter of a century after — smiling, completely without anger. After these memoirs had been written I learned that this episode was mentioned by Tamm in one more conversation with his two closest colleagues. I decided to write about this episode not to prick or belittle anyone but only because it illustrates with outmost clarity the attitude of Igor Evgen'evich to "precedence mania". It was important for him to know for himself that he *could* do it and if somebody extracted pleasure from sharing external recognition without having done anything — God be with him, let him enjoy it, this was just funny. This same story possibly illustrates in which sense Igor Evgen'evich was a proud man.

A separate important issue is on the relations of Igor Evgen'evich with his disciples. Everybody knows that around him there grew a vast school of theoretical physicists, that the many a year pedagogical activity — lectures in Moscow University, then in Moscow Engineering Physics Institute, then again in MSU, his course on electricity theory exerted a big influence on generations of physicists. At the same time, paradoxical as it seems, he had no thought-of system of bringing up young scientists. A brilliant school of theorists created by Landau had in its foundation a detailed plan, worked out by him, of how a student enters science. First, examinations on the famous carefully composed and thought-of "teorminimum", then reports on literature at the seminar and, finally, scientific work (usually on a topic suggested to the student). This system, with no second opinion, gave brilliant results. It turns out, however, that a different approach is possible.

Talking about lectures, Tamm simply chose for reading those courses that interested him. He did not like to repeat the same course many times and it is not difficult to understand him. I first heard him in 1932 at MSU when he read a theory of electromagnetic field. He had already read it many times, a second edition of *Foundations of Electricity Theory* had been printed. As he told me himself he was really bored with this course: "I know this book as an educated Jew knows Talmud: If you pierce the book with a pin I will be able to tell which word is punctured on every page."

Nevertheless during lectures he enflamed and ignited the students. His lectures were loved. Undoubtedly it was simply a charm of his personality that played a significant role in this.

As a rule the work with a student began only after his graduation from the university, sometimes several years later. One could remember only a few cases (three? four?) when Tamm had diploma students. During his education he selected disciplines according to his own taste. A system of teaching at the Moscow University before the revolution and especially at the Edinburgh one in which he began his studies left a significant freedom of choice. Also after graduating from Moscow University, before becoming close to L.I. Mandelstam, he expanded his education himself. Perhaps he expected to find the same independence by his students — even in the conditions of a strictly regulated university programs as is common here. He was once told that one of his students "fell in love" with him already in the third year. Tamm expressed surprise: "I can not work with students at all, I do not give them anything."

In fact what "worked" here was probably mostly an example of attitude towards science — not only the logic of reasoning, the choice of considered examples, a measure in putting together physics and mathematics and of the strict and stable textbook material with the new actual problems, but also a lecturer's interest, activity, his evident joy obtained from tracing a way to the truth. At the same time the relations with those diploma students that one can remember things did not go too well. With one of them who he liked a lot during diploma period he had big difficulties in working together at the PhD stage so that the student distanced himself from Tamm. He did not want another one to continue working with him and later this student became a good theorist.

Usually the arrival of a so-called disciple was marked by bringing Tamm his questions, results and ideas which showed an independence of approach (which was considered to be most important) and an ability to work. Then Tamm was enflamed, there appeared a sympathy. Using the physics terminology, his "response function" was step-like with a very big step. To provide such a person with the possibility of working in science Igor Evgen'evich, in case there arose difficulties, energetically supported the Ph.D. studies or a leave from the factory laboratory if he was not allowed to leave it, etc.

However, even after a young man began working in the department the methods of entering physics were diverse. It was again considered to go without saying that one had to do with an independently thinking physicist, a colleague with whom one just needs to share one's experience. This was also reflected in the fact that according to the custom of the time of his youth Igor Evgen'evich addressed everybody with a name and patronymic even if they became acquainted when Tamm was already a well-known scientist and the new acquaintance was a student. I can remember only one colleague whom after his quarter of a century work in the theoretical department Tamm *sometimes*, in rapid conversation, called simply by name. It is impossible to imagine that he could address somebody, be he one of the youngest, with the familiar "you" getting in response a polite "you" as it happens nowadays.

Probably this constituted a significant element of bringing up an independently thinking scientist and removal of authoritarianism which in our time often leads to the Ph.D. student's "looking into the mouth" of his supervisor (I know that this is by no means universally true and can myself find examples when addressing in familiar "you" and one-sided negligence of patronymic did not influence the independent thinking of a Ph.D. student and, vice versa, when addressing a PhD student by patronymic did not ensure a respect to the supervisor, but I think that this is in general true). Vitaly Lazarevich Ginzburg and myself worked with Igor Evgen'evich for 30–35 years, since our Ph.D. studies. However only during the last years of his life he insistently suggested that we call each other by name. It was however impossible to call him "Gora" — this is how his relatives and friends from his youth called him. We agreed only on the asymmetrical variant and "the suggestion did not pass".

Let us however return to the working relations of Tamm with his disciples. Some began working with him on the suggested topic,; some just got the topic, worked independently and sometimes consulted with Tamm; some chose the topic, sometimes completely independent from the interests of Igor Evgen'evich, themselves, and just discussed intermediate or final results with him, getting advice and critique. I would say all three variants had the same rate of occurrence.

One could wonder how can one describe Igor Evgen'evich's supervision and why can one talk about an existence of the "school". The main things here were his attention, good will and at the same time a totally uncompromising critique, an example of his own incessant work, his great erudition, an example of the ability of combining the physics approach, physical understanding of the heart of the problem with a convincing mathematical treatment, a cultivation of the usage of similar elements in distant domains of physics, cultivation of attention towards the problems that are most actual in each domain, a development of such an attitude towards the works of others, when a respect to an esteemed author (who could be the supervisor himself) goes along with the critical spirit and a circumspection when seeing a new previously unknown name goes along with a serious analysis of his work that admits from the very beginning an appearance of a new talent, finally a creation of such an atmosphere in which a work on applied physics that essentially uses good physics and high professional qualification is not at all rated lower than research on "high" physics topic.

Here is a concrete example. I was a Ph.D. student of Igor Evgen'evich and worked together with him in the department for decades. In all this time we did not publish a single paper together. After the first unsuccessful attempts of joint work Tamm once suggested me a topic. The topic of my Ph.D. thesis. In the end I showed him a quite complete study. Later I was mainly working in the domains that were not too interesting to him (so it was almost impossible to consult on this) and only partially in those more close to but still not touched by him. However, I lived in the atmosphere of the Tamm theoretical department, I was an active participant of this wonderful brotherhood, always had noble examples before my eyes and one should not wonder that I gratefully (and undoubtedly with a reason) consider myself to be a disciple of Igor Evgen'evich.

One has to stress that the critique of Igor Evgen'evich was not aimed at enforcing his own evaluation of the perspectives of working in the given direction chosen by the colleague. There were lots of cases when he doubted something done in the department or even had serious skepticism towards it. This skepsis was however expressed very cautiously. Probably this was a reason for the fact that among Tamm's disciples one finds so many people with distinctly different personalities, with widely different domains of research, with clear distinctions in the schools created by them.

With such a broad view on possible approaches to problems, on estimating the perspective of different directions when a conviction in the rightfulness of one's own choice went together with an extreme delicacy and respect of the position of the other, it was quite natural that a benevolent support of own disciples and colleagues could never lead to an arrogance with respect to the works, views, style of other schools, of "outside" theorists. His behavior did not show even a trace of "belonging to a sect" with respect to research of problems distant from his own interests. He evaluated such research in completely impartial way.

His genuine delight when meeting any interesting and fruitful work was evident and instructive for anybody witnessing it. If his reaction was restrained this meant that not seeing direct errors he doubted the result or was convinced that this research did not have a perspective, considered it useless although it was in principle impossible to prove that he was right. Only the direct anti-science, fake science, caused his furious attacks (still correct in form). One could say that his scientific "patriotism" never became a "chauvinism".

After reading these notes one could wonder what was finally so special in the features of the person described above. Was he simply a truly decent and good person whom, above that, Nature gave the talent of the scientist? Don't we find such features in not only Einstein, Bohr, Mandelstam, but also in many less great persons and those not great at all? Of course this was essentially what was said. This is true. In Igor Evgen'evich these features showed up with rare completeness allowing us to consider him as a standard to be upheld.

Above I attempted to "put onto different shelves" some basic features of the personality of Igor Evgen'evich and illustrate each feature by facts from his life. Rereading the notes one can conclude that this attempt was not too successful. Some facts illustrating his spiritual independence could be selected for illustrating his courage or firmness of principles and vice versa. This is not accidental. The thing is that all of them melted into a character of wonderful integrity and, simultaneously, complexity. A charm of his personality felt by almost anyone meeting him can not be expanded into elements and rationally understood at all.

I.E. Tamm in the process of discussing a scientific problem.

His life went through different stages. At the beginning, in the first quarter of a century, this was a conscious formation of his attitude towards the world, a search for and choice of the way forwards. Then, for 30–35 years, an extremely productive (especially during the first 20 years) period of scientific work was accompanied by the growing recognition of colleagues and simultaneously marked by a number of tragic events that were so characteristic for the social and political conditions of that time. Finally, the last decades were those of wide social, official, scientific recognition and esteem — of something that one usually calls fame.

However, both difficult the times and times of fame only sharpened the basic features of his character, those features of his personality which, like nuclear radioactivity laws, could not change under the influence of external conditions. This caused a deep sympathy, even with people who did not contact him directly and those even incapable of estimating, even very approximately, his contribution to science.

This was a scientist who embodied links with the epoch of Einstein and Bohr and who upheld a standard of decency in science and social life. A person with both physical and mental courage. A powerful and sophisticated theoretical physicist. A delicate tactful supervisor who taught by example and benevolent critique and not by detailed "leadership" and didacticism of the elder. A true friend. A person jovial and serious, charming and perseverant in fighting for something that was in his view of principal importance. A man prompting love and joyful respect from many and generously presenting friendship and joy of communication himself. Steadfast in achieving any set goal. Courageously facing tragic events brought by the epoch into his private life and the ruining of heartfelt ideals of his youth that so painfully influenced him. A good person and an important scientist.

2.2 A Life of a Russian Intelligent[24]

Igor Evgen'evich Tamm was a person remarkable in many ways. One has written about him as a scientist and a person, a participant in the formation of Russian physics. There remains however one side of his personality that until very recent times could neither be sufficiently explained nor understood. One was silent about it. I have in mind his social, political, civil

---

[24]Published in the journal *Priroda* (N7, 1995). The text is expanded.

position, his life as an element of a fate of Russian intelligentsia of nineteenth — twentieth centuries. Formation, development and transformation of this position and his view of the world are characteristic of a particular wing of our intelligentsia.

One can not say that he was a "typical Russian intelligent" because the intelligentsia of that time was not homogeneous. At the same time it had a common feature distinguishing it from intellectuals in other countries — a tense feeling of pain for the people, a feeling of guilt towards them. It was called into life by the particular history of the country, first of all by the long preservation of serfdom (and after its fall an extreme poverty and a cultural backwardness of a vast majority of the population) and also by an absolute power of the monarchy already impossible in Europe. A savage absence of civil rights and culture and the poverty of the lower classes went along with a high culture of intelligentsia, first aristocratic, to which the established term "intelligentsia" is applicable only with reservations, and from the middle of XIX century also, to *raznochintsy*.[25]

This combination of an almost medieval backwardness of one part of society with a high spirituality of another part which increasingly gained a position of one of the leaders of the world culture, led to a division of the country population into completely different parts.

Such polarization inevitably created sparks of liberal and even revolutionary protest. By the end of the XIX century there raged fire. A moving force of this protest was precisely a feeling of guilt toward the people that often led to a revolutionary extremism.

The polarization of society led to a mutual misunderstanding between the intelligentsia and the lower classes as so brightly described in literature (*Three years in the country* by Garin, *Fruits of enlightenment* by Tolstoy, *New dacha* by Chekhov, etc.). The people did not understand and often also despised the intelligentsia for what it was doing; an intelligent always remained for the people a *barin*, attempts of an intelligent to help the people were met with disbelief, sometimes even hostilities. Precisely this disbelief and estrangement led, after the revolution, to the humbled, oppressed position of the intelligentsia which became a social class of the "second sort".

A country estate lifestyle of an educated Russian aristocrat of the first half of the XIX century was usually tied from childhood with the lifestyle of the people (in particular that of house-serfs). A thoughtful, spiritually

---

[25] The teachers, doctors, low-rank clerks, etc.

developed person could not fail to understand the crying inhumanity of serfdom and its absurdity in the period of development of the best features of Western civilization: The progressively strengthening democracy and civil rights development. Nevertheless at first the feeling of guilt towards the people did not arise. It is characteristic that we do not see it even by supersensitive Pushkin with his praising of freedom in general, with his sober understanding of what was possible and what was still impossible in the Russia of his time. He understood that a serf blessed his fate when his corvée was replaced by a tribute, but with him one does not feel the suffering for this serf that Nekrasov and Lev Tolstoy, Dostoevsky and Sofia Perovskaya that soon appeared did have.

However, beginning from the middle of the XIX century when a circle of *raznotchin* intelligentsia began expanding, an understanding of the ugliness and injustice of Russian life strengthened sharply.

In the West the society was for centuries adopting to the co-existence of poverty and richness by developing civil self-understanding and civil rights (let us recall a mill in Potsdam which Friedrich the Great could not take from its owner), with an existence of the vast middle class. The gap in the cultural and spiritual levels between the two social classes was correspondingly smaller. By us only a sharp transition to the Western model in the epoch of great reforms in the sixties and seventies of the XIX century opened up beautiful perspectives. However, this realized itself predominantly for the rich, partially for the middle class and successful former serfs, etc. In other words the gap between the two social classes did not become significantly narrower.

What could the attitude of an intelligent towards the reforms (which grandiose importance he certainly understood) be? Here one can distinguish three main ways, three psychological and social positions.

There were those called Slavophiles. A worry on the destiny of the lower social classes was reshaped by them in a hope that Russia, because of the communal way of rural life and the orthodoxy, possessed its own special way. In the course of further development this Slavophile position, by the end of the XIX century, progressively got moral and philosophical direction seeking the way out in the moral (religious) perfection of personality and society. This movement was represented by many outstanding writers and religiously oriented philosophers.

The majority of the intelligentsia, however, did not share the hopes of this way of Russia's development. It enthusiastically accepted the reforms of Alexander II and saw its purpose in catching up with the West and joining the general process of world development. For this, these people thought, one had to build, cure, teach, introduce zemsky self-government, culture of labor and civil self-consciousness i.e., in essence, one had to create a cultural revolution. This activity brought wonderful results but the goal was so grandiose and the backwardness of the masses so deep that the time that history gave it was insufficient — there began the First World War. Nevertheless the half a century before it was used with great success and significantly changed the face of Russia.

There appeared talented businessmen, splendid lawyers, world-famous engineers. Here it is appropriate to repeat what was already written in the Section 2.1: at the core there was a working middle-class intelligentsia. It was this layer that worked out, in a natural fashion, a non-written moral code defining decent behavior. Igor Evgen'evich was merely one of those accepting this code with particular thoroughness. A feeling of guilt towards the people, a worry for the people went along with this code in a natural way.

The most significant representative of this intelligentsia was Chekhov who understood better than many that there was no way to the "sky in diamonds" other than a long, tortious, filled with hard labor, but unavoidable one. It was exactly the Chekhov moral: bitter recognition of estrangement from the people, and a worry for their destiny reflected the world view of this intelligentsia.

Many, however, did not agree with this position. This same working intelligentsia with all its virtues gave birth to the third layer, that of impatient and revolutionary minded intelligents. It was not accidental that Yuri Trifonov called his novel on narodovoltsy *Impatience*. In essence they were moved by an idea of the immediate equalizing restructuring of the society with a quick transition to socialism or something close to it. The great reforms of Alexander II did not satisfy the impatient.

On the eve of the XX century the new state authorities responded to the people seeking protection from its side by the absurd, unnecessary, shameful war with Japan, the "Bloody Sunday" and the stupid management of the war with Germany.

For this absent understanding of the needs of the people and the country and an unwillingness to continue following the path of reforms the ultra-conservative monarchy paid not only with the life of the weak tsar but with the destruction of the monarchy itself, the country thus facing a cruel revolution and a bloody civil war.

How strong should a feeling of "impatience", a feeling of guilt with respect to the people should have been by Countess Sofia Perovskaya, engineer Nikolai Kibalchich, their followers and the revolutioners that came later, by all the revolutionary feeling wing if their main principle was: "My life belongs to the people, to revolution", and this was not an empty phrase: they readily gave their lives for revolution.

An anticipation and a passionate desire of a revolution that would bring justice were not influenced even by understanding of the fact that without a preceding "cultural revolution" it would be, in the words by Valery Bruysov, an invasion of Huns[26]:

> And you, wizards and poets
> Keepers of mystery and belief
> Let's carry the newborn lights
> To catacombs, deserts and caves
>
> . . . .
>
> Everything could tracelessly disappear
> We knew this all alone
> But you who will erase us
> We greet by the welcome hymn.

Alexander Blok, with the same worry for the people and with the same expectation of a tragic revolution, wrote at the beginning of 1910's[27]:

> Yes, so teaches inspiration
> My free dream
> Clings to abasement
> And dirt and darkness and poverty
> There, there, humbler and lower
> Where another world is more visible
>
> . . . .
>
> To life's impenetrable horror
> You open, open your eyes
> Before great a thunderstorm
> Would wipe out everything in your Motherland
> Let rightful rage ripe
> Prepare your hands to work
> You can not — let depression and ennui
> In you accumulate and burn.

Even the revolutionary-minded intelligentsia underestimated, however, what could be done by the masses, by lower class people having passed through the fire of the terrible "German" war. This war taught them

---

[26] Excerpt from V. Bryusov, *Coming Huns*, translated by A. Leonidov.
[27] Excerpt from A. Block, *Iambic verses*, translated by A. Leonidov.

what an easy thing was murder when extermination of millions of people by other millions was already perceived as something normal, when a the value of life cheapened, the verb "to murder" was no longer describing something terrible and meaningful and was replaced by colloquial "to slap", "to send to waste", "to the wall" and became an everyday customary notion.[28]

The future of the country was determined by the fact that during the revolution the victor was the extreme wing of the impatient — bolsheviks and leftists. It is very important that by that time they had also been morally crippled by the heavy atmosphere of an underground with its inevitable mutual suspect, provocations, with an inevitably arising conviction in a permissibility of any methods of action in order to achieve the great goal. So there grew the "daemons" of Dostoevsky. They managed (and in this, with great efficacy, there helped an already mentioned three-year war) to drag masses of people with them.

In his youth Igor Evgen'evich was a Menshevik, i.e. belonged to the "impatient", but not to the "daemons". After the revolution, however, he could not stay in this position although an influence of the ideas of his youth was important throughout his life.

Igor Evgen'evich was born almost a year before the Khodynka Tragedy when during the coronation of Nikolai II, 1400 people died and 1300 were heavily injured, but not only did the tsar did command a prayer — that evening he went to the ball given by the French ambassador. The name Khodynka became symbolic.

Tamm was nine at the time of the war with Japan (and could not miss constant discussions condemning those who had begun it) when on "Bloody Sunday" a demonstration of workers with gonfalons and tsar portraits was raked with fire with the tsar's consent. Once again the tsar did not even command a prayer. At ten Igor Evgen'evich learned about an uprising in Moscow, a general strike and a new constitution. For several years to follow peasants were setting fire to the landowner's estates and for this were shot (let us recall a drawing by V.A. Serov who, it would seem, was so distant from politics) and hanged in such numbers that the gallows noose was called Stolypin necktie. Indignant discussions of these events took place in all intelligentsia families and in general throughout the country.

---

[28] A bitter recognition of this turning point is reflected in a somber comic story. A young man from a peaceful family was mobilized, got an overcoat and was sent out into trenches. When a battle began he jumped out and shouted, "You are crazy, what are you doing? People are sitting here!"

The parents of Igor Evgen'evich were not revolutionary-minded at all. This was a working intelligentsia family but should one wonder, knowing the passionate character of Igor Evgen'evich, that his political position was determined already during his gymnasium years? This was, of course, the revolutionary one. He read a lot of socialist literature, considered himself a convinced Marxist, had contact with social democrats. He attended meetings of the Marxist group of workers.

"Politics" as Igor Evgen'evich called his desired future activity had already captured him. However he faced strong resistance from the side of his parents who were more moderately oriented and understandably feared for the life of their temperamental son. A compromise was arranged. As was already described, after graduating from gymnasium he would, giving a word of honor of non-participation in the underground activities, for one year go to study abroad. The university of Edinburgh was chosen. In the August of 1913 he left. We will learn a lot on his year-long stay there from the remaining letters to his future wife Natalia Vasil'evna Shuiskaya.

Igor Evgen'evich was prepared extremely well. Therefore much time was left for satisfying other interests. He earned money on the side by teaching Russian at foreign language courses (until he demanded a raise of the ridiculously small payment and, after not getting it, left. He began studying *The Capital* by Marx and "began hating capitalism" as he jokingly wrote in one of his letters). He was totally absorbed by this new life. With a characteristic avid interest he absorbed diverse impressions.

During the first days he made acquaintance with "a dozen of students". Already in September he wrote: "I take most active part in the political life of the country... Yesterday I voted [for] the revolution expressing a strong reprimand to the English government."; "I became a member of a student socialist group (this should not be known at home)... all the activity of the group is in reports and debates."; "I have just returned from a meeting of the Fabian society (a socialistic one, I am a member"). Tamm describes the horrors of children's lives in the slums which "petrified" him.

Christmas vacation was used for a trip to London. He read a lot of literature forbidden at home, talked with Russian emigrant socialists, etc. He wrote,

> How stupid is it that I am studying to "for an engineer". My aspirations are for pure science, I would be a bad practical worker and, most importantly, I will never work as an engineer. Oh, how does my word of honor distress me ... One says this is

> a sacrifice to my parents but one should not sacrifice something that makes sense of one's life to anybody.

Later he wrote:

> I have spent the yesterday's evening very well. An Armenian acquaintance took me to the International Club where 26 nationalities and countries are represented. I listened to a talk on New Zealand but most interesting were discussions and meetings.

So, when learning life, there formed Tamm's internal world. Here are the fruits of his youthful profundity:

> "... how pleasant is it to run across a book in which you find your own thoughts, but deeper and more coherent. ... I have just read one such book, the *Living life* by Veresaev ... Here is a mixture of his and my thoughts. The biggest mistake in the world is that people have elevated and exaggerated thought. It accomplished much, very much and the people began thinking that it is omnipotent and can even find the meaning of life. Just think — the thought that by its essence can only perceive the external, find relations between phenomena and not their cause (he seems to have read a lot of Kant. — E.F.), which gives everything a mechanical treatment, should find the meaning of life! ... But it is not at all where it is looked for. Consciousness is only a small part of a human being, only a film on its surface. Everything most important and meaningful (even life itself is born there) takes place in an infinite sphere of subconscious, where feelings and almighty human instincts are rooted. The sense is in them. Justification and meaning of life is in the life itself. A mighty instinct gives it a meaning. To perceive that you live, move, feel, think, that you are "something" — here is the meaning. In experiencing joy, suffering, love, hatred, in mental in physical pleasures — here is the meaning."

He concludes: "Nonsense? This is, by God, a very interesting thing but I can not express it."

Such are the thoughts of a "philosopher at eighteen", of a man tensely thinking on himself, on the world, on humanity, on God. There formed a hard atheistic position that he kept throughout his life.

An enthusiasm with respect to the book by V.V. Veresaev, a doctor, a writer, a literary critic, a typical intelligent of Chekhov's kind who put intelligentsia itself, the people and their interrelations at the center of his writing, is natural. This witnesses that Igor Evgen'evich himself belongs to the same "class".

At the beginning of the summer of 1914 Igor Evgen'evich returned home and matriculated to the faculty of physics and mathematics of Moscow University. And, very soon, the First World War that shook and changed Russia's fate and the life of the whole world broke out.

During the first two years the students were not drafted into military service. However the views and the character of I.E. Tamm did not allow him to stand aside. It seems that very soon he took a sharply anti-war position. Many clever people, even from the monarchist camp, understood that Russia did not need war. Already in February 1914 a former minister of interior, a member of the State Council, P.N. Durnovo, persuaded the tsar in a detailed letter that conquering new territories was useless and even harmful for Russia. He warned that in any case, be it a victory or a defeat, Russia will face a turmoil of an unprecedented scale. In essence he gave an astonishingly accurate forecast of much of what would happen in three years. He wrote, in particular, that simple folk do not need a political revolution. They need land and work in the factory, that during revolution the folk will first and foremost turn their fury on the same liberals and democrats that are calling the people to it.

However the revolution with which Durnovo threatened and which happened exactly as described was precisely what Tamm and all the "impatient" intelligentsia wanted.[29]

Everything changed after the February revolution. I.E. Tamm plunged into political activity. He gave speeches at numerous anti-war meetings and had success as a speaker. He printed and distributed the anti-war literature. Finally, he was elected as delegate from Elisavetgrad to the First All-Russia Congress of Soviets of working and soldier deputies in Petrograd. He

---

[29]Of course a question of whether it was possible for Russia not to participate in the war is not so simple. Let us recall that only due to Russia that in the August of 1914 sacrificed the army of general Samsonov thrown into the offensive being not completely prepared for it (earlier than the time in the strategic planning) Germany had to send a part of troops advancing to Paris to the Russian front. Because of this Paris, France and, consequently, the "second front" in Western Europe were saved from the defeat. It could be that if this had not happened, if Russia had stayed aside then Germany, after finishing with one enemy in the West, would have then attacked a lonely Russia — as it happened in the Second World War. Although "if" is not a valid argument in historical issues, the history of German wars shows with sufficient clarity that a war into two fronts is impossible for it while at a single one it is victorious. This follows not only from the defeats in the XX century but also from the victories of the XIX one (first Austria, then France — both were defeated).

belonged to the fraction of Mensheviks-internationalists and actively continued his anti-war struggle. When at the famous session on June 19 (July 2) a new offensive in the front was announced and a resolution supporting it was suggested (and adopted) Tamm was the only non-Bolshevik who voted against this resolution.

To which party did he belong is not quite clear. In an earlier letter he wrote: "How do you like a dirty attack of the liberal press on us (Bolsheviks)? ... The students ... surpassed my expectations. You hear only: "provokers, German spies, enemies of the Motherland, Jews"".

From one hand, when involved in political activities in Presnya he acts against a prohibition of the party leadership (it is not clear which party it was, probably the menshevist) and tries to create a unifying social-democratic organization (of Mensheviks and Bolsheviks); from the other hand he summarizes his impressions from the Congress of Soviets by writing:

> All lines and contours of political views became much more distinct and clear. The truth is that my views have not changed, they have only strengthened. They have changed in only one aspect — I have twice convinced myself that Bolshevism exists in the masses only as a demagogic anarchism and unruliness. Of course this does not refer to its leaders who are simply the blinded fanatics, blinded by the truth, the really big truth that they defend but which hinders them from seeing anything else besides it.

However, when the revolution had happened and the fights had begun in Moscow, Tamm was isolated. He lived in a big house near Nikitskie Vorota and, when he tried to go out and join the Red guard, he was arrested by a White detachment. They took his pistol and, together with other inhabitants, forced him to return to the house (shot at from all sides) and blocked it. During all days of fighting (six days) Tamm stayed at home having only fragmentary information on what was happening in the city. Some actions of Bolsheviks sparked his indignation. However, in one of his letters he writes that he was "tensely thinking" on the way out but "it was impossible to sneak to the red and he had nothing to do with the white".

Feeling useless he decided to finish university (he still had to pass through several exams). On November 14 he wrote: "My position is the same, it has only strengthened. I am further to the left in Menshevism as before. I will certainly be active in politics, even more than before, and if I am now busy with the university — this is only to free the hands... All in all I am convinced that one can not mix regular political work with exams, nothing productive will come out on either side."

Therefore he still thought that with the new regime a not purely bolshevist political activity would be possible. Although he managed to graduate from the university, nothing went on as he had thought. A dispersal of the Constituent Assembly, a rapidly strengthening terror, a raging "demagogic anarchism" and the soon-coming civil war tragically sobered many "impatient" intelligents.

After the deeply emotional poem *The Twelve* that still justified an invasion of Huns by their great goal, Alexander Blok, who had earlier been writing dozens of verses a month, wrote only two poems. The first of them (*Scythians*) was in essence revealing a confusion that captured him; in a paradoxical form he still tried to justify the horror that had set in but afterwards probably gave up and was broken. In the three remaining years of his life he wrote only one verse, *To Pushkin House*. At one of literary disputes of that time one enthusiast, pointing his finger at Blok, menacingly said: "Blok, you are dead!" After this when late at night Blok, together with Chukovsky, were returning home, Blok said, "Yes, I am dead". He went towards his death without resistance. At the time when the poet was working on the revolution-praising *The Twelve* Tamm declared that he was "still further to the left in Menshevism", i.e. closer to Bolshevism. Sobriety, however, came to him as well.

After graduation from university, life moved Igor Evgen'evich to different places. In spite of the initial "almost Bolshevism", the events seen and experienced compelled him to abandon politics for good and be "just" a theoretical physicist. However, even seeing that a new Russia was a caricature on the socialist ideal of his youth he kept it in his soul for his whole life. That was why he avidly looked for those elements in the further development of the country that still spoke of socialism. These possibly fuelled his hope that once evil forces faded away, a step to the "bright future" would be made. This was a consolation for many intelligents even when the "daemons" had fully exposed themselves.

What was the basic mistake of the especially "impatient" in their striving for an immediate "equalizing" revolution? In the country that had only recently freed about a third of its population of serfdom and was keeping the traditional monarchy intact; the country with the population just beginning to develop an understanding of the elements of civil rights and other pillars of democracy; the country that had not gone through what is usually called the cultural revolution i.e. everything that took centuries in other countries, the country in which dozens of millions of people from the "lower classes" had additionally been embittered by the world war — in such a country one could not expect anything different from what Bryusov and Durnovo did and what really happened.

Answering a corresponding reproach by a Menshevik N.N. Sukhanov Lenin wrote: Why not *seize* the power first and make a cultural evolution only *afterwards*? The history gave its answer: this is impossible. It was possible to liquidate the almost general illiteracy, then organize general school education and open universities but when doing all this "on the class basis" in the horribly militarized country and with an unprecedented repression of personality it was possible to maintain the people's deprivation of rights only by terror. It was possible to reach "obrazovanshina" (a well-pointed word belonging to Solzhenitsyn)[30] but not the true general (i.e. also encompassing "lower classes") culture inseparable from democracy.

Revolution, civil war and terror turned over all the foundations of the former status and former position of the intelligentsia. The now openly despised "second class" was nevertheless useful to the state that needed "spetsy".[31] The intelligentsia felt the pressure of state power together with all the people, as a rule in a stronger way. Therefore there could be no feeling of guilt with respect to the people that characterized the pre-revolutionary intelligentsia. The pain was now for the whole country and for one's own self.

The three groups of intelligentsia that were described at the beginning of this piece faced a different fate. Those who we conditionally termed Slavophiles, inseparably related to orthodoxy, were either expelled from the country (writers, philosophers and other outstanding people) or crushed, together with the church, by merciless terror or, if staying alive, saved themselves and their culture by going into a silent "internal emigration." What has miraculously survived, mainly through children and grandchildren, is coming back to life before our eyes — along with a simultaneous revival of an inexorably conservative and aggressive church, dreaming of the restoration of a pre-Peter situation when the tsar was afraid of the patriarch whose influence was the same as that of the tsar.

The majority of those belonging to the extreme revolutionary wing of the intelligentsia fully accepted revolution and merged with the authorities. Soon, however, they were exterminated anyway (leftists almost

---

[30] People who are educated but lack true general culture.
[31] People having, e.g. engineering expertise.

immediately, others within 1-2 decades). As foreseen by Durnovo, the lower classes that had not lived through "cultural revolution" pounced upon those very liberals and democrats who had called the people for revolution.

The major part of the working intelligentsia could no longer display any political activity and continued to work and follow the old moral code, but this was very difficult and many turned into "co-followers" that tried, to a bigger or lesser degree, to accept the ruling ideology and find a justification for it.

An existence of the intelligents (in the pre-revolutionary sense) was to a large extent determined by the possibility of finding such a niche in which they could protect their personality by working professionally or, even better, creatively. Of course, no such niche did provide a guarantee that an executioner's axe would not fall upon one's head.

In the new circumstances finding this niche was much more difficult for those working in the liberal arts (writers, people of art) than for engineers, specialists in natural philosophy and other experts whom the state needed at that moment. Initially the ideological press was for the latter less heavy than for scientists in liberal arts. Pasternak, however, wrote already in 1923[32]:

> We were the music in the ice,
> I am talking about the frame
> With which I planned
> To leave the stage and that I'll do
> Here there is no place for shame.

He was writing with sarcasm on those intelligents that were passionately believing in the new regime (Bruysov? Mayakovsky?)[33]:

> In the legend's glaze behind
> A fool, a hero, an intelligent
> In the light of decrees and commercials
> Blazed praising the new power
> That in the corners secretly reviled him
> An idealistic intelligent
> Printed and carried the banners
> On a joy of his dawn.

Still he could not hide an understanding of a certain appropriateness of what was happening and his amazement with respect to the personality of Lenin with whose "voice extract ... screams history itself" on what in it "by

---

[32] B. Pasternak, *High Illness*, translated by A. Leonidov.
[33] *Ibid.*

reality's blood inscribed is" and Lenin was "its sonic face". He acknowledged that Lenin "controlled a flow of thought and thus the country," but[34]

> A *genius* comes as a herald of benefits
> And revenges his leave with oppression.

Years had to pass before this some kind of dualism transformed into a final verdict:[35]

> I was leaning towards poor
> Not because of being high-spirited
>
> . . . .
>
> Although acquainted with gentry
> And delicate public
> I was an enemy of spongers
> And a friend of the poor
>
> . . . .
>
> But I have spoiled myself
> Since time was touched with spoil
>
> . . . .
>
> To those whom I trusted
> I am no longer true
> I have lost a man
> Since everyone has lost him.

In different versions this process of mind transformation gripped almost all the intelligentsia of the pre-revolutionary origin.

For scientists in the field of natural philosophy this was easier. In spite of many losses from terror, of the strengthening ideological press of illiterate "supervisors" they could still breathe a pure air of their science which rapidly developed because the new regime needed it. Scientific institutes and universities were growing like mushrooms. A universal schooling, with all its drawbacks, inevitably gave rise to generations of people capable of thinking. Of course this often merely led to "obrazovanshina" and not to true "intelligentnost".

The regime needed these people but was also afraid of them and therefore kept them in permanent fear. In a satirical story by D. Granin *Our dear Roman Avdeevich*, its character, a typical member of the Brezhnev

---
[34] *Ibid.*
[35] B. Pasternak, *The Change*, translated by A. Leonidov.

politbureau, produces a wonderful statement: "Fear gives rise to conscientiousness."

The fear reigning in the country indeed impelled people to instinctively look for and find the features of the regime that allowed at least partial reconciliation with it. Such features were of course present. An aspiration for somehow blending into the reigning system, to accept the spirit of the time can be found by many poets and writers, even the best ones, but of course not by all (let us recall the unbending Akhmatova and Bulgakov).

An intelligent, a scientist in the field of natural philosophy that had found his niche was more inclined to recognize positive features of the regime than others. It was not accidental that young Landau was loudly proclaiming himself to be a materialist and even marxist — both at home and abroad. It was not accidental that courageous unbending Kapitsa was underlining the positive sides of this same regime. It was also not accidental that Tamm, during his visits abroad in 1928 and 1931, was convincing of this to his foreign colleagues and probably even convinced Dirac.

All this did radically change at the time of the "big purge" at the end of the thirties. In the spring of 1938 Landau took part in writing a leaflet to be distributed during the demonstration of May 1. It declared that Stalin had betrayed the ideals of the October Revolution and established a regime akin to the one of Hitler. He was of course arrested. After a year in jail and suffering during investigation he was nevertheless released (a miracle secured by courageous and wise Kapitsa[36] who gave his guarantees). In these years many people, also physicists, disappeared forever. There took place horrible public processes.

One of the disgusting "habits" of that time was to "discuss" the people "of interest" to the regime at the staff meetings of the place where these worked. Like a kennel, watchful activists pounced upon people having "terrible vices": arrested relatives or friends, "compromising" past, "bad" social origin or relatives abroad. It was expected that those "discussed" would recognize their mistakes and repent of them; would condemn the already 'discussed' friends or relatives, break with them, etc. Most absurd was the fact that it was almost never known what the repressed "enemy of people" had been accused of: The "organs" did not inform about it. There, at these meetings, facts from their lives were brought up in an attempt to prove their

---

[36]See Section 10.1.

guilt, their work for the enemy that the "discussed" had not unmasked and still had to unmask.[37]

Such unmasking was however suitable only for those to whom there referred the following well-known paraphrase from Pushkin's *Evgeni Onegin*: "He could easily speak and write in Marxist, acknowledge mistakes and confess." Indeed, why not renounce, not curse a friend who is already executed and one can neither help nor harm him? For a Russian intelligent this was however impossible and neither Tamm nor many others, including scientists, could do this. He did not lose face in such situations but his emotional experience was very heavy. In spite of direct pressure and threats he did not disavow anybody, did not condemn anybody postmortem and in general remained the Tamm known to everybody. On the surface he stood firm but only those close to him knew what it costed him, personally.

And, then came the war. In the face of Hitler, all of the country, including the intelligentsia, was swept along by a feeling of true patriotism. Although it was clear to many that Stalin as a despot is akin to Hitler, an outstanding role was played by the national idea. After all, Hitler openly declared as its aim enslavement of the Slavs, extermination of the Jews and not just the conquest but the subdual of the country and an establishment of the rule of the "superior race."

As for our intelligentsia, Stalin, with all his disdain and even hatred towards it, demonstrated an unexpected understanding of the necessity of preserving it for the sake of the future of the country. A decision of the State Committee for Defence of September 15, 1941 categorically forbade sending to the front (and, in general, using not according to an area of expertise) of all the teaching staff of the institutes of higher education and scientific workers (both working in natural philosophy and humanities). The patriotic feelings of scientists were realized therefore in passionate work on problems that could be useful for defence. Such work also started in FIAN. Tamm, who had devoted all preceding years to extremely abstract, so one thought, problems of a theory of nucleus and elementary particles, found himself "out of work." He looked for application of his outstanding craft and the things he did were really useful. At his scale, however, these were insignificant subsidiary projects.

In 1943 there started and underwent rapid development the Soviet project on creation of the atomic (nuclear, uranium and plutonium)

---

[37]More detail can be found in the chapter "Vavilov and Vavilov's FIAN" below.

weapon. Although due to the above-mentioned decision of GCD the scientists in this field were available and in the end they completed the project with success, their number was too small for such a great problem. The specialists from the scientific disciplines having something in common with the project and even Ph.D. students from the front were mobilized to the project. They were rapidly learning the new topic.

It seems that Igor Evgen'evich with his broad knowledge of different branches of physics and his brilliant talent of a master was needed. However, at the beginning he did not take part in this super-secret work. A reason can be seen only in his political "unreliability" in the eyes of "organy." In addition to his above-mentioned "vices" a personal enmity of A.A Zhdanov, an ideological boss in everything related to culture, could play its role. This showed up also later in 1946 when new members of the Academy of Sciences were elected.

At those times the lists of candidates were preliminary checked in the party CC[38] and its approval was mandatory. The party members were then obliged to vote for the candidate while many of non-party, although the vote was secret, yielded out of fear (such was a situation up until perestroika). One hour before the vote on the day of election in each Academy branch there took place a meeting of the "party fraction" (the members of the Academy that were members of CPSU) and, in the presence of the CC representative, they were told for who they should vote and for who they should not. One can recall only two cases when the Academy did not yield. During Khroushev's time it rejected Nuzhdin, a close collaborator of Lysenko (Khroushev, in fury, wanted to abolish the Academy) and in 1966 it rejected a head of the CC science department, an illiterate Trapesnikov and the scandal was quite loud (an announcement of the voting results was out of fear postponed for two days).

Thanks to the influence of at that time Academy president S.I. Vavilov in 1946 it turned out possible to get an approval and elect really good physicists. The candidature of Tamm was however removed from the list by Zhdanov himself. Although for the world of science this looked stupid, this "expert in science" who was sharply critical towards quantum mechanics perhaps considered electing a "bourgeois idealist" Tamm to the Academy impossible.

Only in 1946 was Tamm allowed to work on some problems that were "safer" from the point of view of secrecy. This is how there appeared his

---

[38] Central Committee.

work 'On the Width of the Large Intensity Shock Wave' allowed for publication only in 20 years. Same fate was shared by his work on the interaction of accelerated particles in the accelerator (this issue was at that time also related to the atomic problem).

There passed, however, only two years and, be it because Zhdanov died or due to the personal influence of I.V. Kurchatov (a scientific supervisor of the whole problem), things changed. There arose a problem of creating a more terrible weapon, a hydrogen bomb. Although even a principal possibility of creating such a weapon seemed very problematic Igor Evgen'evich was asked to organize an analysis of this question in the theoretical department of FIAN.

Igor Evgen'evich accepted this offer and assembled a group of his disciples and coworkers. It included, in particular, V.L. Ginzburg and A.D. Sakharov who already after two months suggested two of the most important original and beautiful ideas that allowed to make this bomb in less than five years, faster than the Americans. In 1950 Tamm and Sakharov moved to the super-secret town-institute known nowadays as "Arsamas-16". Yu.A. Romanov also followed them, some other collaborators whom Tamm wanted to take with him were not allowed to do so because of the "biographical particulars."

A work on the realization of the main ideas was extremely intense and difficult. It required a solution of many problems from many domains of physics — nuclear physics, hydrodynamics, gas dynamics, etc. In the article 'With the eyes of physicists of Arsamas-16' published in the 3-d edition of the book *Memoirs on I.E. Tamm* and in the magazine *Priroda* (N7, 1995), Yu.B. Khariton (a scientific leader of the institute) and his coworkers tell what a giant role Igor Evgen'evich played as a researcher and as head of a theory group. He was also a participant of the real test of the first "article" in the summer of 1953.

As is well known many physicists distinguished both by their talent and high moral values worked in the institute at that time. This collaboration was remarkable and Igor Evgen'evich perfectly 'fitted' into it. He was one of the recognized leaders. Igor Evgen'evich told me what an extremely strong pressure of responsibility he experienced before the first planned test when, together with another specialist, he studied meteorological data to decide whether it could be made. From the meteorological situation there depended whether an extremely dangerous radioactive trace would form in the right direction so that populated places would not be struck. Finally they gave their approval and were right.

The success of the project changed the opinion of people in power on Igor Evgen'evich and drastically changed his level. His authority in their eyes sharply increased. His election to the Academy was finally approved. Igor Evgen'evich returned to Moscow, to FIAN, and immediately continued, together with his younger collaborators, his passionate and intensive work on fundamental problems of theory of particles and quantum fields. He started frequently going abroad both for scientific visits and to participate in the Paguosh Conferences where scientists aimed at preventing the nuclear war. It does not seem to me that his evaluation of these meetings of liberal scientists — people of great human value, but having to deal with Soviet participants who were under strict and detailed control of highest party bosses was high. Still he did not find it possible to refrain from participating in these meetings.

Still, the general public is constantly asking two imminent questions. First, how could Tamm and other scientists take active participation in creating a horrible weapon that for already half a century cast fear in the whole mankind? Second, how he and our other scientists could create such a weapon for Stalin (like it happened earlier, without Tamm's participation, with the atomic bomb)?

The answer to the first question is relatively simple. Many centuries of scientific progress had led to a possibility of taming the nuclear energy. If not in one country, it would inevitably happen in another, perhaps with a delay of several years. It is thus unjust to blame the scientists, even those having made the last step. They understood the meaning of their discovery themselves. It is still remarkable that countries separated by almost prehistoric wild animosity found strength to reach an agreement and for already half a century the sword of Damocles is above us but did not fall. This is very reassuring, that mankind has learnt to be reasonable and to suppress wild impulses.

It is as senseless to accuse scientists as it is to accuse Prometheus, who had brought fire to people on Earth, that became a great blessing for humanity, but also gave birth to evil. It is unfortunate that a development of science went ahead of the moral and social development of a mankind that was incapable of using the benefits of this grandiose discovery and, at the same time, suppress its evil sides.

It is more difficult to answer the second question. It is necessary to take into account that two factors were of importance. First, our scientists worked not for Stalin but for mankind and for our country. Stalin and his regime were no mystery for very many of them. Of course, in subconsciousness a purely scientific enthusiasm towards the grandiose physical problem played its role. Fermi expressed this soberly and realistically: "First of all this is good physics." However, for the overwhelming majority this was not a primary motivation. This rather was in understanding that there was only one way of preventing the evil — a liquidation of the monopoly of one country and an establishment of equilibrium between two rival camps in terms of nuclear weapons. Then nobody would dare to start a nuclear war in which there could be no winner. I have many times heard from Landau: "Physicists did a good job, they made war impossible."

It is also necessary to recall that already in 1944, before the first test of the nuclear bomb, Niels Bohr (this was, of course, not only his opinion) tried to convince Western politicians that it was necessary to share with the Soviet Union the secrets of atomic weapon in order to prevent possible complications after victory over Hitler. Politicians, however, played with him as if he had been a ball. Roosevelt seemingly agreed but sent him to Churchill. Churchill was outraged and even wanted to intern Bohr. This did not happen but both leaders agreed at once that no such sharing should take place.

When the 'Iron Curtain' was lifted many Western scientists (about 40) took part in a Session of the Academy of Sciences on peaceful usage of atomic energy in Moscow in 1956. I was first astonished how fast did the old contacts resume and new friendly ones appear between the people that had, on the different sides of the curtain, created nuclear weapons for their countries. These meetings took place in an atmosphere of joy. However, soon I understood that they thought they had worked for the common cause, on preventing the nuclear war.

For our scientists another factor was also of importance — an inertia of patriotism born during the war and sharply amplified by the Fulton speech by Churchill which marked a beginning of the cold war. Of course, he understood that the aggressive politics of Stalin did not disappear after the war and one could expect anything from him. The speech was followed by many statements in the USA calling to use the Western monopoly and do away with communism by dropping nuclear bombs on the Soviet Union.

As is well known the monopoly ended in 1949 and Stalin immediately started a war in Korea. This, of course, supported Churchill's vision and disproved a statement that in the situation of nuclear equilibrium war is impossible in principle. It was, however, possible only with conventional weapons, nuclear weapons in Korea stayed unused. At the same time the monopoly inevitably leads to a temptation of using nuclear weapons. This was proved by the bombardment of Hiroshima and Nagasaki which, according to the conviction of many critics all over the world, was from the military viewpoint no longer necessary.

Of course, the threat of nuclear war did not disappear. However, a half-century period of silence of nuclear weapons strengthens the optimistic hopes of scientists. This is, perhaps, an answer to the second question.

Igor Evgen'evich died in 1971. To the end of his days he remained true to the basic principles of the Russian intelligentsia and its ideals (as I wrote elsewhere, a year or two before his death when we discussed, with sorrow, a difficult situation in our country, he said: "Yes, but it is impossible to deny that economy was reorganized on socialistic principles"). The strong sides of the Russian intelligentsia, its weaknesses and mistakes, its woes and joys were, as I have tried to show, resonant in him.

We see that it would be erroneous to think that the intelligentsia was completely crushed, gave up, turned into a humiliated gray mass and lost its previous noble look. The very existence of such unique personalities as Igor Evgen'evich Tamm and other Soviet scientists described in this book proves that this is wrong. The same humiliating opinion towards scientists in humanities and artists is wrong as well. Despite the betrayal of the great culture of some and mistakes and compromises of others they stretched a thread of this culture, although abused and oppressed, to the new days. The hunted Osip Mandelstam was correct to write[39]:

> For the thundering glory of future times
> For the high unity of mankind
> A cup at the father's feast did I lose
> And a joy and an honor of mine
>
> The century-wolfhound jumps on my shoulders
> But I am not a wolf by my blood.
> Push me better, as a hat into sleeve
> Of the fur-coat of Siberian steppe
>
> To see neither a coward nor puffing mud
> Bloody bones in road rut
> So that blue foxes would shine me all night
> In their pristine beauty
>
> Lead me into the night where Enisej flows
> And a pine-tree that reaches the sky
> Because I am not a wolf by my blood,
> And will be killed only by equal to me.

---

[39]Translated by A. Leonidov.

This verse is both a moan of the oppressed intelligentsia and a declaration of belief into the "thundering glory of future times" and a confidence that endured sufferings were not in vain, these are for "the high unity of mankind." And a scream of disgust when seeing a coward, "puffing mud" and bloody bones in road rut. And all this moaning that tears the soul suddenly ends by the two last proud lines: "Because I am not a wolf by my blood, And will be killed only by equal to me." He knows that he is stronger than a giant pack of monster wolfhounds. And life showed that he had been right. One could torture and kill the poet in the Gulag camp in the Far East but new times came and his verses sounded, stronger and stronger, for the "high unity of mankind." The wolfhounds that killed him disappeared, the verses of those quasi-poets that fulfilled the "state order" and prospered when Mandelstam was hunted and he was dying, but it turned out impossible to kill his verses as well as those by other true poets. The above-cited remarkable verse could be thought of as a credo of that intelligentsia of years gone that yielded neither to temptation nor to oppression, that had preserved persons in it.

We have to be grateful to all of them, to Russian intelligents of different origins and generations that were able to "carry" the already weakening great culture into the new epoch in which it has real chance to regain its former power.

2.3 I.E. Tamm and the Formation of Domestic Physics[40]

This is not the place to describe all the important events in the life of Igor Evgen'evich, the features of his personality that showed themselves in university years and in the years of civil war. Let us only repeat[41] that he graduated from the university in 1918 and only at that time terminated his political activity (in 1917 it had still been very intensive). He was considered to be a "menshevik-internationalist", i.e. quite close to Bolshevism. However, the October Revolution significantly enlarged this distance.

Several forthcoming years were "empty" for his scientific work. For some time, however, he taught at Simpheropol University where he established contact with many remarkable scientists (physicist Ya.I. Frenkel, mathematicians V.I. Smirnov and N.M. Krylov, biologists A.G. Gurvich and A.L Lyubishev, and others).

---

[40] This essay includes material requiring knowledge of physics. The corresponding parts can be read by a non-physicist reader "across the diagonal".
[41] See e.g. Section 2.1 in this chapter.

Then in 1921 he moved to Odessa to join L.I. Mandelstam who became his lifelong elder (at 16 years) close friend and, one could say, a mentor. Before that the turbulent years of the civil war had been filled with moves, sometimes very dangerous because of crossing the front lines. In essence his scientific activity began only in 1921 in the cold and starving Odessa.

At this place it is reasonable to pause and say several words on a situation with physics in Russia. Before the revolution, physics was, on the whole, weak. While in chemistry there already were such names as D.I. Mendeleev, A.M. Butlerov; in mathematics, N.I. Lobachevsky, M.V. Ostrogradski, P.L. Chebyshev, A.A. Markov, A.M. Lyapunov; in physiology, I.M. Sechenov, I.P. Pavlov, A.A. Uchtomsky; in physics, after Lomonosov, there had only been V.V. Petrov, soon completely forgotten, then E.Kh. Lenz and B.S. Yacobi, after that there flashed an inventor of radio, A.S. Popov. Among the most prominent physicists were N.A. Umov and A.G. Stoletov, both little known in the West, while the scientific activity of a really remarkable physicist P.N. Lebedev had just begun. His importance was not only in being a brilliant experimentalist who for the first time observed and measured the pressure of light (a Nobel level work). He was a person whose head was always filled with physical ideas, the first one who at the beginning of the XX century founded a Russian physics school akin to the Western European ones. He worked with his disciples (S.I. Vavilov, P.P. Lazarev, N.N. Andreev, V.K. Arkadjev and others). However, in 1911, together with more than hundred of professors of Moscow University he left it protesting against the reforms of Kasso, the minister of education, who sharply reduced a traditional university autonomy in order to fight a revolutionary student movement.

This event (the intelligentsia in Russia did not separate itself from liberal movements, from social and political problems) made the Moscow University bleed white and was completely deadly for the university physics. In a year, P.N. Lebedev died from heart disease at the age of 46. University physics fell into decay and lagged far behind the world level. Igor Evgen'evich recalled that in the university he attended lectures on a theory of electricity of some professor who, having reached the Maxwell equations, declared that this theory is too difficult so that he would not lecture on it. Several more serious young professors (N.N. Andreev, G.S. Landsberg) could not influence the overall atmosphere.

A special situation was in Petersburg where in 1907–1912 there worked an outstanding Austrian theorist P. Ehrenfest who had not found a job at home. A theoretical seminar organized by him exerted a great influence on the formation of a group of young theorists and theoretically educated experimentalists who fully showed themselves later, after the revolution. Other physicists of elder generations were as conservative as their colleagues in Moscow. An exception was the remarkably knowledgeable and sensitive to everything new O.D. Khvolson, whose five-volume course on physics was not only many times reprinted, in modernized versions, but was translated abroad.

One could say that if Russia was "pregnant with revolution", its culture was "pregnant with big science." In physics there was such researcher as A.A. Eichenwald who experimentally proved an equivalence of convection and conductivity currents (the Eichenwald effect); there was E.S. Fyodorov who classified crystallographic symmetry groups; there was A.A. Friedman (who should rather be considered as a mathematician and an expert in mechanics, only in 1922–1923 he made an outstanding discovery in physics by finding nonstationary solutions in Einstein cosmology and thus proved a possibility of the existence of an expanding Universe). A drawback however was in the absence of scientific schools, with an exception of that of P.N. Lebedev, that soon decayed.

Almost all prominent physicists of the pre-revolutionary times (except probably for N.A. Umov) were for many years studying and working abroad, the majority of them in Germany, which before the Hitler devastation of science, was an indisputable world leader in natural sciences. P.N. Lebedev was a student of Kundt, Kolrausch and Helholz; A.F. Ioffe was a student of Roentgen; A.A. Eichenwald, N.D. Papaleksi, D.S. Rozhdestvensky, N.N. Andreev, B.B. Golitsyn graduated from German or Swiss universities or at least worked there for several years after graduating from Russian universities. After the revolution almost all of them created their own schools, organized institutes, etc.

L.I. Mandelstam was no exception. After his expulsion from Novorossijsk (Odessa) university for participation in the student movement, he was, from 1899, a student of Brown and then worked with him in Strasbourg, where he became a professor, got a world name through his works and only in 1914 returned home. He exerted great influence upon Igor Evgen'evich. It was under his guidance when, at the age of 26, there began Tamm's scientific activity. Igor Evgen'evich was already fully prepared for this. In particular, his mathematical skills were at a high level. He proved it when in 1922–1925 there appeared his first three papers.

The first short paper written together with L.I. Mandelstam was published abroad after the other two but in fact was a source of two bigger papers (one only in Russian) published earlier. This was a research on electrodynamics of anisotropic media in relativity theory and was interesting from a general, principal point of view. However, even nowadays we do not literally deal with anisotropic bodies moving with relativistic velocities. It is no wonder that these questions were for a long time afterwards not touched upon in the literature. Only after a quarter of a century Yauch and Watson, not knowing about the papers by Mandelstam and Tamm, became interested in the problem and, in particular, re-derived some of their results.

The above-mentioned papers were basically written in Moscow where Igor Evgen'evich moved in 1922, when after the introduction of NEP, the situation was slowly normalizing. His life was however rather uncomfortable. To make money for his newly born family he had to teach in uninteresting institutes, write popular essays, give popular lectures on physics, translate books.

The overall situation with physics in the country was, however, rapidly improving. In Petrograd-Leningrad there appeared large physics institutes previously absent in Russia. This was, first of all, a physico-technical institute organized by A.F. Ioffe that became a cradle for many institutes branching from it, also into other cities, starting from the beginning of the thirties. In Kharkov, Dnepropetrovsk, Sverdlovsk (Ekatirenburg) and Tomsk there appeared new institutes with the kernels formed by the groups coming from the A.I. Ioffe Institute (the Kharkov and Sverdlovsk ones later becoming powerful scientific centers). Another big institute, an optical institute of D.S. Rozhdestvensky, concentrated on more applied topics (only due to its work we could develop, "on an empty spot," an industry producing a vast range of optical devices), but in combination with research on deepest questions of optics. It was not accidental that in the institute there worked, besides D.S. Rozhdestvensky himself, such physicists as V.A. Fock, S.I. Vavilov, A.N. Terenin. Finally, one should mention a Roentgen and radiological institute of V.G. Khlopin in which later, in 1937, the first European synchrotron was launched.

In Moscow rapid changes began somewhat later. A significant push was a move of L.I. Mandelstam, who became a head of a chair of theoretical physics in the university in 1925. Igor Evgen'evich became a privat-docent of a physical (more exactly, of, at that time joined, physical-mathematical) faculty in 1924 and in 1930 replaced L.I. Mandelstam as head of the theoretical physics chair.

Around L.I. Mandelstam there gathered the best part of professors and young Ph.D. students — A.A. Andronov, A.A. Vitt, M.A. Leontovich, G.S. Gorelik, S.E. Khaikin, S.M. Rytov and others. It is necessary to stress that the invitation of L.I. Mandelstam to Moscow took place after a long fight of these people with the rest of the professorship. A prominent role in this belonged to young people, in particular by those active in various student structures that were very influential at those times. Especially active was the student A.A. Andronov who later became an academician. Among the professors, an important role in inviting L.I. Mandelstam was played by the young S.I. Vavilov.

Physics in Moscow was also developing in a new technical institute akin to the above-mentioned Leningrad one and belonging to the people's comissariat (i.e. a ministry) of heavy industry. In such a way there appeared an electrotechnical institute where, in its physics department, I.E. Tamm worked at the end of the twenties. It is interesting that his most important works on quantum theory of radiation, that will be mentioned later, were published as done in this institute.

Around the same time well-known opticians V.A. Fabrikant, V.L. Granovsky, K.S. Vulfson, disciples of L.I. Mandelstam and G.S. Landsberg, began working in the institute. Other applied physical problems were studied, under the supervision of A.S. Predvoditelev, in the teplotechnichesky institute headed by a thoroughly educated engineer L.K. Ramsin (later sentenced as a head of a mythical "Promparty").

The state, within the limits of the possible, did not grudge in funding the development of science channeling it, as mentioned above, even through the industrial Narkomate. However, the country still did not have an industry producing research equipment. Members of the staff (in 1930 — including myself, at that time a laboratory assistant of the teplotechnichesky institute) went through second-hand shops buying everything relevant: broken amperemeters and voltmeters made by Hartmann-Brown or Siemens-Halske which could still be repaired; lenses from old photographic- and movie-cameras, etc. Physicists traveling abroad would often buy necessary materials from their own money. However, a still poor state bought more and more equipment abroad.

Let us mention that in 1931 disciples of L.I. Mandelstam — A.A. Andronov, G.S. Gorelik, as well as close to them people from Leningrad, M.T. Grekhova, V.I. Gaponov and others — moved to Nizhny Novgorod and organized an institute that nowadays is a powerful complex of institutes well known in the world. It is appropriate to mention that all these moves (of people from Moscow and Leningrad) to the periphery was to a large extent based on the civic understanding of a necessity of developing science not only in the capitals but in the whole country. Traditions of Russian intelligentsia were still alive. These people had to face tremendous difficulties but their cause moved forward.

At the beginning of the thirties the lack of equipment started diminishing. I saw myself how the previously empty shelves in the physics institute of the Moscow University started being filled. For example, there appeared a huge number of mirror galvanometers "FI" produced in the Leningrad physics institute. At the beginning half of them broke almost immediately but their quality improved with every year. They were cheap but they did work. There also appeared optical devices. In the Radium institute in Leningrad in 1932 there began a construction of the cyclotron, earlier in Kharkov, of a Van der Graaf accelerator. Industry fulfilled special orders for these bulky constructions.

The difficulties in the everyday life of scientists, especially the young ones, were still present for a long time. Until 1935 Igor Evgen'evich lived in an "apartment" rebuilt from stables in the university yard. The floor was at the level of the ground, the apartment was frequently flooded and the "conveniences" were in the yard. However, nobody felt shy about these things. Dirac, a friend of Igor Evgen'evich, who twice visited the Soviet Union, stayed in this apartment. To the second of these stays there refers a well-known episode when a short-spoken and exact Dirac said, replying to the embarrassed Igor Evgen'evich who glibly and rapidly apologized that nothing had improved after the first visit, "Why, nothing changed? Earlier one had to go there with a candle, now there hangs a bulb."

After this lengthy detour let us however return to the works of I.E. Tamm of that period. Having begun his scientific activity very late, he worked with remarkable intensity. After the first three papers, there followed several works, still done within the old Bohr quantum theory (before the appearance of quantum mechanics), that were of minor significance. In the same years he worked on his course 'Foundations of a theory of electricity', first published in 1929. This course, of remarkable physics clarity, became very popular. In the lifetime of Igor Evgen'evich there were many editions, almost always expanded and improved. It was also reprinted after his death. If one recalls the above-described level of teaching the theory of electricity in Moscow University before the revolution it is not difficult to imagine what a refreshing phenomenon this book was. This book is valuable nowadays as well.

One could consider that a breakthrough came in the year of 1928 when Igor Evgen'evich, already an established scientist, went abroad for almost half a year. He visited Born in Gettingen but spent most of the time with Ehrenfest in Leiden. Here he really mastered the newly born quantum mechanics and established close relations with Ehrenfest, who was of a high opinion on him, and Dirac who came to Leiden.

After returning home Tamm finished a work on the classical electrodynamics of a spinning electron that he had begun in Leiden. It is now not easy to understand why was it necessary to develop in 1928–1929, with the quantum mechanics and Dirac theory of electron with spin existing, a complicated relativistic non-quantum theory of the electron magnetic moment. However, the mere fact that the most part of this work was done in Leiden and the paper contained acknowledgments to Ehrenfest and Fokker "for many useful discussions" shows that the best physicists of that time considered the problem of interrelation between the Dirac electron spin theory and the classical picture of a rotating charge to be an important one. Igor Evgen'evich worked in physics with an all-consuming passion. When at that time he developed a great interest to the unified field theory by Einstein he devoted to it five (!) publications within 1929 (two of them together with M.A. Leontovich) aimed at understanding the behavior of the Dirac electron in this theory and, in a broader sense, as he wrote in the first paper, attempting to show that "a new Einstein theory possesses certain quantum mechanical features."

However, the fate of this work that consumed a lot of energy and time was not a lucky one. Einstein unified field theory in which its author tried, for three last decades of his life, to unify electromagnetic field with gravitational one similar to the way in which electromagnetic field unifies electric and magnetic ones did not solve the problem.

Nowadays we know why it was doomed: In the unified field theory the electromagnetism should be joined by weak (and also strong) interactions that at that time were unknown. Practically the whole world of physicists of that time considered Einstein's perseverance as an eccentricity pardonable for a genius. It took several decades for his *idée fixe* to reappear at a new level and become a universally accepted fundamental problem.

Evaluating this period of the first eight years of Igor Evgen'evich's work in theoretical physics showing his high technical professionalism and broad knowledge, one should admit that it did not bring him real success. Of course, one should not forget his remarkable course on the theory of electricity, but the significant scientific results were practically absent.

At this point Igor Evgen'evich makes a sharp turn in his scientific career. From the consideration of most general problems (relativistic electrodynamics of anisotropic bodies, applicability range of the correspondence principle, unified field theory) he turns to studying concrete phenomena within the framework of quantum mechanics and, within several years, achieves very significant results.

His close relations to L.I. Mandelstam, his almost exact coeval G.S. Landsberg and a younger friend M.A. Leontovich, had a direct influence on the choice of a topic of the first work of this period. These physicists were engaged in in-depth studies of light scattering in solids, both experimental ones and the ones within classical theory in which this phenomenon was treated as a scattering of a light wave on elastic oscillations of a crystal.

Igor Evgen'evich gave a quantum theory of the process but the importance of his work goes far beyond simply giving a consistent theory of the particular phenomenon. The thing is that he quantized the elastic oscillations in the same fashion as the quantization of the electromagnetic field was done by Heisenberg and Pauli. As a result the collective oscillations of the lattice particles appeared as a gas of "elastic quanta", quasiparticles, each including the motion of all particles in the lattice. Ya.I. Frenkel suggested to call them phonons. The concrete formulae obtained contained certain deviation from the results of the classical approach which found an immediate experimental confirmation in the experiments of G.S. Landsberg and L.I. Mandelstam.

The main thing here was, of course, that *for the first time in physics a motion of many interacting particles was reduced to the gas of quasiparticles*. An importance of this step is difficult to overestimate. Nevertheless the quasiparticles of various sorts, in particular phonons, have since that time gained such firm ground in physics, have become such a habitual notion, that if one looks into some physical or special encyclopedia, the articles "phonon" or "quasiparticle" would not even mention that these were introduced into physics by Tamm.

Right after this there followed another important paper by Igor Evgen'evich on light scattering on free electron, i.e. a theory of the Compton effect. This is again a particular process but the results are again of principal importance.

Igor Evgen'evich reconsidered the Compton effect problem using a consistent Pauli-Heisenberg field quantization, i.e. a second quantization formalism. The resulting formula coincided with that obtained a year before by Klein and Nishina who had used the correspondence method. So did Tamm just confirm their result and the work belonged to the class of "Verklaerungen" and "Neubegruendungen" (clarification and new substantiation) so despised by Landau? Not at all. The thing is that second quantization introduces intermediate states and Igor Evgen'evich found that intermediate states in which the Dirac electron is in a state with the negative energy play a fundamental role. Even in the limiting case of the scattering of long (infrared) waves which results in the classical Thomson formula these negative states are completely essential.

It is necessary to recall that at those times a presence of the negative energy states in the Dirac theory was a "headache", because all real electrons having positive energy would have fallen onto the level with the infinite negative energy.

There is a private episode related to this paper that is, perhaps, worth a digression. There exists a letter from I.E. Tamm to P. Ehrenfest dated February 24, 1930 from which one sees that at first Igor Evgen'evich sent the paper containing a small mistake. Here is an excerpt from this letter:

> Dear Pavel Sigizmundovich, You have perhaps received my telegram. I am awfully, awfully ashamed. When writing, I have checked my calculations three times before sending you the letter. Then I began finalizing the paper for publication and, as I always do, did all the calculations anew, not looking into the earlier calculations. And here it turned out that at the very beginning I have always written a wrong sign for sinus! If doing everything correctly at the end there is no deviation from the Klein-Nishina formula!
>
> This story is especially vexing to me because now I managed to reshape the calculations into an elegant form that I like. With a slight change one can, for example, calculate the probability of spontaneous transition of an electron in a state with positive energy into a state with negative energy. I am working on this right now and will finish this work in a few days.
>
> I am terribly sorry that having sent you a paper for publication for the second time I have again made such mistakes (last year I did not symmetrize the wave equation)...
>
> Yours Ig. Tamm.
>
> P.S. Of course everything told on the dominating role of transitions through negative energy states remains true. Ig. T.

In my memoirs on Igor Evgen'evich (see Section 2.1 in this chapter) I wrote that he did not publish any erroneous paper and that the papers were submitted for publication only after he diligently checked them out many times. It turns out nevertheless that there were moments when he saved himself from publishing an erroneous result at the last minute. Let us note that, as it follows from the text of the letter, the corrected paper was expanded, in particular by the formula for electron-positron annihilation cross-section.

Tamm published two more notes on the Dirac theory of electron in 1930 (on light scattering on two electrons) and in 1934 (in which in the place where one gives the name of the institute in which the work was done there stood "Teberda-Caucasus", i.e. this one was done during a vacation in mountains).

In 1931 Tamm went to Cambridge to work with Dirac whom he called a genius and whom he admired. On the way he visited Ehrenfest in Leiden and on the way back he visited Jordan in Rostock. Relations with Dirac developed into a real friendship. In a letter to L.I. Mandelstam, Igor Evgen'evich wrote: "In Cambridge I am feeling very good... Scientifically the most interesting thing is a new work by Dirac finished "before my eyes"... He shows that quantum mechanics admits existence of isolated magnetic poles." Igor Evgen'evich adds that in connection with this work he wrote a "mathematical paper, a study of amusing properties of electron eigenfunctions in the field of the magnetic pole."

Dirac taught Tamm to drive a car and they went to Scotland where Tamm, in turn, taught Dirac mountain climbing (he writes in the same letter: "... before that, as an introductory course, I climbed trees with him.") After that, when Dirac visited the USSR, Tamm went with him on mountain-climbing expeditions to the Caucasus many times.

Let us however return to physics.

After the works on the scattering of light and Dirac theory of electron Igor Evgen'evich switched to a new field, the quantum theory of metals, which at that time was in its infancy. Here three important problems were studied.

Perhaps the most important paper of this cycle is Tamm's work in which he discovered a possibility of the existence of particular surface electron states in metals. In such a state an electron can neither leave the metal nor penetrate inside it. These "Tamm levels," as they have been called since then, turned out to be exceptionally important for the physics of surface phenomena and, in a quarter of a century, for transistor technics, etc. By the sixties this field became so large that there appeared a monograph by

S. Davison and J. Levin, *Surface (Tamm) States* (it was published in Russian in 1973). Besides that were published, together with Tamm's disciple S.P. Shubin, the fundamental paper on photoeffect on metals and the one on photoelectric work in metals.

Completing the discussion of this period one should mention that Igor Evgen'evich published, for the first time, a paper on methodological aspects of physics in a philosophical journal. Defending new physics, the theory of relativity and quantum mechanics, from ignorant attacks of party philosophers and conservative physicists that had declared themselves marxists, he tactfully explained to them the true meaning of this new epoch in the development of science. The only result was a furious hatred towards him that solidified his reputation as a "bourgeois idealist" that would in the future bring him many dangerous problems.

Let us pause for a moment and consider a situation in physics at that time of its decisive transformation both in the world and in our country.

In the world this manifested itself in the discoveries of positron (brilliantly confirming the predictions of Dirac theory and strengthening the authority of quantum mechanics and theoretical physics in general) and neutron (after which it became clear that nuclei consist of protons and neutrons and opened a new epoch, that of the physics of atomic nuclei). In April of the same year Kokroft and Watson did for the first time split atomic nuclei at the high-voltage apparatus designed by them. All this as well as other achievements in physics changed, the atmosphere of research in general, its scale experiencing a rapid growth. This transformation is well described in the well-known novel by Mitchell Wilson (a physicist by education) which was due to some reason published in Russian under the title *Life in Darkness* (the true English title, *Life with Lightning*, appeared only in the new edition of 1990).

In our country this was a period of physics transforming into a developed science and getting international reputation. By this time several discoveries at the Nobel Prize level scale had been made. For example, in 1928 G.S. Landsberg and L.I. Mandelstam discovered combinational light scattering in crystals, the Raman effect, which is named this way because Raman, in India, had observed it in liquids and had sent a telegram to the journal *Nature* three months before G.S. Landsberg and L.I. Mandelstam sent their paper for publication, and thus Raman got the Nobel Prize.[42]

---

[42] In fact G.S. Landsberg and L.I. Mandelstam observed the effect even earlier than Raman, see chapter 1 in this volume.

This retardation turned out decisive — despite of the fact that Raman's understanding of the observed effect was at lower level than that by Landsberg and Mandelstam who clearly saw its physical meaning.

One could also mention the discovery of D.V. Skobeltsyn. In 1927 he found that cosmic rays when observed at the Earth surface consist of high energy electrons and thus could not be explained by, e.g. radioactive admixture in the atmoshpere and in 1927 showed that they are in fact electron showers. This discovery can be considered as a beginning of high energy physics. The Nobel Prize went, however, to the German physicist W. Bothe who confirmed this result by another method (a system of simultaneously acting counters and not the Wilson chamber as by D.V. Skobeltsyn). It is interesting that the Nobel Prize was given for this particular method. From 1926–1927 N.N. Semenov (at first working with Yu.B. Khariton) discovered chain chemical reactions (in a quarter of a century he received, together with S. Hinshelwood, a Nobel Prize for this discovery).

Finally, somewhat later, in 1933, S.I. Vavilov and P.A. Cherenkov observed an unusual radiation of electrons in media that is now named after them (published in 1934). In a few years it found its explanation in the work by I.E. Tamm and I.M. Frank. For this discovery and its theoretical explanation the Nobel Prize was given 25 years later as well.

Besides this many other remarkable works were done. Let me give only a few examples. M.A. Leontovich and L.I. Mandelstam theoretically discovered that quantum mechanics leads to a possibility of tunneling effect and studied its basic properties. Gamov theory of nuclear alpha-decay was based on this work. In the Kharkov physico-technical institute a Van-der-Graaf accelerator was built on which (in the same year of 1932 as D. Kokroft and E. Walton) A.K. Valter, G.D. Latyshev, A.I. Leipunski and K.D. Sinelnikov (all of them young people) reproduced an experiment on nuclear fission. One could mention other significant works as well.

All this proves that by the end of the thirties, in just ten years, there appeared a new generation of theorists (recall the still young L.D. Landau) and experimentalists that had made themselves names by works done at top international level. Even more works appeared in the following years. Schools created by scientists of the older generation mentioned at the beginning of this chapter gave rise to new schools. Simultaneously, as was already mentioned, there grew an industrial production of scientific equipment. New institutes proliferated rapidly.

While previously, scientific papers had been traditionally sent to German journals *Zeitschrift fuer Physik* and *Physikalische Zeitschrift* and had been rapidly published in them and only with a time lag (up to one year) appeared in Russian in the journal of Russian Physical and Chemical Society, in 1931 there appeared an international journal *Physikalische Zeitschrift der Sowietunion* based in Kharkov which rapidly published papers in German, English and French. It is true that it was "international" only to some extent, but papers by foreign authors, even by Dirac, did appear in it. The main thing was that Soviet papers found a direct way abroad and the journal was being read.

In 1934 in Kharkov an international conference on theoretical physics was organized. The conference was attended by about 30 participants (at that time this was a usual conference scale in other countries as well which is difficult to believe in nowadays). These included, however, Bohr, Rosenfeld, Waller, Gordon — in total, eight foreigners. An interest in and attention to our science grew.

Few years before that time a mere fact of the existence of research on physics in our country was not known to the wider public. I can refer to my own experience. I grew up and went to school in the center of Moscow. Feeling an attraction to physics I went to Roumyantsev (later Lenin) library to read books describing the works by Einstein, Bohr, Rutherford. I was convinced that only geniuses, not ordinary people, could do research in physics and after school tried to become a student of a chemical institute where, at least, there was a specialization on electrochemistry. When in 1930 I learned about the existence of the physics faculty one kilometer from the school and became its student, some acquaintances of my parents asked: "This is what, physical culture?"

Very soon such situations became impossible. The news on nuclear fission in Kharkov was published, in the form of a report to comrade Stalin on a significant victory of Soviet science, on the front page of *Pravda* as the main news of the day. Information on physics was in general spreading rapidly.

The scientific breakthrough of 1932 also changed the direction of the research of Igor Evgen'evich. He switched to nuclear physics.

Of interest is a fate of his very first work (together with his Ph.D. student S.A. Altshuler) in which, based on an analysis of experimental data on nuclear magnetic moments, it was stated that an electrically neutral neutron has a magnetic moment. This was already described in Section 2.1. Nowadays it is very difficult to understand why this idea faced such a strong condemnation at the above-mentioned Kharkov conference on theoretical physics in 1934. Igor Evgen'evich, however, did not see convincing arguments in these objections and insisted on his point of view. After a short period of time it was recognized as being correct.

Of special significance was however a voluminous and extremely important work of Tamm on the nature of nuclear forces. The proton-neutron structure of nuclei did with a necessity lead to a question on the nature of forces binding these particles together. Let us stress that at that time no forces beyond electromagnetic and gravitational were known. A mere assumption on an existence of some other forces was an incredibly bold fantasy. As Igor Evgen'evich writes in his paper, "immediately after the discovery of neutron in 1932 Heisenberg put forward an assumption that an interaction of proton with neutron is due to an exchange of electric charge." Please note — again only the electric one. One could not say anything on the mechanism of this exchange though. However, in 1934 Fermi gave his remarkable theory of beta-decay as of a process in which a nucleon emits an electron-neutrino pair. This was sufficient for Igor Evgen'evich to immediately come up with an idea that nucleons (this word was of course not yet used) interact through exchanging such pairs and anti-pairs. This was an assumption of an existence of new forces beyond gravity and electromagnetism.

He immediately started calculating, performed them at nights during the Kharkov conference and obtained a discouraging result: this exchange did give new kinds of forces, even of the required short-range type, decreasing with a distance r as r^{-5}, but many orders of magnitude weaker than it was necessary to explain the stability of nuclei. Igor Evgen'evich published the formula for the interaction potential in *Nature* and was srongly distressed by this failure.

During the following two years Tamm was, without success, considering other variants of the realization of his idea. In 1934 Igor Evgen'evich concluded an article with the melancholic sentence: "A further examination of a vast number of possibilities of this kind without knowing some general so far unknown principles appears senseless." I remember this mood of him very well. He forced him to send the paper for publication.

He was, however, wrong. The Tamm theory of beta-forces describes the really existing weak forces between the nucleons (these forces were after several decades discovered experimentally) but they are not responsible for the stability of nuclei.

Very soon, in 1935, Yukawa, directly referring to Tamm's work, put forward a bold idea according to which the origin of nuclear forces is in exchanging a new, at that time unknown, particle with a mass of order of one third of that of a proton — a meson. Now we know that this is a pion. A starting point for Yukawa was an idea of Igor Evgen'evich on interparticle forces due to an exchange of massive particles.

So, in some six-seven years Tamm published three cycles of papers: on scattering of light on solids (introducing a notion of quasiparticle, the phonon) and electron (including electron-hole annihilation and proving the necessity for Dirac negative energy levels); on quantum theory of metals (predicting an existence of surface "Tamm levels" and explaining the photoeffect in metals); and on nuclear physics (putting forward a statement on the existence of neutron magnetic moment and, most importantly, giving a theory of nuclear beta-forces).

This did at once bring him recognition and respect in the world physics community. As B. Pontekorvo told me, Fermi highly esteemed Tamm. Dirac was his friend and worked with him. Ehrenfest suggested him as his successor at the chair that, before him, had been occupied by Lorentz. He was recognized by Russian physicists as well: In 1933 Igor Evgen'evich was elected a corresponding member of USSR AS, he established good relations with L.D. Landau and V.A. Fock, an old friendship tied him with Ya.I. Frenkel. However he also received a reputation of "bourgeois idealist" among local marxist philosophers and reactionary physicists.

It was quite natural that when S.I. Vavilov organized (after the move of USSR Academy of Sciences to Moscow in 1934) the P.N. Lebedev Physical Institute (LPI) and invited there the best Moscow physicists, Igor Evgen'jevich was among them. He organized and headed, until his death, the theoretical department now bearing his name. Very soon he transferred there from the university his weekly seminar and in general made the institute a center of his activity.

At this time there intervened a new episode. As was already said in 1933 a Ph.D. student of Vavilov, an extraordinary diligent and attentive experimentalist P.A. Cherenkov studied, following Vavilov's proposal, luminescence of uranium salts dissolved in liquid under the influence of radium gamma-rays and did with horror observe that the liquid emits light, with almost the same intensity, in the absence of uranium salts. This made his studies of uranium salts senseless (he was also convinced of that). S.I. Vavilov, however, did at once became interested in the "parasite" radiation and did quickly understand that this is a new previously unknown type of radiation emitted by fast electrons torn from atoms by gamma-radiation. For a number of reasons this seemed impossible and caused a lot of mockery. However, in 1937 Tamm and Frank showed that Vavilov had been right and gave a physical explanation and a theory of this "Vavilov-Cherenkov" radiation (in the West one simply says "Cherenkov radiation" neglecting the important role of Vavilov in this discovery). As has been already told, it was precisely this discovery for which Tamm, Frank and Cherenkov received a Nobel Prize (Vavilov had died a long time before and this prize is not awarded posthumously).

It is funny that in the scientific biography of Igor Evgen'evich this work awarded with such a significant prize was in essence an episodical deviation aside. In these and following years his main topic was nuclear physics and elementary particle physics.

In this main field of his interest there deserves mentioning one work. The issue is that when muons were discovered (in cosmic rays) and their decay was observed one erroneously thought that their spin was equal to one. Tamm found that an exact complete system of wavefunctions of such particles in the field of a Coulomb center necessarily leads to falling of this particle onto center. A way out of this situation is possible if one ascribes to these particles finite size — as suggested by Landau. Landau thought that the radius of a spin one meson should be $r_\mu \sim e^2/m_\mu c^2$ where m_μ is its mass.

This was, of course, a crude estimate. Nobody knew how to introduce into a local theory a relativistic particle having finite size. Nevertheless this estimate did, together with the conclusion of Igor Evgen'evich concerning the falling onto center, serve as a ground for a hypothesis by Landau and Tamm that such a meson falls onto center until it reaches a distance of order of r_μ. At such a distance an energy of electromagnetic interaction of meson and proton is of the order required for nuclear forces. According to this hypothesis a neutron, for example, consists of a proton and a negatively charged spin one meson. This work of the two great theorists distinctly shows in what thick a fog physicists paved their way to understanding the nature of nuclear forces, how distant from the truth were the hypothesis that they put forward. Before the discovery of pions the situation was indeed a dead-end one.

Let us however remark that in the period from the middle of the thirties up until the war the activity of Tamm resulted in only three papers. First, there is the Nobel work on the theory of Vavilov-Cherenkov radiation which, with all the high professionalism required for its accomplishment can, paradoxically, hardly be considered as especially outstanding among the other works of Igor Evgen'evich although it was, of course, one of the best. The second one, done together with S.M. Belenky, was on improving a theory of electromagnetic showers in cosmic rays. The third paper was on charged spin one particles in the Coulomb field.

Is it a lot or is it not? It is of course evident that these works are completed by a high-class theorist. One could however say that one could have expected more from the creator of phonons and the theory of beta forces. What was then the reason? It can not be explained as an age-related decline in creative activity. At that time Igor Evgen'evich was 40–45, according to the ancient — acme, a man's heyday. His later activity also serves as sufficient disproof of such a suggestion.

One could however wonder how Igor Evgen'evich did all this. These were the years of horrible Stalin terror. Friends of Igor Evgen'evich and his colleagues were blotted out, sent to Gulag camps. There followed, one after another, horrible mendacious court processes. As already described in Section 2.1 this led to truly tragic events in the family of Igor Evgen'evich.

It may seem strange that in this atmosphere people kept their personalities, even worked quite well. It is still difficult to understand how a brain squeezed by horror and ideological pressure could at the same time contemplate in an independent and creative way in its professional field. The answer perhaps is that the work was a salvation, a kind of internal emigration, provided a chance of keeping one's personality. How could one otherwise understand, for example, that after a year in jail and under cruel methods of investigation characteristic for that time L.D. Landau could publish in the following one and a half-two years a dozen papers including a fundamental theory of liquid helium (for which he later would receive a Nobel Prize).[43]

During these years Igor Evgen'evich was squeezed, exhausted from persecution, threats, disappearance of people close to him, understanding that the regime that had promised socialism (his life's dream from his younger years) transformed itself into a despotic merciless dictatorship. He was however a man of strong will and did not display this state of mind before others in the institute. For such a creative person a clean atmosphere of honest scientific work was that gulp of clean air that allowed to survive "with a loop

---

[43] I automatically recall how during the several years of his terrible deadly illness a half-paralyzed Igor Evgen'evich was nevertheless continuing active scientific work. When somebody expressed his astonishment with respect to this in the presence of M.A. Leontovich he said: "What else could one do to save oneself in such a situation?"

on the neck." One should add that this looked the same for the majority of people that devoted themselves to science.

The war brought theorists and experimentalists in nuclear physics into a state of embarrassment. To the common horror and grief there added an understanding of the otiosity of one's work. When on Monday, June 23, at the next day after the beginning of the war,[44] Igor Evgen'evich gathered a few, at those times, members of the LPI Theory Department, the depressed state due to this understanding was obvious. Perhaps only M.A. Leontovich who had in preceding years been working on a theory of propagation of radio waves knew that for him a transition to defence themes was natural (radar!).

The main direction of research in the department was however on the theory of elementary particles, nuclear forces, etc. Now nobody needed this any longer. Therefore there began a feverish search of the actual narrowly applied topics. Let us recall that in the Leningrad physics-technics institute the research on nuclear and particle physics was immediately terminated. I.V. Kurchatov, under whose guidance there had already begun a research on the cyclotron in the radium institute, the first in Europe, that had been just launched into operation, moved to working on ensuring the security of military ships with respect to magnetic mines joining A.P. Aleksandrov and some other scientists from the Leningrad physics-technics institute that had been already working on this for a long time. In FIAN, soon evacuated to Kazan, D.I. Blokhintsev began working on lowering the noise level of aviation engines (at that time when radars were still not available the anti-aircraft defense was based on noise detection) for which a serious progress in acoustics was needed. I.Ya. Pomeranchuk who at that time also worked in FIAN worked on this topic as well. Others chose narrowly applied topics. For example during evacuation the starving M.A. Markov began constructing anti-tank shells with better aerodynamic characteristics (this was, of course, somewhat naive — constructors in defense institutes were better professionals here), etc.

The passionate desire of being of some help to the front was however deep and sincere. I was lucky. Leontovich advised me to solve one problem of great practical importance in radiophysics that had waited its theoretical solution for more than 20 years. It is astonishing, but although freezing,

---

[44] On June 22, 1941 Germany invaded the Soviet Union and there began the Great Patriotic War.

starving and periodically bedridden from tuberculosis, I solved it. Only in two-three years did an acute need in nuclear physicists arise again.

In such an atmosphere there appeared the work by Tamm (made together with V.L. Ginsburg) on a foliated core that was needed for radiophysics (it was used by N.D. Papaleksi who was working in FIAN) and the one on variations of the magnetic field of the Earth. As was already mentioned Igor Evgen'evich calculated, at a request of A.P. Aleksandrov, complex magnetic fields of ships, etc.

Nevertheless during the whole stay in Kazan he continued an intensive research on problems in theory of particles and nuclear forces. As Ginzburg, along with the work of practical importance (on propagation of radiowaves in the atmosphere), continued working on the theory of particle that can have states with different spins, they began collaborating. This resulted in the work in which a relativistic equation for particles with varying spin was suggested. The paper contained some details that did not satisfy the authors. These were later clarified when this approach was generalized by I.M. Gelfand and A.M. Yaglom who suggested their own equation. Igor Evgen'evich was not quite happy with the results and postponed the publication of this paper until 1947.

So, what was Tamm in fact doing during the four wartime years? We have mentioned some insignificant, on the scale of this outstanding theorist, narrowly applied works. To these one could add some participation in the work of the laboratory of G.S. Landsberg which helped in organizing a production in Kazan, within the Academy, of steeloscopes, the devices for the express analysis of the structure of metals. These were of great need to the front for the fast sorting of metals from broken equipment (so that the noble sorts of steel would not go into a general remelt). Igor Evgen'evich helped in calculations on optical systems, etc.

We have seen that besides all these "trifles" on which he worked with a particular passion only one work (together with V.L. Ginzburg) was done — that on particles with higher spins. What else? I can only witness that Igor Evgen'evich was intensively working all the time not paying attention to the surrounding circumstances. It is worth quoting from the memoirs of V.Ya. Frenkel, at that time a boy, whose parents were close friends of Tamm. He describes one of the evenings in Tamm family:

Igor Evgen'evich sat on some small children bench, Natalia Vasiljevna, his wife, was busy with household matters and her father, a very old man with a full beard, was repairing shoes.

At that time there was nothing unusual in this in the families of scientists. When my mother and I came in, Igor Evgen'evich jumped up, greeted us, said a few words and sat down again on the bench with a notebook on his knees. My mother asked: "Would it disturb you if we talk?", "No, no, please talk and do not pay attention to me!", "Gora (this is how Tamm's wife and the friends of his youth called him) can completely switch himself off", clarified Natalia Vasiljevna.

In other memoirs one reads that in the corner one saw a heap of potatoes, a characteristic detail typical for the difficult life in evacuation even for most prominent scientists.[45] (At the same time, in September of 1943, FIAN returned to Moscow where life was less difficult. At least the food coupons were not just pieces of paper, one really got food for them, albeit in quite moderate quantity.)

One could think that this work resulted in a big and very important paper published right after the war (submitted for publication on August 27, 1945). In this work Igor Evgen'evich suggested an approximate method of effective consideration of nuclear phenomena in which there participate pions. Igor Evgen'evich himself called this method, which in the literature is called the "Tamm-Dankow method" (because five years later it was rediscovered in the USA by Dankow), that of cut or truncated equations. Igor Evgen'evich returned to it much later in two papers published in 1952 and 1955.

Having reached this place the reader would probably wonder why after formulating the method in the paper published in 1945 Igor Evgen'evich turned to its applications only seven years after. What happened during this interval? In fact, a lot happened.

Among the more or less prominent Soviet theorists Tamm was one of the very few who was not participating in the work on an atomic problem from the very beginning. He did not deserve confidence. At the same time he was an expert in the theory of atomic nucleus and, as we have seen,

---

[45] In the remarkable novel *Life and Fate* by Vasily Grossman, who spent the whole war on the front and apparently had little understanding of life at the home front, he describes a certain prominent physicist who was, together with his institute, evacuated to Kazan. In one colorful scene there enters his daughter who brings from the special shop for outstanding scientists two kilograms of butter. This is, at least with respect to Kazan, a completely improbable fantasy. Apart from Tamm's family I knew another family of the corresponding member and can assure that there was no special shop for them. All of them lived with the same hunger as other scientists and certainly never saw such luxury as butter. Additional privileges for scientists (of all specialities including humanities), writers, composers and others appeared only in 1945.

performed research in widely varying fields. He solved concrete practical problems easily and in the "atomic problem" a quantity of such problems seemed countless. Igor Evgen'evich, however, stayed aside.

Time passed and, as already described in another chapter, in 1946 he became involved in discussing some questions, not those of major importance, on the "classified" topics of the atomic problem. When there arose a new problem, that of constructing the thermonuclear weapon then, as already discussed, I.V. Kurchatov probably convinced the "proper people" in the necessity of using Tamm's talent. Igor Evgen'evich was requested to organize in the FIAN Theory Department a group of "supporting" or "checking" the theoretical research already conducted on this problem by the group of Ya.B. Zeldovich. The FIAN group included V.L. Ginzburg, S.M. Belenky, A.D. Sakharov who had just finished his Ph.D. studies and the Ph.D. student E.S. Fradkin and soon, after their graduation from MIFI, Yu.A. Romanov and V.Ya. Fainberg. For all of them the problem was completely unknown and new. It could be that precisely because of this, having not been influenced by the approaches developed in the American group of E. Teller and our group of Ya.B. Zeldovich, they obtained a result that turned out to be fantastically unexpected: Instead of the "support" Sakharov and Ginzburg put forward, already after two months, two decisive completely new ideas. The simplicity of the main ideas did not save from solving many physical (not mentioning technological) problems requiring extremely intensive research. This was the work to which Igor Evgen'evich gave his forces and his time.

In March 1950 Tamm, Sakharov and Romanov were detached to the nuclear research center nowadays known as "Arzamas-16" headed by Yu.B Khariton.

Here it is impossible to fully describe the works of Igor Evgen'evich on this "main research topic" of those years. From the article by Yu.B. Khariton, V.B. Adamsky, Yu.A. Romanov and Yu.N. Smirnov *Looking with the eyes of the physicists of Arzamas-16* (see p. 99 in the book mentioned on page 61) we learn that Igor Evgen'evich played an outstanding role as a leader, "conductor" of a big collective who inspired others, went into every detail and at the same time was himself solving the continuously arising concrete problems. The spectrum of these problems ranging from subtle and complex physical to in essence almost organizational is astonishingly wide. The group by I.E. Tamm, as well as the group by Ya.B. Zeldovich which worked in parallel, had to solve the permanently arising new difficult theoretical questions.

Working with such great intensity Igor Evgen'evich did nevertheless find time to follow the literature on principle issues in physics. When coming to Moscow he took part in the general seminar of the Theoretical Department and, moreover, together with the young staff members V.P. Silin and V.Ya. Fainberg worked on new development of his method of "truncated equations", the "TD method". This, however, still does not fully characterize an intensive scientific activity of the already 55-year-old Igor Evgen'evich. It included two more directions.

First, there was research on the idea of controlled thermonuclear fission with the help of the so-called magnetic thermonuclear reactor which is nowadays called Tokamak. The idea appeared in 1954. It is customarily called the Sakharov and Tamm idea and when Sakharov in the seventies and eighties was in strong state disfavour his name was completely omitted from the literature. Igor Evgen'evich however always stressed that the idea itself belonged to Sakharov. I remember that when he heard the words "The work by Tamm and Sakharov" he jumped from his place and exclaimed "Sakharov and Tamm, Sakharov and Tamm!" emphasizing by intonation the name of Sakharov. N.N. Golovin recalls how at the first meeting of the high committee chaired by Beria who was in charge of all the works related to atomic and thermonuclear problems, Sakharov, after describing briefly the essence of the proposal, said that main calculations were made by Tamm. Golovin writes: "Tamm, in agitation, asked for the floor and began emotionally explaining that the main ideas belong to Sakharov and that the main credit should go to him. Beria waved impatiently the hand and interrupted Tamm with the words: "Nobody will forget Sakharov." Sakharov however was also right — the bulk of extensive volume of calculations requiring new specific ideas was performed by Igor Evgen'evich.

About half a century has passed and we see that much is still to be done to realize this idea but the efforts of physicists from different countries have led to significant progress. As we know, currently a project of a gigantic international reactor (Russia, USA, Japan) is being discussed. In relation to this I would like to recall one story by Igor Evgen'evich on the psychological state of the physicists involved in these problems.

At that time everyone was hypnotized by the remarkable success of scientific predictions. When constructing the uranium and plutonium bombs scientists worked through mountains of extremely complex nuclear, gas dynamics, chemical, mechanical, metallurgical and other scientific problems as well as purely design issues and everything worked perfectly well in the first tests both in the USA and in Russia.

One started working on the thermonuclear bomb, faced a new mountain of scientific and technological problems, and again everything combined into a "device" that worked from the first time or, as practical people say, "from the first demonstration." One was sure that the controlled thermonuclear fission would be soon successfully realized. The first attempts were undertaken by L.A. Artsimovich and ... a miracle! After creating a strong current gaseous discharge and obtaining a so-called pinch in the discharge filament experimentalists discovered a flux of neutrons coming from the filament.

Everyone was happy. Artsimovich himself declared however that these were "wrong" neutrons. Igor Evgen'evich told me that he was for two weeks persuading Artsimovich that this was an awaited success. It seemed that everything was in agreement with theoretical estimates. Nevertheless Artsimovich stood his ground. Finally he persuaded others in his joyless truth as well. There began a period of large scale incessant work on the problem by us, in England and in the United States that continues nowadays. I remember that a well-known Indian physicist Bhabba told me during his visit to the Soviet Union in the fifties that he had made a bet. He bet that the problem would be solved within twenty years and would be solved in our country. His premature tragic death did not allow him to pay his loss.

I have however said that during this period at the turning point between the forties and the fifties Igor Evgen'evich had, besides the main work on the hydrogen bomb, two important fields of activity and have so far described only one of them.

The second one was of completely different character. This had no relation to the bomb and was related to fundamental particle physics. This was an idea of nucleon resonances, "nuclear isobars" as Igor Evgen'evich himself called them at that time. As has been already described in Section 2.1 Tamm, on the basis of Fermi pion-nucleon experiments, put forward a daring hypothesis on the existence of the unstable particle, a baryon with isotopic and mechanical spins equaling 3/2. He called this particle an "isobar" decaying into nucleon and pion.

A verification of this hypothesis required extensive calculations which, with technical possibilities of that time (the usual ancient mechanical adding machine, only at the end of the work one used electrical adding machines "Mercedes"), were extremely difficult. These efforts were however rewarded by the fact that it turned out possible to give a good description all numerous experiments, albeit with one unpleasant feature of the result — the energy of the isobar level was only somewhat bigger than its width.

This feature caused deep skepsis among many Moscow theorists (including L.D. Landau and I.Ya Pomeranchuk who were personally friendly towards I.E.).

But Igor Evgen'evich had a good "feeling" of the calculations, was inspired by the result and turned out right. The resonances (a modern name for Tamm isobars) now enjoy full rights of belonging to a vast family of known elementary particles. The particular level (3/2,3/2) discovered by Tamm and collaborators is now a well-known resonance $\Delta(1236)$. However, like it was with the phonons, the literature does not mention that a "resonance" particle was first predicted by Tamm.

Reviewing the after-war period of 1945–1955 of the scientific activity of I.E. Tamm it is difficult not to admire the boldness of the intuitively guessed ideas and the wide spectrum of different domains of physics, both fundamental and applied, that he covered within his own research and by organizing a joint work of the talented collective of theorists first in Moscow and then in Arzamas.

We see that the after-war decade was scientifically extremely productive for him. Although an important place belonged to applied physics this was the physics of an immense scale both by its practical importance and the width and diversity of physical problems covered. What was done there by Igor Evgen'evich was really something that only a physicist of an extremely high level could do.

Turning to his works on principal, fundamental issues we should first of all recall a paper that has not been mentioned before, although it had been written in 1944 and was published in 1945. We are talking about a joint work with L.I. Mandelstam (who soon passed away) devoted to the uncertainty relation for time and energy. In this paper the relation $\Delta E \cdot \Delta t \sim \hbar$ becomes a formal consequence of quantum mechanics.

As we have seen in fundamental physics Igor Evgen'evich did two more important things during this period. First, this is a creation of the method of truncated equations (Tamm-Dankow method) and its application to interactions of nucleons. Second, a bold idea of nucleon "isobars", i.e. "resonances," the excited states of nucleons decaying with emitting a pion. Igor Evgen'evich had to defend this idea in a series of papers (written together with young coworkers) and argue with the same passion and with equally high authorities as it had been in his younger years when nobody had believed him that the neutron possesses magnetic moment.

This period is also characterized by the flood, for the first time in his life, of honors and awards, prizes and titles. The death of Stalin relieved the country from a large portion of horror in which everybody had been living. Igor Evgen'evich became *persona grata*. He remained however the same deeply democratic person. He began traveling abroad on different occasions, to GDR; to Sweden for participating in the ceremonial procedure of obtaining the Nobel Prize; to the USA and England for Paguosh conferences (4 times); to Switzerland to the International conference on peaceful usage of atomic energy and a meeting of experts on nuclear disarmament. To India, France, Japan, China ...

For the first time in his life his passion for traveling, learning new countries and new people could be satisfied. This was a happy period also because the country was becoming a little happier. His purely scientific work — at home, at the desk, did practically stop for some time. Of course the passion for science still possessed him. I.M. Frank tells in his memoirs that even in Stockholm amidst celebrations accompanying the Nobel Prize ceremony Igor Evgen'evich, having heard about some new interesting experiment, spent nights working and trying to understand its theoretical significance (he was not successful — fortunately, because the overheard rumor turned out false).

He was still not feeling well because of the absence of real scientific work. The main thing was the absence of new ideas (with his character he needed big ideas, not the rank-and-file which he could continue producing). He wrote articles devoted to the memory of his friends, fought for their rehabilitation. Using his increased authority he fought with fake science, especially with lysenkovshina. He was captured by new ideas in molecular genetics. He studied these works and tried to do something himself, to decipher the genetic code, but Gamov was faster. And finally in 1964 such an idea seemed to come and capture him.

At that time there appeared a lot of "crazy" theories aimed at replacing the problem-ridden quantum field theory, the basis of all the particle theory. One could mention the work by Heisenberg (1957) based on nonlinear Dirac equation. It was however not accidental that Bohr called this theory not crazy enough. This theory had to be rejected. Among the rivals were also a method of axiomatic S-matrix, nonlocal theories, etc., but the goal remained unachievable.

Igor Evgen'evich, captured by his own "crazy" idea, worked avidly, also when a terrible incurable illness fell upon him. In Section 2.1 we have already described that in 1967 the USSR Academy of Sciences awarded Igor Evgen'evich the M.V. Lomonosov Gold Medal and how he, being chained to the respiratory machine, decided, as required by the Charter, to present his talk to the ceremonial General Assembly of the Academy. He did prepare this talk. This talk was read at the meeting by A.D. Sakharov. The breadth of view, the definiteness of the viewpoints, the optimism piercing the text were astonishingly characteristic of Igor Evgen'evich. He concluded his talk with the words: "I hope that we shall live up to a new stage in the theory whatever it will be."

He did not. Igor Evgen'evich passed away on April 12, 1971.

A. D. Sakharov

Chapter 3

SAKHAROV, Andrei Dmitrievich (1921–1989)

3.1 For a Future Historian[1]

In 1996 the I.E. Tamm Theoretical Department of FIAN in which Sakharov worked for so many years released a collection of memoirs on A.D. Sakharov [Keldysh (1996)]. At the beginning of my contribution I asked a question: Why had the memoirs collected in that volume been written? Subjectively, first of all, to express one's admiration at the remarkable phenomenon of Nature that Sakharov was and, by using examples of communicating with him, to show what caused this admiration. There was however a more important objective goal. Time will pass and there comes a sage and perspicacious writer who will manage to grasp, think over and understand Sakharov better than we, his contemporaries, still imprisoned within a complex mixture of emotions and opinion clashes and not yet free from the heritage of living through a difficult and as yet still not overcome epoch. Such writer would first of all need the facts from the life of Sakharov and people that surrounded him.

Therefore one should hasten to collect everything that we remember, fix everything not yet forgotten and, within the limits of possible (this is a very difficult task that can not be accomplished without failures and mistakes), not distorted by imperfections of memory and biases of the author. We are, with our common characteristic inattention to events, to seemingly insignificant details that, however, often become very important for the history,

---

[1] Written for the most part in 1991. Published in the collection of memoirs on Sakharov [Keldysh (1996)]. The present text contains insignificant changes. The journal variant of the article is published in the journal *Novyj Mir* (5, 1994) under the title 'Sakharov in FIAN'. A less complete collection of memoirs on Sakharov was also published in English: *Andrei Sakharov. Facets of a Life* (Paris: Editions Frontieres, 1991).

in many respects already late. We are feverishly sorting occasionally-kept, offhand notes, trying together to reconstruct dates. We have too literally understood Pasternak, his two lines,[2]

> One should not organize an archive,
> And keenly keep one's manuscripts.

These lines have been understood as an unquestionable law of modesty and decency. Such comprehension led to big losses for the history of culture.

Let us thus try to recall everything we can, fix everything in the verity of which we are sure — not deifying Andrei Dmitrievich, however strong our love and admiration for him is, remembering the great responsibility for what we write, the main most important goal of our "memoirs." One should write about his grandeur and his illusions, his astonishing insights and mistakes. About "illusions"? About "mistakes"? Who am I to judge a great person? First, great people also make mistakes although sometimes these become clear only a posteriori. The great Napoleon, the ruler of "half of the world," went against Russia and lost everything — half of the world and freedom.

Second, I have spent my life in such a scientific community in which a right of judgment equally belongs to everyone from a young Ph.D. student to a famous scientist. The true science, that without "generals" inflated with self-importance, is perhaps the most democratic (if not the only truly democratic) system in the world. Sakharov himself was to the highest degree true to this tradition and a colorful confirmation of this will soon be given below. Also, in general, is it possible to give "witness testimonies", provide "facts" fully separating them from one's attitude to them, from one's relations with Andrei Dmitrievich, from estimating their consequences? Particular features of an "observer" are necessarily influencing the "observation" results (the selection of the facts does already bear this seal). Others would possibly tell something very different about him, show their own different attitude to him. These estimates are themselves the facts that a future historian will take into account.

My contacts with Andrei Dmitrievich are cleanly divided into two periods — the first, from his appearance in the FIAN Theoretical Department in January 1945 until his leave to the "object" in 1950; and the second one, after his family returned to Moscow in 1962 when he would more and more often come to the weekly seminars in the department (and also to

---

[2] Translated by A. Leonidov.

our home), officially again becoming a staff member in 1969, up until the tragic end.

3.2 The Years of 1945–1950

When a slender, lean, black-haired attractive young man, almost a youth, appeared in the old and so comfortable old building of FIAN at Miusskaya Square we did not know yet with whom we were dealing with. He arrived from Uljanovsk, probably at an official invitation from the founder of the department, Igor Evgen'evich Tamm (this was wartime and free entry to Moscow was not allowed). The invitation itself was, according to plausible rumors, sent at the request of the father of Andrei Dmitrievich, Dmitri Ivanovich. He knew Tamm very well through teaching together with him a long time ago in the second MSU (now the state pedagogical university). According to the same rumors (I did not bother to check this with Tamm when it was possible) Dmitri Ivanovich reportedly told him: "Andruysha is of course not as talented as your Ph.D. student N (he named a university friend of Andrei Dmitirevich to whom A.D. (as he was commonly called) paid a lot of attention during his lifetime; after finishing his Ph.D. studies he was not even left in FIAN but later made, however, significant works in a different domain of physics), but still — please talk with him".

In 1988 I asked A.D. why did his choice fall on Tamm. He answered: "I liked his publications I have read, his style." (This was probably, first of all, the university course 'Foundations of the Theory of Electricity'). He added that when after the University he had been working at a plant where he had written four small papers on theoretical physics (he wrote in his autobiography: "... this gave me confidence in my abilities which is so important for the scientific work") and had sent them to Igor Evgen'evich. I dare to assume that I.E. did not reply — he was not very accurate in "small" things, even told that if he would have replied to all letters there would have been no time and energy left for the main cause (see the memoirs of V.Ya. Fainberg in [Feinberg (1995)]).

Anyhow, on the day Sakharov came to FIAN and talked to Tamm in his office I was coming through the corridor. Igor Evgen'evich, in a state of extreme agitation, rushed from the office, bumped into me and blurted out: "You know, Andrei Dmitrievich has guessed himself that within uranium boiler (N.B. this is how the reactor was called at that time) uranium should not be placed homogeneously but rather in blocks." (This means that the paper of A.D., one of the four, was not known to him.) Tamm's excitement was understandable — this important and involved principle which made real a construction of uranium-graphite reactor with natural uranium had long been known in America, England, Germany and by us, but was everywhere a secret. A.D., however, understood this sitting in Uljanovsk without having any contact to physicists and having probably read only the well-known "pioneering" paper by Ya.B. Zeldovich and Yu.B. Khariton on the chain reaction in the system uranium-moderator (at that time they had not yet known of this principle).

Quite soon all of us began to understand that we were dealing with a very talented person. His quiet confidence, based on uninterrupted work of thought, his politeness and softness that went together with firmness in the issues he considered to be important, an unobtrusive feeling of self-confidence, an inability to insult anybody, even somebody hostile towards him, and his extreme frankness and honesty became clear very soon. I am convinced that in general he never said anything not agreeable with what he really thought and felt at that particular moment and never did anything that would have contradicted his words, thoughts and conscience. And, at the same time, already then was he persistent, more exactly, pertinacious in following his chosen route. These features of him are now known to everyone because of his social and political activity of the last decades.

In the very last years television showed a gray-haired, almost completely bald, stooping (but in a way beautiful) man; firm, fearless and extremely active. In his younger years he did not, of course, behave in the same way as on the congress rostrum, but the wonderful features of his character did very soon completely naturally cause a feeling of sympathy both in his fellow Ph.D. students and by those of elder generation. The "seniors" were from 29 (V.L. Ginzburg) to 37 (M.A. Markov) years old, the "patriarch" I.E. Tamm was 50. The elder did however address the younger with patronymic while within each age group shorter names were also used. This old-fashioned style has only partially survived nowadays and it seems strange to many. Nevertheless neither age nor hierarchical difference hindered intensive communication. The democratic tone was of course in the first place determined by the personality of Igor Evgen'evich Tamm.

I was 9 years older than Andrei Dmitrievich (when he began his Ph.D. studies I was already a doctor and professor). Nevertheless neither in scientific discussions nor in our relations was there a trace of hierarchy. Like me, he had a little daughter and when meeting in the corridor (we were working in different offices) we began reading, with pleasure, verses for children.

When in 1949 I had to be transported home from the hospital after a broken leg and surgery (I was still awkwardly getting about on crutches) my wife, Valentina Josephovna Konen, asked A.D. to help. This was completely natural, they already knew each other well.

To elucidate this I can, making a huge jump in time, tell another story referring to the year of 1970 when Sakharov did again become a staff member of the department. He was already an academician, three times Hero and in general an extremely highly revered scientist but was at the same time already "bad" being the author of the famous 'Reflections'. He returned to pure science, published a number of wonderful papers on gravity, cosmology and quantum field theory but gaps in the knowledge of scientific literature on physics of particles and fields that had arisen during the 20 years of all-consuming work on applied physics were still felt.

At that time Igor Evgen'evich was already in bed, terminally ill, chained to the "respiratory machine," and I was, instead of him, in charge of the seminars of the Theoretical Department on the above-described range of topics. Once A.D. told me that he had finished another paper devoted to a further development of his idea of "null lagrangian" (or "induced gravity," I will discuss it below) and would like to discuss it, as usual, at the seminar before submitting it for publication. I was not an expert on the subject of the paper but it seemed to me that everything was fine. I asked one of the "seniors" who was closer to this topic to read the paper and he confirmed my impression. The talk was set, an announcement made.

Soon after the beginning of the talk a young talented and passionate probationer, I.A. Batalin, who had just graduated from the university but had been already accustomed to our atmosphere, began asking questions. A.D. replied. The questions became more frequent and more aggressive and transformed into an argument. Batalin confidently said that at such-and-such point the calculations should be done differently, that something had long been known, etc. As a chairman I tried to stop him (even caught his hand having sat behind him) to allow A.D. to finish the talk, but it was clear that his remarks were on point and competent and his rage was unstoppable. One should add that the whole argument referred to the "technical method" used in the paper but not to the principal important idea put forward by A.D. in brief form in 1966 and having had its development in his big paper of 1975. At last, after long debates, Batalin indignantly said that one of the principal points of the talk, the above-mentioned "technical method," was not new and had been already developed by Schwinger (a well-known American theorist and Nobel Laureate) about 20 years ago.

This was already clear to other people present at the seminar as well. A.D. was confused, somehow finished the talk and the seminar ended. Everybody left the hall. The sorrowful A.D. sat in the now empty hall, his elbows leant on the chair arm, the cheek on the palm. I approached him and said: "Andrei Dmitrievich, if you, without knowing it, did the same thing as Schwinger at the time when you were overly busy with the bomb you could only be proud of that."

Indeed, in the course of this work Sakharov did in fact invent an extremely important proper time method (and the regularization with respect to this variable, I will not explain what this means) having wide application in the physics of quantum fields. It can be traced to the prewar paper by Fock. When I had read the manuscript of the talk I did not understand that A.D. had invented it and had considered it new. In response to my propitiatory words he, turning away, only waved with his characteristic gesture with the wrist of one hand and did not say anything. In a day I visited Igor Evgen'evich and learned that A.D. had visited him and, having told about his failure, had summarized: "In all this story the single good thing is that we would possibly write a paper together with Batalin."

Everything here is characteristic of the atmosphere in the department: The honesty of scientific discussion, equal rights that a young probationer and a celebrated academician enjoy, the fact that this episode did not influence their future relations and mutual respect. Nowadays Batalin (still a staff member of the department) is not just a doctor of sciences but has a solid international reputation. He regrets his unreserve at that seminar very much, still sincerely admires A.D. and the fact that he invented the above-mentioned proper time method himself. It is worth adding that this public talk by somebody who was not safe to approach was attended by people who did not understand anything in science. They sat at the end of the hall. One of them afterwards launched a rumor that at the seminar there was a special troublemaker (reportedly even a gebist[3]) aiming at disrupting the talk of the "undesirable" Sakharov.

I have described something that happened many years later after Sakharov's return to FIAN but in the forties the basic principles of the life of the department were the same.

During those years Sakharov's behavior was still somewhat restrained (this was never true for private, home relations. Here he was somewhat

---

[3]Somebody working for KGB.

restrained only at the very beginning but afterwards always gentle, natural and even more attractive.) In a scientific discussion it was sometimes difficult to understand him right away. His explanations were brief and looked like a dotted line with intermediate links that probably seemed evident to him omitted.

Once a grey-bearded man who taught English to Ph.D. students climbed to the fourth floor to the department. He came to ask what kind of a person Sakharov was — "he, unlike others, somehow thinks in his own way." It is true and because of this there arose paradoxical situations. Here is one of them. A.D. had to pass a Ph.D. examination in his special field. As required there was formed a commission including I.E. Tamm (chairman), S.M. Rytov (a theorist from the laboratory of oscillations) and myself. A topic for the report was formulated and one certain day we began listening to his talk. A.D. liked to use colored pencils. He drew, on a large sheet of paper, blue electrons, a green Dirac electron background, a red hole (a positron), in it and began the talk in his usual style (the topic was related to the problem of the electron electromagnetic mass). At some point I.E. interfered: "You are telling something wrong." (I do not remember the essence of his objection.) A.D. was silent and then pronounced one short phrase. I.E. was surprised and displeased. The statement of A.D. seemed wrong to me as well. As usual, Igor Evgen'evich talked very fast and A.D. dropped short phrases that only strengthened our bewilderment. When after answering questions on other topics Sakharov went out, Tamm, at a loss, said: "What should we do? Not to give Andrei Dmitrievich "five"? This is unimaginable — he and "four"? However, we have to be honest. Nothing could be done." This ended up in us giving him "four". In the evening Sakharov came to Tamm and explained to him that he (Sakharov) had been right.

Another episode is related to a short period of his teaching, at Tamm's recommendation, in the Moscow Energy Institute. For many decades one heard stories that students did not understand him. They complained to the dean's office that Sakharov did not know his subject, demanded a new lecturer. This was much later confirmed to me by Valentin Aleksandrovich Fabrikant who at that time headed the chair at which Sakharov gave his lectures. However, when in 1988 I asked Sakharov about this he was enraged: 'These are myths made about me to picture me as a special personality. I am a man like others, did normally read two courses in two semesters, took tests and exams and left already in 1948 when there began my "classified" things.'

The issue remained unclear until one found out that Nina Mikhailovna Nesterova, who listened to these lectures herself, still worked in FIAN, in the Laboratory of Cosmic Rays. She said that Sakharov had read not two but three semester courses (I think that it was Sakharov who was wrong with respect to their number). It was indeed difficult to listen to him but, as through a miracle, when afterwards one used notes in preparing for the exam everything turned out to be logical, consistent and quite understandable. Students did indeed complain about him. A lady from the office of the head of studies would often call Tamm but he replied (probably in irritation) that he can not recommend anyone better than Sakharov. M.L. Levin, Sakharov's friend from his youth, recalled the words by Sakharov that he "... learned to speak clearer when having to explain a lot to high administrative bosses and generals."

The life of Andrei Dmitirievich at that time was difficult. With his wife and a newborn daughter he lived on his Ph.D. stipend and did not have a permanent place to live. The house in which he lived with his brother and parents before the war had been destroyed by a bomb. He rented a room, sometimes in a wet cellar, sometimes in a better place, sometimes outside Moscow. His financial position did only somewhat improve through teaching. The above-mentioned V.A. Fabrikant told me that when Igor Evgen'evich had recommended Andrei Dmitrievich for teaching he had been surprised that he was in need of money. Indeed, his father was an author of published textbooks, collections of problems, popular books on science, was well-to-do and could help. I know him well, said V.A. "But you do not know Andrei Dmitrievich," replied Tamm.

In all circumstances Sakharov was working systematically and persistently. When in 1980 he was sent to Gorky I took from the drawer in his desk in FIAN some left-behind miscellaneous scientific notes. Among them there was a large notebook in which he had made notes on journal papers of interest to him. On the first page he wrote "Bibliographical Reference Book". Then, in a column, the contents: "Part A. Elem. particles. Nuclear forces. Cosmic rays. Experiments at high energies. Part B. Decay, conversion. Miscellaneous. Isomers and spectra. Structure of heav. nucl." and so on up to the eighth, "Part H. Hydrodynamics." These titles, as well as those of "Part C. Nuclear reactions, cross sections" and "Part E. Astrophysics", were underlined. One easily sees that these are precisely the domains in which he would work or had worked before (in the candidate thesis). For example, the title "Part F. mathematical literature" was not underlined (and was just empty). He was probably reading almost exclusively the main foreign journal *Physical Review*, sometimes *Proceedings of the Royal Society*, occasionally *Nature*.

On the second page of the notebook there followed, in three columns, "The used literature." In the first column one saw '74.1; 74.2; 74.3', etc. Two similar columns followed nearby. Only some of these numbers were however encircled — these probably marked the articles that he considered as those completely worked through. This referred to the whole volume 75 of *Physical Review* and issues 1–6 of volume 76.

One summarizing line was sufficient for him. The rest he probably remembered. The author's name was usually not written. For example, in the part "Astrophysics" we read:

> 75.1.208 and 211 Earth magnetism.
> 75.10.1605 Polarization of stellar light due to absorption.
> 75.10.1089 Remarks on expanding Universe.
> 76.5.690 On charge related to mass (cf. 73, 78 and the work by Rumer).

And so on. The first number gives the volume, the second is the number of the issue within the volume followed by the page number. In some cases he wrote down the values of constants of interest to him (e.g. the Hubble constant). The notebook refers to the years of 1948–1949, he had just begun filling it. Then form the *Proceedings of the Royal Society* (he writes P.R.S.):

> 195.1042.323. Classical electrodynamics without divergences.
> 365. Relativistic invariant quantum field theory.
> 198.1955.540. To the theory of S-matrix.

From this one sees that his main emphasis was on fundamental questions of field and particle theory and that these were of main interest to him at that time. He did however sacrifice them (at the very age most beneficial for any theorist) for something that he, at that time, considered being most important for his country and that really interested him. At the same time I do not remember him participating in the permanent passionate discussions of these problems in which other Ph.D. students and young staff members (V.P. Silin, E.S. Fradkin, Yu.M. Lomsadze) were deeply involved. In general, A.D. seldom participated in common unorganized discussions. He was rather listening. Even at the seminars he was not active but always attended and listened very attentively, sometimes making short remarks or asking short questions.

In the later notes in the same notebook one notices a particular interest to the reactions of light nuclei (deuterium, tritium) and, in general, to subjects that were later useful in the work on the bomb. It was at that time when he was included into the group gathered in the department by Tamm for the study of this problem and the open notes in the notebook were naturally no longer made. There also stopped, for a long time, the mention of the literature on fundamental problems of quantum field and particle theory that interested him.

This was a time of the great "quantum-electrodynamical revolution." Quantum electrodynamics was formulated and in "Part A" we see an orderly list of the main papers on field theory:

> 75.7.1079 Quantization in unitary field theory.
> 76.1 Magnetic mom. of nucleons (after Schwinger).
> 65.5.898 Schwinger. Radiative corrections in the scattering problem.
> 76.6.749 Feynman. The theory of positrons.
> 76.6.789 Feynman. Exposition of his method.
> 76.6.818 Application of Schwinger to nucleons.
> 76.6.846 Fourth order corrections to electr. magn. mom.
> 75.8.1241 Schwingerian (Weisskopf and French).
> 75.8.214 Ma. Vacuum polarization.
> 75.8.1270 Schwingerian.

One sees that he studies a question and not just lists; he returns to earlier articles. Then, after listing papers on particular physics issues ("nuclear cross sections for n-p in the int, 95–270 MeV"; "polarization effects in nn and pp scattering",; "on magn. poles", etc.) related to the volume 76 there comes again:

> 75.3.460 The work by Welton.
> 75.1321 The work by Welton.
> 75.3.388 The Lamb shift (Lamb and Kroll).
> 75.3.486 Radiat. theories by Schwinger and Feynman.
> 74.1070 The paper by Wentzel.
> 75.3.651 Schwinger II.
> 74.1439 Schwinger I.

In the spring of 1947, after two years of Ph.D. studies, A.D. Sakharov presented an excellent Ph.D. thesis (by that time he had published three papers on completely different topics, one of them being a short summary of the thesis). At the end of the eighties a well-known English theorist, Dalitz, visited Moscow. He had heard that one of the paragraphs of this unpublished thesis coincided with the topic of his own Ph.D. thesis that was written much later and was very glad when he got from me a copy of Sakharov's thesis [Dalitz (1991)].

He could however defend it only in the autumn which distressed him a lot because, correspondingly, this postponed the improvement of his financial position that a scientific degree should have brought. The reason was that he could not pass a candidate exam on "politsubject." Here one should not look for political reasons. At that time Sakharov was quite loyal with respect to official ideology. Probably the examiners could not understand the logic of his reasoning. At the exam he surely spoke in his non-standard fashion.

This period of our communication was sharply cut off in 1950 when Igor Evgen'evich and Andrei Dmitrievich (and also Yuri Alekseevich Romanov) moved to the "object" (the now known to anyone Arzamas-16 which, before the new epoch, was forbidden to be named other than as the "object"). Later Tamm said that he insisted that I should have been on board as well but the "terrible" CV of my wife (her family spent the twenties in the USA, she was an American citizen until 1934, her father died in imprisonment) not only hindered my participation but led to my general dismissal from the classified research on reactor physics which I had been doing since 1944. (Since that time we had evidently prepared a sufficient number of young physicists.) I was glad because, from the scientific point of view, this work was no longer interesting for me and I had only been doing this governed by a feeling of duty and under the request of S.I. Vavilov (nevertheless I had been doing this quite intensively; it seems to me that any domain of physics, when deeply studied, becomes fascinating). I think that for secrecy reasons A.D. could not continue contact with me and my 'suspicious family'. This is why our contacts resumed in 1962 or somewhat later.

3.3 A Return to "Pure" Science and the Beginning of Social Activity

As Sakharov himself writes in his "memoirs" it was perhaps starting from that "push," the now famous toast by Marshal Nedelin, that for A.D. there began a tortious work on rethinking his social and political position. This referred to relation with authorities, the true meaning of deeds of the rulers, our social system as a whole and the generic problems of the mankind. The attempts of A.D. Sakharov to stop excessive testing of "superbombs" carrying deadly harm to the health of dozens, even hundreds of thousands of people, and the hard reaction of Khrushev to these attempts in 1961, marked the first stage of this process. He began to understand the cynicism of authorities in its relations to scientists (you have done your job, have given us a bomb, but in the issues of its usage we will not require your opinion. This was in essence what was said by M.I. Nedelin at the banquet just after the successful test of the first hydrogen bomb) and to the living people dying from these tests.

He could still play a significant role in signing the treaty on the test ban in three media[4] but the moral aspect of the situation that was so important for him caused a growing feeling of protest. On the one hand he delved deeply into studying and rethinking of the factual state of the people and authorities in our country and their interrelations with the rest of humanity; on the other he began returning to pure science, to something that was his main passion and that had been abandoned for 15 years so as to give all his stregth to reaching an equilibrium in the world as a guarantee from the nuclear war. At that period he still believed in the necessity and inevitability of the fall of capitalism and the rule of socialism.

At the beginning of the sixties, A.D. began more and more frequently attending our weekly seminar and in 1966–1967 published brilliant papers on cosmology and theory of gravity. This was, first of all, an explanation of the world's baryon asymmetry in which he tied into a unified system, hadrons (i.e. the nuclear particles — protons, neutrons, etc.) and leptons (electrons, neutrino, etc.) which according to his theory could transform into each other. For example he predicted that a proton can decay, albeit with a small probability (he even calculated its lifetime).

This seemed so fantastic and crazy to me that when he presented me with a copy of the article with a nice dedicatory inscription (to which he later added a frequently cited jocular quatrain understandable only for physicists working in this field: "From the effect of Okubo ...", etc.), I said to myself: "Of course, Sakharov can allow himself everything, even such fantasies." (In ten years I told him about this and we laughed.)

However, only ten years have passed and development of particle theory has completely, independently of him, led the best theorists of the world to the same concept but from a completely different viewpoint and searches for proton decay were declared the "experiment of the century." Efforts of several experimental groups have so far not been successful but this is interpreted as a drawback of the currently used concrete variant of the

---

[4] Earth, air and water.

theory, the idea of the unity of all particles (and, correspondingly, of the possibility of proton decay) still capturing the minds of the physicists.

Another work of this period refers to the theory of gravity. Sakharov explains the nature of mutual attraction of two bodies by the fact that their presence changes the zero quantum oscillations of the spatial metrics. This deep idea was developed by many theorists and is called the "induced gravity." One could also mention in this connection Sakharov's work in which inhomogeneity of matter in the Universe (stars, galaxies) is also explained by the fluctuations of metrics.

This remarkable beginning of the new stage of his scientific work had its continuation in other papers by him but went into contradiction with another direction of his activity. In these years he would often visit me at home at Zoologicheskaya Ulitsa and, after discussions, put on his non-dandyish (softly speaking) coat, put on galoshes on his feet, took a simple "avos'ka" from the pocket and went to a nearby Tishinsky market to buy some products that were not available in the Sokol district where he lived.

I remember how once he came, in a state of extreme agitation, with a thick blue folder in his hands, a manuscript by Roy Medvedev 'Before the trial of the history.' It contained facts on "Stalinshina" (RedTerror) and on Stalin's way to power. Most of these facts were published during the Khrushev Thaw, but combined in one place with new material added and re-thought over made a very strong impression. This was probably in 1967. I think so because when I thought we had discussed everything I began telling him about my work in physics published in 1966 which I considered interesting, but he stopped this discussion almost immediately and returned to the previous topic (I was naturally not happy about this).

This process of evolution of his social and political views, of intensive work of thought and mind, discharged in a publication (abroad) of his famous article 'Reflections on Progress, Peaceful Coexistence and Intellectual Freedom' in which he appeared as an outstanding social philosopher (I like this expression by which his daughter Tatjana Andreevna characterized him). Here a great idea of a convergence of two systems was declared. It seems all the more remarkable to me because an outlined way of saving the country turned out to be inseparable from one of the salvation of humanity as a whole.

In the atmosphere of those years the idea of convergence was, for the official ideology, monstrously heretical and, even for the hardest critics of our system, utopian and unreal. One had to see something that everybody does understand now: capitalism had come a significant part of this way. It is already not the one of the times of Marx and even of that when imperialism was declared to be its last stage. Indeed, one could say that this was the last stage of classical capitalism and that it was gradually becoming different. An idea of the responsibility of society and state for the welfare of each citizen accepted in practice even in the USA in Roosevelt's times did, after the war, become along with democracy the main principle in the life of capitalist countries (dole, pensions, free medical help and free education did, in different forms, realize this idea in the majority of states; one should not forget that all this was to a large extent achieved due to the fierce struggle of masses). One could not but wonder about the fact that the idea of convergence was put forward by the man who had overcome his earlier confidence in the inevitability of a collision of two systems that should have ended in the defeat of capitalism.

Andrei Dmitrievich never said a word that he was preparing such an article (this was in general his manner when preparing scientific publications as well). I heard it in the summer of 1968 on radio (BBC?) being on vacation far from Moscow. Everybody's impression was that of a bomb's explosion. Before Sakharov we have heard only critics of the regime, some harder, some softer. Here there was not only critique but a great constructive program which the country began to accomplish in only 17 years.

Sakharov immediately became "bad" in the eyes of the authorities. He was, of course, for a long time still protected by the recognition of his outstanding service, high awards, etc. He was however immediately debarred from the works at the "object." In reality, according to my understanding, an uncertainty persisted until the end of 1968. I guess (and may be mistaken) that scientific leadership tried to somehow keep him for the further work on these topics but failed.

At the end of the year he was formally fired from his work and from the beginning of 1969 remained a "private person," "simply an academician" in Moscow. No external manifestations of special measures with respect to him was to be noticed. I remember once when he visited me I asked: "Do you think you are followed?" "Of course", he answered, "it should necessarily be one. They should be sure that I do not go to the American embassy". However, even if it had been the case (I agreed with him — he should have been followed), it was done professionally and he, not feeling any real guilt, did not watch this closely, did not try to notice this and did not worry in this respect.

At the very beginning of the autumn of 1968 there took place an event which I, at last, have to describe — an acquaintance of Aleksandr Isaevich

Solzhenitsyn with Sakharov. It happened in my home. Both Solzhenitsyn (in *The Oak and the Calf*) and Sakharov (in the *Memoirs*) briefly describe it but, for understandable reasons, do not mention my name and the circumstances of their meeting. Time has come to describe it.

A close acquaintance of my wife and I, Tamara Konstantinovna Khachaturova (at that time she worked in FIAN's library), was also a close acquaintance of Solzhenitsyn. She told me that Aleksandr Isaevich wanted to meet A.D., and had, through her, written him a letter and mentioned where he wanted to meet. Sakharov, having had agreed to meet, insisted, according to the words of Tamara Konstantinovna, that the meeting should take place in my apartment. At the set day and hour Solzhenitsyn was the first to come (in the previous version of this article I erroneously wrote the contrary — Aleksandr Isaevich has recently corrected me). I tried to send my daughter and niece on their way but they lingered at the front door when then rang the bell and, when I opened the door to Solzhenitsyn, they were about to leave. Aleksandr Isaevich quickly came in. He was irritated and was sweating (the day was hot) and, seeing the girls, gave them an angry look.

Before that Solzhenitsyn had once visited us and we three had had a very interesting (at least, for us) discussion at dinner. I understand however that I deceived his hopes (or expectations), turned out to be a "wrong man." And although we parted quite friendly and then met once again at Igor Evgen'evich's place when Solzhenitsyn came to him with his first wife (and Tamm also invited V.L. Ginzburg and myself and our wives) Solzhenitsyn, apparently, was disappointed by Tamm and lost his interest in me (although when we occasionally met at the Belorusskaya train station he was quite nice again).

My wife and I did not quite understand what the procedure of the meeting should have been. Be it as it may we decided to lay the table "with food" in the (only) big room. When Aleksandr Isaevich saw all this he said, being more than displeased: "This is what, a reception?" It was clear our choice of style was wrong. I showed him where to wash hands and then appeared Sakharov with Tamara Konstantinovna. I, with my wife and Tamara Konstantinovna, went to the small room leaving Sakharov and Solzhenitsyn at the laid table. Nevertheless I felt myself somewhat awkward.

I understood, of course, that A.I. had come only to meet Sakharov and nobody else. Still the previous contact and the feeling of being the master of the house made me enter, a couple of times, the room, once for bringing tea. Each time, after staying for a minute, I felt from A.I.'s mood that I had to leave, and I left.

It would be naive and wrong to consider such behavior of Aleksandr Isaevich as simply impolite or unfriendly. One has to remember that at that time he was consumed, carried away, possessed by his deed and this went along with the all-consuming purposefulness of his clear-cut actions (a truly "American business approach and Russian (counter)revolutionary scale"). Everything secondary was swept away.

I was a witness and participant of his three attempts of finding himself a worthy ally among Academy physicists having attractive social reputation. He met, from their side, a sincere admiration, a willingness to help (say to retype on the typewriter his unpublished "awful" works) but for him all this was "not that." Now he came to first meet a man who was from the same environment but had already made a great deed, "crossed the threshold." Therefore everything else was inessential, could only hamper.

They talked sitting half-turned by each other. Aleksandr Isaevich, leaning on the table with one hand, did edifyingly drum something to Andrei Dmitrievich. A.D. pronounced separate indolent phrases and, as usual, did more listening than speaking.

I do not remember how long this conversation lasted, probably, about two hours. (According to Solzhenitsyn, in *The Oak and the Calf*, they were sitting for "four evening hours" but in the same place he writes on his "two-hours foolish critique" of Sakharov's reflections. I think nevertheless that the length of the conversation was closer to two hours: three of us, sitting in the small room, would in four hours have been completely exhausted, and this was not the case). Finally their conversation was over. We talked a little bit all together and then the "guests", one after the other, left.

In *The Oak and the Calf* Solzhenitsyn writes that this conversation took place in not too good conditions, "they were not always left alone," but he was astonished how attentively, without being offended, did Sakharov listen to his critique. Solzhenitsyn probably did not know that this was characteristic for his behavior in the Theoretical Department or in general.

Sakharov notes, with surprise, that first of all, before the conversation started, Solzhenitsyn had curtained the windows. He also mentions that according to the opinion of A.I. their meeting did not stay unknown to the KGB. A.D. doubts that. In particular he writes that for some reason when he went out in our forgotten street, where only one side was occupied with houses, there stood a taxi (T.K. Khachaturova is convinced that Sakharov forgot the circumstances of their leave. In fact, two taxis, one after the other, were called for by phone, whose cord had first been cut following the wish of Aleksandr Isaevich and then, with difficulty, reconnected. Sakharov's driver asked him: "You have had a meeting?"). I agree with Sakharov, I am of better opinion than Solzhenitsyn on the professionalism of KGB people.

I also remember (possibly, not very accurately) how A.I. himself told me before that living officially in Ryazan he had spent much time in Moscow and, to have a still hideous place for work, had bought a log cabin in the village about 80 kilometers from Moscow in which he had hidden himself for work being sure that nobody had known about that. However once he had received an envelope without an address but with his name on it. Thus the "secrecy" of his shelter was fictitious.

This meeting of two outstanding people of our time marked a beginning of their relatively short-term personal contact. Already in 1974 after the quite correct but hard critique of Solzhenitsyn's position expressed in his 'Letter to the leaders' (see *Znamya*, 1990, 2) by Sakharov there came, if I understand it correctly, a certain cooling with the mutual respect totally kept.

In general, after the publication of 'Reflections' Sakharov became a focus of attraction for many of those who had an opposition attitude, did not want to make it up with humiliation, lawlessness and oppression coming from illiterate and cynical authorities, who were ready to participate in the heroic human rights movement or had already participated in it, had "crossed the line." The activity of Sakharov as a "social philosopher" (the 'Reflections' were followed by other publications abroad following the same line) was inevitably expanded through his activity as a factual leader of the human rights movement. I remember how this began forming.

Once in November 1970 we were leaving a building of the Presidium of the Academy of Science where we had together participated in a meeting. While we were walking in the yard along the giant flower bed Sakharov told me he had organized, together with two like-minded persons, a group (this had been exactly a week before our conversation) on a problem of defending human rights with an aim of providing consultation on this issue and develop it based on the UN Declaration on Human Rights. It was clear that Sakharov joined a movement that was later called a "dissident" one (he himself did not like this word and preferred to use "human rights defense". We had the following conversation (it belongs to those that have been deeply engraved in the memory, here I am describing only such cases):

> 'Andrei Dmitrievich, please tell, when do you think we had most free, democratic time during the last more than 50 years?'
> 'One can tell it exactly: from the twentieth congress to the Hungarian events, February–October 1956.'
> 'Right, I also think so. Did it come as a result of protests from below, an opposition movement?'
> 'Of course not.'
> 'Andrei Dmitrievich, good social reforms in Russia were always made from above: the reforms of Aleksander II, NEP, Khrushev's Thaw. Do you really think that with an existing merciless apparatus of oppression one could achieve something?'

A.D. answered with something indeterminate that I have not remembered but I know for sure that he did say with conviction: "Nevertheless this should be done."

Time has passed and one can judge that in a narrow pragmatic sense I was right. In 10–15 years the authorities managed to disperse and suppress this movement, sometimes acting with extreme cruelty and completely neglecting the indignation and protests that their actions caused in the whole part of the civilized world that was not under their control. The positive results achieved such as a creation of the 'Chronicle of Current Events,' documentation of placing healthy people into psychiatric hospitals, etc. were of course important but, in comparison to the sacrifices made, not too big. (I was, in general, not of a high opinion on the pragmatic value of such actions. The recent experience of devastation that happened in mathematics in 1968–1969 [I will describe it in more details below in Section 3.6] was a good example).

Once (probably at the end of 1970) when we were together, walking out of FIAN, I told him:

> 'All of us are "walking under Almighty." Although, as you know, I do not participate in the dissident activity it could be that I will be imprisoned as well. It could then happen that you and Igor Evgen'evich will begin a campaign in my defense with signing collective letters, etc.'
> 'A.D. answered ardently, "We will surely do this!"'
> I objected, 'So, I am begging you not to do it in any case. This will not help me but an understanding that because of me blows would fall on those people who would participate in this would make my state psychologically unbearable.'

'Well, we'll see,' said A.D. softly.
'Let us hope that this question will not arise.'

I was however deeply wrong in considering all this activity hopeless. Of course he suffered heavily himself but a movement of courageous unsubdued people got a leader who turned out to be an embodiment of high spirituality, purity, courage and love. I could not imagine at that time what a scale could an influence of this ennobling commencement would reach. Once V.L. Ginzburg drew my attention to the newly appeared book under the title, it seems to me, "*Lev Tolstoy and the Tsar Government*". This book showed that Tolstoy became a second government in Russia, a spiritual and moral one. Of course, a very different one from A.D., one could say — with a different ideology, but their role in the country and the society was in many aspects very similar. It turned out that we had a courageous and incorruptible person whose morals and activity raised the spiritual level of the people, helped some to straighten up, and helped some to begin a non-violent struggle. With special clarity this showed itself in the days of his funeral when his very death made masses of people better and purer. Already in the seventies many sought help from him, intercession, and wisdom. The people said: I see but do not dare to say but Sakharov has dared.

Once my wife and I visited Andrei Dmitrievich and Elena Georgievna at their place. When we were leaving, A.D. accompanied us to the landing. On the stairs there stood, in indecision and likely, for already quite some time, a girl with a pale frightened face. She was likely not a Muscovite but rather from the provinces. She asked timidly: "Are you Sakharov?" "Yes". "Can I come in?" Friendly A.D. said: "Come in, please". How many of these sorrowful faces enlightened by hope did he see!

During those years the belief in him showed up in funny forms, for example in the anecdotal (but quite possible) phrase reportedly overheard in the liquor store: "Brezhnev again wanted to raise the price of vodka but Sakharov did not let him do it."

Of course one should not separate him from his heroic but less famous colleagues with whom he was completely united and tireless in the struggle of alleviating the fate of the persecuted. Once I told him: "I think you are playing a no-lose game: If your ideas will be accepted, this will be your victory; if you will be imprisoned you will be glad to suffer like your comrades." He laughed and agreed.

The whole essence however was that here there was nothing from calculation or play. He just defended others for purely human reasons. Once he interfered in favor of three, unknown to him, Armenians accused of organizing an explosion in the Moscow metro and sentenced to execution (in newspapers only a little, barely understandible information was published). This interference was of course used in the process of defamation that Sakharov was exposed to at that time. When we met in FIAN I reproached him that he involved himself in an unclear matter. "What else could I do? There come three girls, crying, kneeling, trying to catch my hand to kiss. I sent Brezhnev a telegram asking to postpone the execution and carefully reexamine the case. This is what he himself did a few days ago". Indeed, the Pakistani president Bhutto, who was loyal to our government, had just been overthrown and sentenced to death. Brezhnev had sent the new Pakistani rulers exactly the same telegram as the one concerning terrorists that Sakharov had sent him (Elena Georgievna, having read this in my memoirs, noted that she did not remember the episode with three girls but the main thing was that this action of A.D. was preceded by his long fight for an impartial and just consideration of the whole case. Thus it was not just an impulsive action.)

When abroad, I told my friends asking me about Sakharov, "He is not Christ and is not Albert Schweitzer, but is made from the same material."

Let us however return to his life after he was fired from the work at the 'object' in Arzamas-16.

3.4 A Return to FIAN

The beginning of 1969 was marked for Andrei Dmitrievich by a personal tragedy. His wife Klavdia Alekseevna, the mother of his three children, died in March from a lately diagnosed cancer. He experienced a deep shock. In his memoirs he writes that at that period he was living and acting somewhat mechanically.

In the same memoirs Sahkarov writes that Slavsky (the minister in charge of the "object") sent him to FIAN. This is not exactly true. His memory tricked him. He mixed up two events which I know for sure because all the administrative procedures related to them went through me: From 1966 until 1971 I was a deputy of I.E. Tamm as a Director of the Theory Department. This is what happened in reality.

I would think that in the year of 1966 or 1967 when Tamm was still healthy and Sakharov was not yet "bad", A.D., who attended our seminars quite regularly, told me after one of them (when we, together with Tamm and Ginzburg, did as usual come to Igor Evgen'evich's office to talk) that he had permission to have a half-position in FIAN (apparently from Slavsky, nobody else could have given such a permission). I was glad: "What is the matter then? Please sign an application." I quickly wrote the text and gave him the paper and a pen. I am seeing it even now — how he is standing in the middle of the room, holding the paper in his somewhat stretched right hand and reading it with the pen in his left hand (he often wrote with his left hand). After thinking for some time he suddenly said: "What do I really need it for? No, let us not do it, I would feel myself freer, let everything stay as it is ..." he returned the paper and the pen to me. All my persuasions (I wanted to affiliate him with us very much) were in vain. It is possible that if he would not have refused then additional later complications could have been be avoided.

Now, in 1969, when he, in sadness, was staying at home, one had to do something. After consulting with V.L. Ginzburg and other "elders" in the department I went to Tamm, who was already chained to the respiratory machine, and got his warmest approval. Then I went to A.D.'s place. I found him very sorrowful. I told him: "Andrei Dmitrievich, I do not know what are you going to do but this can not continue further. If you would like to return to the Theoretical Department all of us would be very glad." If I am not mistaken he agreed at once and wrote an application. I delivered this application to the director of FIAN of that time, D.V. Skobeltsyn.

Everything turned out to be not so simple though. The direction (I think most probably because of the resistance of Partkom[5]) could not dare to appoint out-of-favor Sakharov as a senior scientific researcher. It very much wanted to get an approval from above, say, from the President of the Academy who at that time was M.V. Keldysh or from the scientific department of CC. As far as I could understand they did not say anything, just told us to decide by ourselves. I went to D.V. Skobeltsyn again — no progress. Then I invented a "move" — I wrote a draft letter from Tamm to Keldysh. I myself especially liked the phrase: "I would worry much less on the work of the department if I knew that young staff members could have a possibility of listening to an opinion of such a remarkable physicist." I went to Tamm, he immediately signed the letter (this was on April 26, 1969) and it was sent. Of course, Keldysh could not just turn down such a respected man as Tamm who was terminally ill (he died in less than two years). Still Keldysh probably needed a sanction from higher authorities. Only in two months after Tamm's letter, on June 30, Sakharov

---

[5]Institute's Party Committee.

was hired. It is possible that Keldysh needed the sanction of Slavsky as well.

In the memory of A.D., the two episodes, the permission for having a second job given by Slavsky before 1968 and the hiring to FIAN in 1969, merged into a single one. The memory of Andrei Dmitrievich was very good but not ideal. I have firmly established this in one case. In 1976 I wrote my article to the collection of memoirs on I.E. Tamm [Feinberg (1995)] and included into it one colorful episode which I have already described in the essay on Tamm (see Section 2.1).

In 1956 there took place an election of the President of the Academy, A.N. Nesmeyanov, for a new term. All sections apart from that of physics and mathematics quietly approved it. Tamm, however, persuaded his section to demand from Nesmeyanov that he should first present a program of revival of scientific biology (and with bureaucratism in the Presidum of USSR AS). This caused a storm at the general assembly of the Academy. It was finally agreed to reelect Nesmeyanov without delay but in three months to have a new meeting to discuss his program. This was realized, again in an atmosphere of hard argument and mutual attack. I knew it only through stories told by others though, I was not a witness of this.

Naturally I decided to check the credibility of these stories with the closest friends of Igor Evgen'evich and his companions-in-arms in the fight with lysenkovshina academicians M.A. Leontovich and A.D. Sakharov. To my chagrin, all of them, having shrugged their shoulders, said: "Nothing like that did happen." The episode was however so interesting that it was regretful to throw it away from the essay. I went to the Archive of USSR AS, asked for the shorthand records of the general assemblies in 1956 and found out that everything had happened as I have described it here. Most surprising was that in the attendance lists of both meetings (at those times academicians signed them personally) I saw the well-familiar signature of both of them! Not only were they present, but Tamm did surely discuss everything with them in advance. Nevertheless they completely forgot everything.

Since that time and until the end of his life Sakharov remained an active staff member of the department. He always (save from the period of Gorky exile) attended weekly seminars in his field, the official one on Tuesdays and an unofficial on Fridays (later it became a separate seminar of E.S. Fradkin). I chaired both of them and if due to some reason A.D. could not attend he called me and explained the reason. He began with the words: "Here's Andrei speaking." He pronounced his "r's" in the French manner saying "Andhei"; I should add that up to the last days the greeting postcards

I got from him were usually signed "Andrei" or "Andrei Sakharov," later "Lyusya, Andrei". This was an explicit hint that it was finally time to stop using patronymics. I could not however overcome the habit established already in the forties later amplified by my growing admiration of Andrei Dmitrievich that did not allow familiarity from my side so it remained "Andrei Dmitrievich" to the end.

With the same punctuality did Sakharov attend (usually once a month) meetings of the scientific council of the department (and the less frequent ones of the scientific council of the institute), participated in discussions of even insignificant interdepartamental problems. Only his intensive political activity after Gorky began increasingly impeding this. Nevertheless here as well he did inform that he could not come. When after exile he became a member of the presidium of AS which at that time held its meetings on Tuesday mornings he was often coming to the seminar even without having had lunch. Sometimes he napped at the seminar (I usually sat by him but did not wake him). This meant that he was working at night (his scientific, literary, publicistic and practical human rights defense activities did tear him apart). After ten minutes, however, he shuddered, with a characteristic gesture passed his hand over the face upside down and resumed listening.

Let us however return to the year of 1969. At that time he lived with his younger daughter Lyuba (she was already 20) and son Dima (he was 12). As usual he was simple in the questions of everyday life. Once I asked him: "Andrei Dmitrievich, how do you manage all this? Now, when Lyuba left for vacations who does take care of you?" He answered: "Everything's fine. With Dima we make a tour of the nearby cafes and each day have lunch in a new one. We have a rule: each meal is only one dish but in large quantities. Sometimes my cousin comes and makes breakfast for us but often this turns out unnecessary."

On May 21, 1971 Andrei Dmitrievich turned 50. On that day there was a Friday seminar. There came all the members of the department and some other staff members of FIAN. I delivered a speech in which I said, in particular: "We, the members of the department are glad and, I can not find another word, proud that Andrei Dmitrievich did choose our department in his youth and after 20 years when he returned to his favorite domain of physics."

The situation around Sakharov was at that time already very tense. His fellow academicians that knew him personally when meeting him during a general assembly of the Academy in the foyer of the House of Scientists did in most cases rapidly greet him and come by. Those passing him by did not turn. More and more often at the meetings of the general assembly (especially after 1973) after having entered a crowded hall I saw that chairs around him remained empty. More and more often (I think this was done on purpose) did he sit in the last row which usually was half-empty. A sight of his isolation was distressing and I sat next to him. Our department did feel a malevolent attention of authorities, faced some attacks and was of course "under surveillance" so in the above-mentioned speech I was choosing the words carefully but told about his political activity as well.

After the seminar I did, just in case (if some problems arose), write down my speech. This was two and a half typewritten pages long. A solemn and warm greeting was naturally concluded with some humor. As usual he was laconic but, I think, satisfied. I recall nevertheless some shadow of sadness in the way he looked. I do not know whether I can allow myself such a guess but I think that few of those that became close to him during an intensive work at the 'object' did allow themselves to congratulate him.

I should note that I did have a right to tell about the pride felt by the members of the department. During all the years, including those of Gorky exile, all of them from the senior scientific leaders to typists did deeply feel their perhaps weak complicity in the fate of Andrei Dmitrievich and were proud of this. Everything that had to be done for A.D. was done immediately and with love.

At that time there took place important events in Sakharov's personal life: He met Elena Georgievna and soon they married. There came together, tightly and for a long time, two persons with the characters and formation history so different! He was a concentrated clot of traditions and moral norms of the Moscow intelligentsia. She was a daughter of a convinced bolshevik who was fighting establishing a Soviet power in Armenia and until 1937 (when he was arrested and executed) was a member of the executive committee of Comintern where he was responsible for the human resources department. Sakharov did not even attend school until the eighth form and grew up in a family where there dominated softness, kindness and complete trust. She grew in a family living in the Comintern dormitory, communicating with people for whom the possibility of death was close, both because of Stalinist terror that demanded numerous victims even among the communists — political emigrees, and because of their devotion to "highest party and class interests" that forced ones to send and others to readily go on mortally dangerous and often cruel missions in fascist countries and in foreign countries in general.

This difference did not hinder their close relations and may have led to working out new features in their common activity in the human rights movement and life in general.

The strength of A.D.'s feeling is now clear to everyone from the two published volumes of his memoirs. But even at that time one saw how his internal life changed. He was shining. He looked different even in the details: He began coming to the institute with his haircut done in time, there disappeared the unshaven halo on his neck, etc. I first met Elena Georgievna in the April of 1972 at a small scientific conference in Baku (A.D. had of course told me about the marriage a long time ago but I had fallen ill, and been in surgery). I remember when the participants of the conference spent a free day out of town on the seashore, we were lying with Andrei Dmitrievich facedown next to each other on a warm slanting rock fragment while Elena Georgievna easily and merrily jumped from one big stone to another which was not safe. A.D. raised himself a little, leaning against his stretched arms and cried in full delight: "Lyuska, don't do it!" (he himself writes about this episode in his memoirs but with more restraint and of course does not reveal this delight). In general, afterwards it was clear that this marriage gave him a feeling of personal happiness that was so necessary for him. It amplified his two-year history of practical participation in the human rights movement in which Elena Georgievna had been active for a long time.

Young A.D. Sakharov

3.5 A Hunt

The tension around Sakharov increased. In September 1973, after the first interview which he gave to a foreign journalist, there finally came an explosion. This began with a newspaper campaign, with the memorable first letter of 40 academicians. In FIAN there of course also began a collective signing of a protest against the political position and activity of A.D. In spite of the immense respect he enjoyed among scientific staff there was a lot of people who signed the "protest" with ease or even with pleasure. There were, as everywhere, those simply befuddled with propaganda under the influence of which they had been living for their whole life, from the kindergarten, and thus were sincerely indignant.

Most important however was simply the influence of the ingrained fear. To the many that refused to sign, the institute partkom had to apply a usual "arm twisting" to make them surrender. In one big laboratory the director, partorg[6] and proforg[7] did decide: "OK, let us three take the shame and sign but other staff members should stay aside." A scientist from another laboratory was exposed to a long "processing" in partkom. To his remark that in 1937 many had also been accused and were later acquitted they replied: "Think what you are saying, you have children." On the next day he came to partkom and signed the "protest". There were other similar cases as well.

In total there were, if I am not mistaken, about 200 signatures. Of course, with the Theoretical Department there was a misfire. All the threats to the party group in the department, all the pressure did not help. The atmosphere around the department was however thickening. The document, the text of the "protest," was kept in partkom and used for blackmail in any convenient case. At that time, for visits abroad and for sending the defended thesis to HAC (Highest Attestation Commission), the decisive document was a characteristic approved by partkom. A senior scientific researcher of the Ddepartment, the party member G.F. Zharkov, was invited to a conference in Paris. He did however refuse to sign the protest and because of that was denied a characteristics. A corresponding member of USSR AS (later an academician), E.S. Fradkin, a most prominent internationally known scientist who had been at the front for the whole war and had been wounded by Stalingrad, had become a party member at the front and had been decorated with battle orders, received a prestigious invitation to give a talk at the Nobel symposium. He was called to the meeting of partkom and directly told "You sign — you go." He refused. A talented young theorist (nowadays a corresponding member of RAS) did brilliantly defend a thesis.

---

[6]Secretary of Party committee.
[7]Secretary of trade-union committee.

Before sending it for approval to HAC he was given a "bad" characteristics. I checked it myself — with such a characteristics the approval would be denied. He consulted with me, somebody not close to him, on what should he do. I told him: "Decide it yourself, but I understand your situation. Even if you yield to this shameful pressure the attitude towards you in the department will not change." He did yield. He was the only one from about 50 members of the department who did sign.

For several more years this document was used in the partkom as a "litmus paper" and used during new splashes for the hunt of Sakharov. In one such case a secretary of the party group of the department, the above-mentioned Fradkin, was summoned to the secretary of raikom[8] who, in the presence of specially attending CC staff member, shouted at him and threatened him — nothing helped. Later, during the Gorky period, when it was even more difficult, I once told A.D. that I admired the behavior of the party members of the department (at that time there were five of them) because for them all this was especially hard. He agreed with me.

There were also more unpleasant, even worrisome episodes. At one all-institute party meeting the participants were vehemently condemning A.D. and those having conciliatory attitudes towards him. The atmosphere became so heated that from the back rows there came forward a worker from the workshop, a powerfully built man, who told something like "... give it to me and I will immediately — "and made a gesture with two hands showing how he would turn off the head. There was noise, he was stopped (later he got a party reprimand). B.M. Bolotovsky was a witness of this. The episode made us worry and either V.L. Ginzburg wrote a letter to the assistant director on "regime" (such a position existed because in some laboratories a "classified" research was conducted) or V.Ya. Fainberg went to talk to him (we could not remember this exactly later) but I do remember his answer very well: "Do not worry, on the territory of the institute not a single hair will fall from Sakharov's head."

One episode from this period of hunt is colorfully characterizing A.D. himself. One of the party leaders of the institute, a good physicist himself, did in that period sincerely fall under the anti-Sakharov madness; he did in the sixties-seventies on the whole become, as some say, "very party." During the period of the hunt he came to the Department (where he had once been a Ph.D. student) and said indignantly: "How at all can you treat

---

[8] District party committee.

him friendly! You should not give a hand to him!" He received very hard replies (but still remained dopey for a long time).

Soon this physicist was being approved, at the scientific council of the institute, in the position of a head of a sector. The council members from the department did naturally vote "against" although they understood that from the scientific viewpoint he deserved this position and that he would anyway get a necessary number of votes (this was just a plain demonstration against his behavior). This indeed happened. After the meeting when we gathered together I said perplexedly: "Why did he get only five votes "against"? There should have been six." A.D. interfered: "I voted for". "You?" "Yes, I considered it awkward to vote "against" without having openly expressed my opinion before the vote."

I think that his behavior is a good lesson for many and that party activist has paid a lot for his folly — his behavior became widely known in the physics community. One cherished his scientific results but his reputation in the eyes of his colleagues fell very low. For decades he was not elected a corresponding member of the Academy of Sciences and only recently when, probably, he has himself understood his shameful folly, he was "pardoned" and elected. In terms of his work, he long deserved it.

As a whole the authorities did nevertheless leave the department alone. It quasi-recognized that this was a special case and confined itself to a general, watchful, unfriendly observation and to constantly creating problems in inter-institute matters, in particular related to trips abroad, hiring Ph.D. students, etc. I would say that I, personally, experienced very few problems.

I think that despite this atmosphere of hunt, there played a role a respect to and even admiration of Sakharov's personality both inside FIAN and in the scientific circles outside it. Here is an example.

In one of the years of the Gorky period at a meeting of the AS presidium there reportedly happened the following episode. The president A.P. Aleksandrov did on some occasion suggest to adopt a resolution condemning some statement by Sakharov (this could possibly be related to his letter to S. Drell that caused a splash of anti-Sakharov statements). A late member of the presidium, academician N.A. Pilyugin, asked: "Why should not we expel him for good?" Aleksandrov quickly replied: "This question is not on the table."

This story became known in the academic circles. I am describing it "as heard" (recently I have heard another version: Pilyugin asked his question at a party conference) but it is confirmed by something told to me by my friend (a man, far from the Academy, whose words can be trusted with certainty). Once he was with some business of his in the Council of Ministers and was a witness on how Alexandrov was asked the same "Pilyugin" question. Anatoly Petrovich replied: "This can be done but there will be so many votes against the expulsion that this will create a bad impression." If he really replied like this then he was right. This is proved by the following fact.

Shortly after this episode at the meeting of the presidium there took place a routine re-election of its members for a new term. Pilyugin did not enjoy a wide publicity: He was elected at the specialization "automatic control" which was probably related to the classified areas of missile technique. Several minutes before the vote I saw the name of Pilyugin in the list of candidates. I reminded the story to one academician, met occasionally in the foyer, who I knew (I myself was at that time not a full member of the Academy and did not participate in the votes). He reacted emotionally and when this story was made known to several other academicians the information spread like a chain reaction. As a result of a secret vote Pilyugin was elected but got an unprecedented number of votes against — it seems, 49. One could understand the meaning of this number only if one takes into account that usually the number of votes "against" in such votes did not exceed five, seldom more. Only one member of the presidium who at that time was thought of as having chances of becoming a president, while many feared this because of his very resolute character, got about 25 negative votes. It was clear that if the story with Pilyugin had been reminded earlier he would have got even more "black balls."

The meaning of this vote was understood in the Academy presidium at once. Here is proof. Soon after that my wife was visited by an acquaintance, a colleague who was a wife of an influential member of the presidium (both were specializing in liberal arts). When I came into the room she asked me at once why had Pilyugin get so many votes "against". I answered: "Well, he was the one who demanded the expulsion of Sakharov". She said: "Yes, of course, I also thought so." She surely asked this question by request of her husband. Never before had we discussed academic matters.

Let us however return to the "pre-Gorky years." Sakharov was continuing his activity with the same or even growing intensity. Once in 1976 I told him: "You know, Andrei Dmitrievich, some people that I know revering you and Aleksandr Isaevich think nevertheless that neither you nor him should give recommendations on particular issues of economy and politics because you are not professionals in these fields." I do not know what would the answer by Solzhenitsyn have been but Sakharov immediately replied: "Of course I am not a specialist, of course I am making mistakes but what could one do when others do not dare to say a word?"

Once at the very beginning of the human rights activity of Andrei Dmitrievich, when we were walking after a seminar (as it often happened) I told him: "They will probably hit not you yourself but those dear to you." "It is possible", he agreed. Finally it thundered over him too.

3.6 Arrest and Exile to Gorky

The turning-point date of January 22, 1980 caused a shock in our department. The reason was not only in the generic fear embracing at that time everyone that of course did not miss us as well. During 12 years we got accustomed to that in spite of the baiting in mass media, A.D. himself was untouchable, that, as I had told him a long time ago, one would hit his entourage, "those dear to him," but not himself.

It became clear: If the authorities did after all choose such an exceptional course of action that would cause indignation and protests around the whole world community they also would not later restrict themselves to half-measures.

Of course in a minimally democratic country one would have to first express a collective protest. By us however the authorities were always particularly afraid of such "collectives" and violently prosecuted them. Exactly in the sixties and seventies, in the course of a process of liquidating the remnants of Khroushev's Thaw, this lead to cruel prosecution of "signers." A bright example that corresponds quite closely to our case (if only and incomparably "softer" one) was a pogrom committed in 1968–1969 in the mathematical community when 99 mathematicians from MSU and other institutions came out with quite a mild letter addressed to the Ministry of Health demanding to free a practically healthy mathematician and human rights activist A.S. Esenin-Volpin who had been forcedly placed into a psychiatric clinic. This resulted in the firing of a number of scientists that had signed the letter. A professor of the Lenin Pedagogical Institute, a well-known mathematician Isaak Moiseevich Yaglom, at that time an author of extremely popular textbooks and educational books on mathematics, was fired and could teach only in the district center Orekhovo-Zuevo. A talented professor of MSU, Sergei Vasil'evich Fomin, had a massive heart attack (this was not the only case) that probably caused, in several years, his premature death.

There were many fates like those. The main thing however was that there followed a special decision of CC on this matter and the brilliant faculty of mechanics and mathematics of MSU, where many stars of our mathematics ensuring an atmosphere of free creativity impartiality and objectiveness were working, was practically demolished. Its leadership was replaced and from scientific oasis it turned into a citadel of preconceived orthodoxy. Chair of academician P.S. Novikov in the Pedagogical institute was ruined in the same fashion. Similar measures were taken to the editorial boards of mathematical journals and publishing houses.

It is not difficult to imagine what a collective protest in the politically much more acute case, against the Sakharov's exile, would have led. One did not need to make assumptions or consider theoretical constructions. The precedent was at hand. I am reminded of this because A.D. himself, with his reckless adherence to the ideals of democracy, considered collective protests as something highly desirable and himself participated in such actions (that only led to new repressions). He thought that a doctor degree[9] provided sufficient defense in such cases (see the memoirs of V.I. Ritus in [Keldysh (1996)]). This opinion did, as we have seen, contradict "the experiment." It was not without reason that the most clever P.N. Kapitsa, who on many occasions stood up for the repressed (and saved, in particular, V.A. Fock and L.D. Landau — which was almost a miracle), achieved his goal by sometimes very harsh and filled with self-respect personal letters to Stalin and Molotov that were kept, however, in total secrecy. Everybody learned about them only after his death. This required however his immense authority.

As could have been expected there began attempts of immediately firing Sakharov from FIAN. The director turned out to be (or pretended to be) ill but evidently somebody "from above" was already calling him and the direction was in search of a formulation for the reason for the firing. One suggested as a reason the fact that Sakharov did not live in Moscow but for some members of the direction it was not convincing (one could not impose such conditions upon an academician). Attempts to learn in the presidium of the Academy of Sciences which decision or special document should one refer to were fruitless. The numerous demands of Sakharov's firing were possibly resulting from the fear that the very name of Sakharov induced. It is probable that there was no definite document. One tried to pass the decision to the Theoretical department but its leadership refused to do it.

---

[9]Russian doctor degree is roughly equivalent to Habilitation in Germany.

Then a deputy director, S.I. Nikolsky, who was in charge of our department, called the scientific department of CC. After indeterminate words like "you understand it yourself, don't you?" he was finally told: "Act according to the law." This was interpreted in the favorable sense. One did not have to fire. Everything was left as it was but in a very unclear and alarming status, "in suspended state."

This could not last long. We, the "elders" in the department, worked out a special program containing three points. (1) Sakharov officially remains a staff member of the department; (2) as an outstanding scientist he gets all possible assistance for the continuation of his scientific work; (3) in the framework of realization of this assistance members of the department will regularly visit him for mutual consultations and discussions of scientific issues. Of course the third point contained an evident implication — a wish to somehow soften an isolation of A.D., his separation from colleagues and close friends.

V.L. Ginzburg, as a head of the department, went to the scientific department of CC and managed to persuade an official who talked with him in the rationality of our suggestions. The official promised to forward these "upstairs." Only on April 9, after negotiations of V.L. Ginzburg with A.P. Aleksandrov, did the latter send an instruction to FIAN, was the problem solved exactly in the framework of our program (for more detail see 'The Gorky File' by I.M. Dremin in [Keldysh (1996)]). This decision was also possibly influenced by our parallel "propagandistic" activity.

It was necessary to lay bare a widespread lie that Sakharov was no longer a scientist, that "he in the last time distanced himself from scientific work" (as written in the *Big Soviet Encyclopedia* and in reprinted editions of the *Brief Encyclopedia*) and did not any longer represent anything as a scientist. (This version remained alive for a long time. In December 1990 one solid newspaper that marked the one year anniversary of Sakharov's death by publishing an excerpt from his Nobel speech wanted to accompany this text by additional material and asked me to answer two questions. The first was: "When did Sakharov stop his scientific activity?" My answer that this never happened caused a surprised comment: "There is however a widespread opinion that he turned to social and political activity because he had run dry as a scientist." Truly: "Slander, slander, at least something may remain." And this remained even when, it seemed, everything on him was already known. It is easy to imagine what it looked like in those days...).

Of course one can refute this nonsense by the simple fact that when Sakharov was detained on January 22, 1980 the department typist was retyping the manuscripts of three new papers that he had left. With this, by the way, there was connected a funny episode. On January 22 I was in a sanatorium. On the very first day of my return to FIAN I got a stern question from above on why, when A.D. was being flown to Gorky by plane and was concerned by something, Elena Georgievna comforted him: "Do not worry, Evgeni Lvovich will do everything." Who was Evgeni Lvovich and what did he have to do? It was clear that A.D. worried about the fate of these papers. But in our department we have people that do not need me present to understand everything. The papers were appropriately prepared for publication, S.I. Nikolsky was not afraid of sending them to the journal, our central *Journal of Theoretical and Experimental Physics* (JETP), where the editor-in-chief was P.N. Kapitsa and his deputy, who took care of all current matters, was E.M. Lifshitz, so everything went smoothly. The papers were published in the nearest summer issues and the "terrible name" appeared on the pages of printed matter. It was of course also important that Glavlit did not forbid the publication. At that time a decision of not hindering Sakharov in his scientific work had been probably already taken.

This was however not all. I began compiling a 'Reference', an annotated list of A.D.'s papers written during the period when he had again become our staff member (at first, in 1966–1969, unofficially). The 'Reference' was completed with the help of D.A. Kirzhnitz and A.D. Linde (see the article by I.M. Dremin 'The Gorky File' in [Keldysh (1996)]). This required some effort in collecting the reprints of A.D.'s papers that some of us had. Due to our usual negligence and the above-mentioned excessive trust to the Pasternak words "One should not organize an archive..." two papers turned out to be missing. One of them was a big important paper of 1975 (published only in Russian) in which Sakharov developed his idea (suggested 17 years before that time) of what in the West is called "induced gravity." It was in particular developed by a well-known American theorist S. Adler (in one of the years of Gorky exile he visited our department for two weeks and brought a big talk on this topic. When we showed him the paper of A.D. [by that time we had recalled it] he was astonished — this contained many results later independently obtained by foreign theorists).

The "Reference" listed 13 papers (in 14 years) including those containing the most important cardinally new ideas: An explanation of the baryon asymmetry of the world and a prediction of the proton decay, an above-mentioned induced gravity, a theory of "many-world Universe" (in essence a consideration of the consistent model of the Universe before the beginning of its expansion which has been happening during the last 10–15 billion of years) and many others.

The "Reference" was typed in about 20 copies and some of the elders began dispatching it to the influential scientists — the president and some of the vice-presidents of the Academy, P.L. Kapitsa, etc. It is probable that V.L. Ginzburg, when going to CC, could already hand it over (none of us remembers this for sure but according to the dates this looks possible).

In the course of my conversation with one of the vice-presidents that was proceeding in an atmosphere of understanding and compassion he said something that I did not expect: "If it were only his protest against the Afghan events. Worse is that he, being a carrier of important state secrets, has visited, together with his wife, an American ambassador and has had a long conversation with him." "This cannot be!" I exclaimed and told about my old above-described conversation with A.D. (when he said: "I have to be watched. They should be sure that I would not go to the American embassy," but this was 12 years ago). My interlocutor repeated regretfully: "Yes-yes, they have been there." When I first met Elena Georgievna after that I asked her about it. She confirmed: "Yes, we went there. What is so special about this?"

It is up to now unclear to me whether this was not a mistake by Andrei Dmitrievich. Of course if their conversation was about the weather or even human rights there was indeed nothing special about it. But if the conversation was on, say, disarmament or the international situation, something at first sight insignificant could slip away. One clever acquaintance of mine who was for many years working with A.D. at the 'object', loved Sakharov and enjoyed his respect (for too clever a reader I should say that this was not Yu.B. Khariton), told me: "You certainly understand that corresponding American experts will study the tape of this conversation "under the magnifying glass".

Meanwhile, I was still not convinced in the rightfulness of our approach and decided to discuss this with one physicist (not from our department). He took me to Lenin Hills and there, during a stroll, I explained to him our plan. He thought for some time and said: "You will necessarily contact with "organy," will draw into this contact and, without noticing it yourself, will become their agent, will get entangled." He was right that this danger did exist, we understood it ourselves and always remembered about it afterwards. Nevertheless we could not leave A.D. alone and, contrary to the advice, resolved to proceed with realizing our plan. With a pride for our department I can say that this worked out. Of course we had to accept restrictions that were from time to time imposed by "organy" but nothing like becoming their "agents" did ever happen.

As seen from what V.L. Ginzburg tells in his article [Keldysh (1996)], at first A.D. himself did have such fears and, for example, protested against any chance to be visited "only by people sent by KGB and selected by it." From the above description it is clear that there was no question of somebody being "sent", that this was our initiative that, as already discussed above, was far from being morally innocuous for us. These dangers were however overcome by our (elders in the department) mutual understanding, full mutual trust, diligent preparation of every step. As is known, in total, 17 members of the department did go to Gorky (the majority of them many times — up to 6). We told the direction whom did we want to send on the next trip then, "behind the scene," there took place some approval process and we, with an exception of one case with V.Ya. Fainberg (I will describe it below), got an agreement or the trip was postponed, sometimes for reasons clear to us (during A.D.'s hunger strikes or just after them) or completely unclear ones (for example days close to November 7 or something of this sort). Of course in some cases our actions conformed with KGB goals (for example when I was passionately dissuading A.D. from his intention to go for a hunger strike; this could be done in usual letters which, of course, were inspected, see below). This should not have been taken into account though.

We had to wait for an approval for the first trip for a long time. It was obtained suddenly, no doubt that in two days one expected a visit to Moscow of the president of the New York Academy of Sciences, J. Leibowitz, and one had to be able to tell him that everything "was not so bad" with Sakharov. The director of FIAN, N.G. Basov, told me, smiling: "You have to go to A.D. immediately." I could however not: I had to deliver a scientific talk at the presidium of the Academy of Sciences. "In this case V.L. Ginzburg should go." So, Vitaly Lazarevich did go [Keldysh (1996)]. Their meeting was very joyful for both of them.

3.7 Trips to Gorky

I will not give a detailed description of trips to Gorky. Others have written a lot about them. I will just tell about some episodes that are, in my opinion, of interest.

When in June 1980 I first visited Andrei Dmitrievich and thought that he was depressed I did, in order to cheer him up, recite a distich by, if

I am not mistaken, Kaisyn Kuyliev: "Patience is an arm of a hero, When another arm has been snatched." A.D. was outraged: "What patience?! The struggle goes on!" This was very different from that "social philosopher" whom he had been in 1968 when he had begun his activity in the social and political arena. Then he told me his ironic quatrian that is now known in several variants:

> On the stone face of the state
> The in the shine of world's glory
> Moves ahead without stumbling
> There are invisible pock-marks.

Sherbinka (in English a 'pockmark') was a district in Gorky in which A.D. had to live. He was quiet and vivacious and in quite good physical shape. The hunger strikes were still ahead. I came together with our staff member of a younger generation, O.K. Kalashnikov. In general, without a special discussion of this, we decided to go in twos. I think the reason was in our, by that time automatic understanding, that "organy" would not have let people go alone. According to the calculated psychology of the "watchers" there should have always been a third somebody capable of informing if in the course of communication with Sakharov something inappropriate happened. That the members of the department could have another psychological motive was probably considered impossible. One did not have to worry though. Nothing illegal ever happened.

However, during one of the visits, there took place an incident. After returning to Moscow, V.Ya. Fainberg was excoriated for wrong behavior (he tells about this himself in his memoirs [Keldysh (1996)]). The thing was that although we understood very well how thoroughly everything that was happening in the Gorky apartment was listened to, we nevertheless acted freely and felt ourselves free. The few things that we wanted to convey privately were written on paper covered by the hand from the assumed objective of hidden camera (we did not bother to search for it, it is possible that there was no camera at all).

I guess that once V.Ya. got too much out of hand and discussed politics, etc. This would not still have been a problem. But A.D. decided to demonstrate how a radio "jammer" installed personally for him jammed undesirable foreign radio stations in a radius of 60–90 meters from his apartment. He switched on a transistor radio and everybody heard a first phrase from a Russian BBC program. After that a mighty "Zh-zh-zh ..." was to be heard from the radio — the jammer was in operation. There followed a burst of laughter (this episode shows, by the way, that an observation was performed not only through taping with subsequent analysis but continuously — as one says, "in real time"). I think that such behavior of V.Ya. did very much offend the "guard" who just did what he had an order to do. This was their service and a mockery on it was of course taken as an offence. As a result a report that was sent to Moscow was very much blackening for V.Ya.

There arose an unpleasant situation and future visits were endangered. Then I volunteered to go to improve the situation. When we (together with O.K. Kalashnikov) arrived then, after first joyful greetings, I told approximately the following: "Andrei Dmitrievich, I would like to tell you something serious and practical. We have to take into account that our visits have a well-defined purpose — mutual scientific consultations. These are very important and enjoyable for us and, we hope, for you as well. These can however continue only when this particular goal, not something unauthorized, will be pursued. Do you agree with this?"

During this admonition Andrei Dmitrievich was sitting in an armchair while I was sitting on the chair in front of him. I was talking distinctly and loudly enough for the eavesdropping device to tape everything correctly. Andrei Dmitrievich did of course understand all this and was sitting smiling. I could not allow myself to smile because this would have reflected on my voice. He knew me however quite well and of course answered with something affirmative. The incident was over. (After this V.Ya. Fainberg went there several times. KGB however reported this to the Moscow party gorkom, and as a result V.Ya. Fainberg could not go abroad until 1988). It is possible that after the "admonition" (or before it?) I wrote to A.D. that there had been problems after the visit of V.Ya. Fainberg but I do not remember this for sure.

I also cannot forget my last visit together with E.S. Fradkin in December 1985. This was after the third hunger strike when Elena Georgievna had already gone to the USA where she was to have heart surgery. Andrei Dmitrievich opened the front door to us and having said: "I have influenza, kisses are cancelled, put on gauze bandages — these are prepared for you in the dining room." He was very lean ("I have recovered 8 kg, half of my lost weight," he told, floor weights were standing by the bed) and looked bad. After that I felt his pulse — there were many extra systoles (if I remember correctly ten and more per minute). The influenza was not heavy but he did not allow us to approach him often. Having had breakfast in the kitchen and having put all the food we brought in place we returned to the bedroom and E.S. Fradkin began telling about his last very important work on string theory, the most difficult and "fashionable" domain of theory of particles and fields. On the wall there was no blackboard and it was very rarely allowed to approach Andrei Dmitrievich to show him some formula. E.S. was monotonously walking back and forth along the room and A.D. was grasping everything "from the voice" inserting questions and comments, discussing particular points.

I was astounded by the force of his mind. These problems interested him very much at that time, he was listening and listening. This went on for four hours! Finally Andrei Dmitirevich told: "Enough, let us have dinner and then have a rest. Please warm cottage cheese for me. A teflon pan hangs in the kitchen on the wall." (As is known he liked to eat everything only in the heated form). Having had rest (while me and Andrei Dmitrievich were sleeping Fradkin went to the nearby Institute of Chemistry and appropriately registered our trip) we again returned to science. I began describing something from my topics but soon saw that did not interest him. He was carried out with strings and there again began, for almost three hours, a lecture-talk by Efim Samoilovich. And again without writing formulae. There came a time to leave. Right before our leave there was another important episode that I will describe elsewhere. When we were going back E.S. said that he was also astonished by the understanding of most the difficult science that Sakharov had shown. Shortly before this E.S. had been abroad at a conference and had told the same things to experts in this field. They had understood much less.

3.8 Hunger Strikes

I am now turning to a very uneasy and heavy topic, to the hunger strikes of Andrei Dmitrievich. As is known there were three of them in Gorky: in 1981, 1984 and 1985.

At the end of 1981 in Moscow there spread the news that caused horror and perplexity in many people: Sakharov had declared a mortal hunger strike. Afterwards one learned that his wife participated in the hunger strike as well. What did he demand? A release from the exile? Interceded for some dissident? No, he demanded permission for the fiancée of the son of Elena Georgievna, Liza Alekseeva, to emigrate to the USA. Who was she? Permissions to emigrate were in general given very rarely so the legendary Sakharov was ready to die for that? Many did not understand anything. I myself knew this lovely and clever girl who was really close to E.G. and

A.D. Some thought that he wanted to save her from prosecution for the closeness to his family like he had earlier saved the children, son-in-law and grandson of Elena Georgievna by securing their emigration to the USA. Many thought however (although this did not seem probable) that Lisa was not a reason at all. One thing was important: Sakharov protested by giving a dare to the authorities and could die. Many thought that he could not risk his life when so dear to the people. There were however people by us and abroad who thought that it was remarkable that the great citizen was ready to give his life for the happiness of an ordinary girl. More important than all argumentat however was: Sakharov could die. This was awful.

On one of those days I was told that Liza had on the previous day been trying to be received by the president of the Academy, A.P. Aleksandrov, but had been for the whole day refused a pass. I took this as an occasion for doing something, decided to go to him myself and persuade him to receive Liza. This was an absurd idea but not doing anything was unbearable. Anatoly Petrovich knew me well and had shortly before demonstrated his good attitude, I knew well his assistant and I simply went haphazardly. Of course I did not achieve anything positive but learned something important — this is why I am describing this visit.

When in the morning I came into the president's waiting room his assistant Natalia Leonidovna Timofeeva told that he was having a regular meeting with vice-presidents and the chief scientific secretary G.K. Skryabin and that I had to wait for its end. I began to wait and was talking with Natalia Leonidovna about Sakharov whom she had known since his younger years.

At that time a still quite young and energetic academician, R.Z. Sagdeev, literally flew into the waiting room shouting on his way: "Comrades, do you understand what is going on? Can you imagine yourself what will happen if Sakharov dies? All our international scientific programs, all connections will go to hell, nobody would want to deal with us!" Having learned that the president had a meeting he ran somewhere then, returning in about ten minutes, informed: "They are discussing precisely this question. A KGB representative declared that one should in no circumstances retreat, the situation is under control and there is nothing to worry about. He said: If we yield they will "completely sit down on our head." And, still excited, ran away again. Soon he reappeared and said: "The vice-presidents are urging Anatoly Petrovich to go directly to Brezhnev and he is jibbing." Indeed for A.P. to address Brezhnev above KGB's head meant to enter into a direct conflict with this dangerous and powerful organization. It was easy to understand Aleksandrov — it was not easy to decide to make this step.

At this moment a prominent physicist-theorist academician, M.A. Markov, came to me and began telling about his idea related to the transformation of the Universe during the first several minutes after beginning of its expansion. I am in general not an expert in these questions but at that moment, like it was during last days and nights, I was internally very tense and in my head there tossed something disordered and depressing. I did not comprehend anything and only mechanically squeezed out of myself "A! ... Yes! ... Interesting ..." "So," said Markov, "in two weeks I will go abroad to the Paguosh Conference and in two months to England to discuss all this with Hawking" (Hawking is the biggest expert in the field of relativistic cosmology). Here Sagdeev, unable to stand this any longer, interfered: "You could possibly predict what happens with the Universe after three minutes from the beginning of its expansion but you do not understand anything in what will happen in two weeks. If Sakharov dies you can surrender your passports, you will not go anywhere." Markov fell silent and, embarrassed, left.

Finally Natalia Leonidovna said that I could come in. Anatoly Petrovich sat at his desk irritated, frowning, even angry. I began telling him that I understood the difficulty of his situation, could not advise anything decisive but begged him to receive Liza — this could help to find a way out, some compromise. All in all, having had no clear cut plan before I spoke, not being sure that this would lead somewhere. I just wanted to help Liza. To insist on him addressing Brezhnev seemed, after I had learned that the vice-presidents had pushed him towards this, senseless.

I had one "back" thought though: a personal contact with somebody always has a positive effect bigger than all discussions. (In the middle of the XX century the following experiment was performed in the USA. A hundred owners of road motels were asked whether they would accept a black guest. More than two thirds [I quote the yield as I remember it] gave a negative answer. But when real black visitors were there only less than one third of the owners refused to accept them. This was considered as resulting from a beneficial influence of personal human communication. In the case with A.P. Aleksandrov one could also count on this.)

Aleksandrov began angrily telling me that he could not do anything, that this was out of his reach. "You see — all these are telegrams of protest from abroad —" he pointed at his desk (that was fully covered by accurately piled telegrams) and awry, added: "You understand, the Academy has more than 7000 people in its staff only in Moscow and all of them have some family matters, I cannot interfere in these". I objected: "I understand that you have enough trouble from this but how many Sahkarovs a century does one count?" He repeated: "I cannot do anything." I left having said again: "Think it over. Perhaps after receiving Liza you will get some idea." All in all as I have already told the visit was absurd but what I heard and learned, especially on the position of KGB ("situation under control"; "one can not retreat") was probably worth communicating further to take this into account in our own activity.

It was clear (and I know it for sure) that a pressure on Alexandrov was exerted not only by those who were afraid of the breaking of scientific contacts but also those for whom Sakharov was dear as a unique personality, simply as someone provoking love and admiration. Sometimes the words on possible breaking of contacts were only a "rational cover" for more personal feelings. I do not know for sure how it happened but Anatoly Petrovich did finally overcome himself and went to Brezhnev who resolved the issue: "Let her go." The life of Andrei Dmitrievich was this time saved without a big effect on his health.

A second shock came in 1984 when Sakharov again declared a mortal hunger strike, demanding this time a permission for Elena Georgievna to visit the states "to see her mother, children and grandchildren and for medical treatment." The last argument was understandable: By that time E.G. had suffered a heavy heart attack (possibly more than one) and had been already sentenced to exile in Gorky. The particularity of her situation as explained to me by Elena Georgievna and Andrei Dmitrievich was in the fact that to save her eyes that had suffered from contusion at the front she needed medication that was at that time counter-indicative for her heart (and medication needed for the heart did damage to the eyes) so for surgical treatment for the eyes she had already three times gone to Italy; this was of course a quite unusual phenomenon in the seventies. A motivation "to see her mother, children and grandchildren" which Andrei Dmitrievich placed first in the list of reasons was not understandable for many.

Of course it was characteristic of A.D. with his unblurred attitude to simple human values. A daughter, mother, grandmother does perhaps have a right to see the people most close to her for the last time in her life. At that time, however, everybody knew: If you are seeing off abroad even somebody close to you — this is a separation for good. This is why so many tears flew in the Sheremetjevo airport. Therefore such motivation of A.D. was not understandable for many. Leaving again all these considerations aside the people knew, however, just one thing: Sakharov is on the verge of death, he is protesting again.

Still, in any case for me, who knew the KGB formula "Situation under control, there is nothing to fear of and if we yield they will sit down upon our head," as well as for many others, at any rate for the majority of members of our department, a hopelessness and therefore the inanity of this hunger strike were clear. We did not know yet what the KGB "control" meant. We later learned about this from the letter of A.D. Sakharov addressed to Aleksandrov (see an article of V.L. Ginzburg in [Keldysh (1996)] or the magazine *Znamya*, N2, 1990). An unheard of brutality of the "control" confirmed my very first impression from the arrest of January 22, 1980: If the authorities had chosen this way of action they would not stop at anything.

A storm of indignation of world public opinion and statesmen (for example, Mitterand) could of course exert no influence on the Brezhnev-Suslov leadership (in 1984, during the second hunger strike, that of Chernenko). It was at that time conducting a criminal war in Afghanistan in which tens of thousands of our young people died and many more were physically and morally crippled. About one million of Afghans died. The whole world was storming. The United Nations unanimously (with an exception of our vassals) condemned us. All this was much more important than a noble struggle of world community for Sakharov (which did indeed provide a big moral support for Andrei Dmitrievich and Elena Georgievna).

The leadership of the country was continuing the horrible Afghan war not paying any attention of the indignation from the whole world. And, at the same time, it systematically and knowingly oppressed a heroic struggle of the few participants of human rights movement. Some went to prison, camps, exile, psychiatric hospitals; others were sent abroad, sometimes in a way that looked like as a concession to the world community (Pluysh, Aleksandr Ginzburg and others); some were forcibly sent away (Solzhenitsyn) or simply deprived of their citizenship (Rostropovich and others). Was it naive to believe, with all this at hand, in the success of the hunger strike? In Moscow, for those for whom A.D. was dear as he was, with these illusions, each day of the hunger strike was painful. When he gave up and stopped the hunger strike this was a relief for us. But not for him. Members of the department that went to visit him saw an exhausted and older-looking man depressed by an understanding that he could not endure the hunger strike. Unfortunately he decided, right at that time, to start all this anew.

One could of course understand him. An adored wife whose health was in critical condition was a sufficient reason. A readiness of putting one's life "at a card" could cause bitter feelings and even condemnation from others but then one should also condemn Pushkin who understood very well what he meant for Russia but nevertheless did die, defending his and his wife's honor from all these dolgorukij's[10] and the like, from the hand of a nobody — Dantes. The lack of prospects of Sakharov's struggle, an obvious hopelessness, did however press and torture. Those who tell that, irrespective of a reason, the very fact of his protest was in some sense a struggle for all of us, as well, are of course right. I, however (like probably many many others) did not want him to fight for me in this way. Better I faced difficulties (everything was after all not so bad) than he becoming an old man before his time.

Up to now I cannot understand how this clever man (and also many clever people close to him that approved his decision) did not understand a simple thing: Deportation to Gorky and other repressive measures related to it was above all a retribution for his protest against the Afghan reckless war. His exile was but a gleam, an echo, a minuscule part of an immense crime taking place in Afganistan. It was to a highest degree naive to hope on the effectiveness of the support of the world community. One says: an American Congress passed a special resolution in defense of Sakharov. This same Congress did however also take a multitude of resolutions in defense of Afgan fighters and sanctioned transferring them billions of dollars and providing them an immense amount of weaponry. This, however, did not move at a hair's width a granite grandeur of stupid and cruel authorities that undertook this reckless war. Exploiting an analogy of Andrei Dmitrievich himself one could say that his exile and the exile of Elena Georgievna as well were just pockmarks on this monumental granite.

The extent to which the authorities did not pay attention to foreign protests could already be judged from the readiness with which they threw out of the country dissidents and oppositionally minded, even slightly, people — writers, journalists, actors, etc. All of them, being abroad, fostered public protests and the exposure of the evil doings of our authorities. This, however, did not influence them at all: "there" do whatever you like.

Nevertheless the Sakharov's decision on the new hunger strike in 1985 was unshakable. We knew about it, were horrified and, like before the hunger strike of 1984, dissuaded him from it (I shall not forget the last "discussion" with him in Gorky before the second hunger strike. This was, of course, on paper, not oral, this was on April 4, 1984). Before the hunger

---

[10] A family participating in Pushkin's baiting.

strike of 1985 he wrote a letter to A.P. Aleksandrov containing a declaration in which he wrote that if his petition on allowing a trip of Elena Georgievna would be satisfied he would concentrate on scientific work on controlled thermonuclear fission, otherwise he would declare his leave from the Academy of Sciences. It was clear that he would also resume a hunger strike. All this did increase our unrest. I decided to write to him the following letter (in what follows I have omitted and replaced by omission marks only those parts of letters in which his purely practical requests were discussed, in particular his intention to sell his dacha).

> Dear Andrei Dmitrievich!
>
> Reproaching you with the actions that I consider wrong would be inhuman (taking into account your sufferings during the last years) and unjust (as you base your decisions on insufficient information: Even when Elena Georgievna was going to Moscow all information came from dissidents and foreign journalists and this is quite biased a source). I do not however consider it possible not to tell you what is, in my opinion, the truth, however uncomfortable it be.
>
> Your "menace" to leave the AS if E.G. would not be allowed to have treatment abroad is, I am convinced, accordant to the strong wishes of very many from the AS leadership. To fulfill this dream of theirs it is sufficient to read at a meeting of presidium one paragraph from your letter to A.P., not even the whole one, and a vast majority would sigh with relief having got rid of the constant annoying burden (in reality the reaction turned out to be more sophisticated, see below) (the resulting noise in the West does not make sense at all and could easily be confronted: Your free desire has been fulfilled). Thus by your "menace" you have practically blocked E.G.'s leave. This is however not the main thing: Having left the AS you potentially put at risk the continuation of your scientific work: AS does not have to provide working conditions and care for a non-member at all.
>
> It seems to me that you underestimate two circumstances. First, Western scientists are nowadays most of all concerned by the menace of nuclear war and the arms race. In January a delegation of the US AS was in Moscow and held talks on scientific cooperation. These went very smoothly and neither you nor other dissidents were even mentioned. One of the leaders of the delegation explained in an unofficial but public discussion that the members of the US national AS exert pressure on the leadership demanding cooperation and putting aside everything that could hinder it.

It is of course quite possible that something is being done through non-public channels (not to irritate the self-esteem and prestige), as is quite common, but I am not aware of any confirmation of this.

Second, a demand on allowing E.G. to have medical treatment abroad is very unpopular. 270 million people have their treatment in the USSR and such a demand is, in the eyes of many, very non-democratic and does not go along with your image as a fighter for justice and democracy.

I urge you to immediately withdraw your declaration on leaving the AS. It is necessary to send telegrams to A.P. and V.L. and I do not know who else (it is possible that these will not be delivered as it has already happened) and a letter or at least make an oral statement.

I apologize for having said such unpleasant things, but nobody else would have done it. Therefore I had to do it.

I wish you and Elena Georgievna everything good — possible and impossible. I regret that my visit to you was postponed again and again (until March 1?) as I have caught flu. Otherwise I would have come.

24/II 85 Yours E.L.

It is evident that the letter was written in despair. I am not sure whether, had the situation been now repeated, I would use the above-described "second" arguments against the demand of allowing Elena Georgievna a trip to the USA. A.D. responded to me with a letter showing his extreme anxiety (canceled and replaced words, additional words and phrases added above lines, etc.).

Dear Evgeni L'vovich,

I am not basing my decisions on information from dissidents or the West. radio. My decision of striving, at any price for myself, for the trip of E.G. is based on how I understand my duty towards her who gave me *everything*. I am very well aware of the concern of Western scientists on nuclear war, I am also concerned about it. This concern does only opportunistically (only sometimes) contradict the defense of Soviet scientists. I cannot, however, calculate chances, I do not have a choice. A demand of giving E.G. [an opportunity] to have medical treatment abroad is not a caprice, her situation is different from that of the 270 million people, the citizens of the USSR, because of the KGB hatred towards her. You could not fail to understand this. She should have a right to see those close to her — this is also not a caprice. You suggest that I withdraw my statement of leaving the AS. I will not do it. I am sure that without this *threat* (not only towards AS but towards KGB as well) Aleksandrov would not be able to do anything at all in this matter.

P.S. And if AS does really want to get rid of me, then it would sooner or later happen anyway and it is better to slam the door. I would rather prefer to die of hunger than to share the company that dreams of getting rid of me.

A.S.

Tragedy is a conflict in which both sides are right. As I, however, wrote on a different occasion, in the tragedy they are right in a different way: One side through its rationality and calculation, the other by its non-calculated, not "weighting chances" humanity. The horror was that like in an antique tragedy, only this time in our real life, the conflict could only be resolved by the death of the human and non-calculating hero. This gave no rest and forced me to give Andrei Dmitrievich unpleasant and sometimes harsh reasons.

I do not remember how this letter exchange took place. I was sick during these months (for the first time it was my heart) and could not go to Gorky.

It turned out that we both could not imagine what a clever possibility of leaving everything safe and quiet would be used by the president in the question of leaving the AS (see below). Sakharov wrote a new letter and intended to pass it with one of A.D.'s children (they could go whenever they wanted). However, they did not want to go for private reasons and Lyuba (his daughter) said that such a letter could have been sent by mail. And indeed I understood that my actions did conform to the "organy's" wishes. This was precisely the case that I could factually be considered as their "agent." Here is this letter. (One should always take into account our letters were always written with an eye towards the "sideways reader" — without innuendos, thus excluding a possibility of undesirable interpretation.)

Dear Andrei Dmitrievich,

As it is completely unclear when the next visit of people from the Theoretical Department to you will take place I decided to ask one of your children to bring you the medicines you have asked for. As you see this is still not everything you need."[11]

---

[11] It somehow happened that I was responsible for supplying A.D. and E.G. with medicines. Andrei Dmitrievich sent long lists and we found the medicines either in Moscow or abroad. Once a required rare medicine was sent by Heinrich Boell. I collected everything together and this was sent; see Section 3.10.

> I am using this opportunity to inform you that, as V.L. Ginzburg has already telegraphed you, your letter was handed over to A.P. Aleksandrov in due time. In relation to this I would also like to inform you on the following:
>
> 1. Your declaration on leaving the Academy of Sciences will not have consequences. The AN Charter does not contain the corresponding paragraph and nobody is going to expel you. You will remain an academician as before.
>
> 2. Your declaration on the possibility of a hunger strike causes great concern. With your present state of health this is mortally dangerous. At the experience of the previous year you know that no noise abroad would bring you the desirable result. This is a pure shake-up of air. This year even that would not happen because nobody in the whole world would even know about your hunger strike. Therefore I strongly advise not to take such dangerous and, with respect to your goals, clearly useless steps.
>
> I do not know whether you have already received my letter sent through FIAN. There I wrote in more detail on medicines, in particular on the normal dose of Nootropil which is not 3 but 6 capsules a day.
>
> I wish you everything good, and first of all — health and balance.
>
> 9/IV 85 Yours, E. Feinberg

Alas, this letter (which of course would not have stopped Andrei Dmitrievich) did come late: On April 16 he began the third hunger strike (our letters, for obvious reasons, went for two weeks).

In her article 'Who needs the myths?' (*Ogonyok*, N11, 1990) Elena Georgievna quotes the second paragraph of this letter (beginning with the words "Your declaration ...) but without the last phrase ("Therefore I strongly advise...") and does not mention my name as the author. She expresses a sharp irony with respect to these lines. "So! .. The "shake-up of air" did however always help. Before I was locked in Gorky everything that Sakharov wrote there was published," etc.

It is true that a heroic activity did in the first place make possible the rescue and publication of what Andrei Dmitrievich wrote but what relation does it have to completely fruitless demands of the world community (on freeing A.D. and allowing E.G.'s trip), to the horrors of the senseless hunger strike? I remain of the opinion that what I said in my letter was right and do not see grounds for irony. In addition, in this article by Elena Georgievna there were plenty of unjust words on other people as well. She herself began it with the words: "There arose a strange situation. I am constantly offending somebody." (This is not surprising, her article was written soon after the sudden tragic death of Andrei Dmitirevich. In these horrible sorrowful days Elena Georgievna demonstrated rare courage and firmness. Her state, however, did not allow this courage and firmness to be boundless and her statements to be just without exceptions.) Myself, I was not offended because I thought (and think) that in this issue I was right.

To the story of this letter, to the same illusions on the importance of the protests of the world community, is related one difficult episode about which I have promised to tell when describing our last trip to Gorky with E.S. Fradkin on December 16, 1985. Let me remind that this was after the third and last of Sakharov's hunger strikes, when Elena Georgievna had already left for the USA. This was the only episode during the seven-years-long Gorky epic (and also during all the 45 years since A.D.'s arrival in the Theoretical Department up until his death) that caused serious difficulties between me and V.L. Ginzburg (and several other members of the department) on the one side and A.D. Sakharov on the other. We are talking about the story with a "package" described in detail by V.L. Ginzburg in his recollections in [Keldysh (1996)] (see the recollections of D.S. Chernavsky in the same volume).

At the very end of our visit with E.S. Fradkin, late in the evening when we were hurrying for the train and Fradkin had already left, Andrei Dmitrievich learned from me that we did not fulfill his wish which he, unlike us, was considered to be very important. A common (of four persons) decision to act in this way (we thought that otherwise a serious danger would arise for at least two completely innocent families with children) was not an easy one for us. Nevertheless we were convinced that it was the right one. Of course in those exceptional circumstances it was difficult to avoid some way of action allowing completely opposite evaluations. It was rather wonderful that this was the only such case. Later Sakharov wrote, dryly and shortly, in his memoirs: "I have understood (but have not accepted) the reasons for the disappearance of one of my documents" (see [Sakharov (1990), p. 15]). His immediate reaction at that moment was sharply emotional. It was first reflected in his letter to me written on the next day. This included the lines that are hard for me to bear up until now:

> Dear Evgeni L'vovich!
>
> I am sending a copy of my paper for sending it [to print — E.F.]. I forgot to hand it over on Monday.

> I have to write to you that I experienced a shock for our conversation in the last minutes of your visit. I asked my question just in case thinking that the answer would surely be quite different. The fears which you mentioned[12] seem imaginary to me (I will try to motivate it when having a chance, but what you have told me seems to me being an insufficient reason in such a vital thing). The decision you took practically put us (or could have put) on the verge of death and you could not fail to understand this. I could probably never (or for a very long time) be able to put aside the feeling of disappointment and bitterness. I would ask you to show this letter to Vitaly Lazarevich.
>
> 17/XII 85 Respectfully, A.D. Sakharov
>
> P.S. I hope that you and Fima did not catch flu from me. This would grieve me a lot! 18/XII 85

One sees that the letter was written after a night filled with heavy emotion. Still he did not send it at once. The addendum added on the next day does already point out some softening. After two more days he sent me and my wife a New Year postcard:

> Dear Evgeni L'vovich and Valentina Dzhozephovna!
> Congratulations with the New Year! I wish you happiness and health. All's well that ends well.
> 29/XII 85 Yours, A. Sakharov

Later neither Sakharov himself nor anybody from us returned to a discussion of this issue with him so that he could not "motivate it when having a chance" and we did not clarify him anything. Our first personal contact after this episode took place only in a year when (as I describe below) on the day of his return to Moscow Andrei Dmitrievich came to the department and spent with us such warm and joyful six hours. Nobody wanted to stir the old disagreement up. I myself, of course, consider his sharply emotional statements like "put us or could have put on the verge of death" being imaginary like he (at least at that first moment, in addition not having full information) considered as imaginary our considerations that were explained to him in a very brief and incomplete way. Still the inner bitterness from this episode remains with me until now.

Let us however return to the beginning of the third hunger strike on April 16, 1985.

---

[12] Of course we did not talk. We wrote on a piece of paper. A.D. was afraid of infecting me, everything was cut short and I limited myself with two phrases and could not explain the situation in sufficient details.

It is not difficult to understand what this new step towards the physical death would end up with. However, at this time there happened a miracle. After the death of Chernenko there took place a famous April plenum of CC at which the leadership of the country was handed over to M.S. Gorbachev. He was at that time completely unknown to the broad masses and still had to establish his authority among the people and in the power structures. According to quite reliable rumors, the election of Gorbachev was a difficult one and was possible only because one managed to avoid the participation in the plenum of such Brezhnevists as Sherbitsky and Kunaev. On April 23, Gorbachev gave a program speech at a CC plenum in which such unusual words as "glasnost," "social justice," "perestroika" were used. These words caused wonder but at that time nobody paid attention to them. However, on May 31 a high-ranked KGB officer visited Sakharov in Gorky. From a conversation with him Elena Georgievna concluded that "Gorbachev instructed KGB to clarify our case. But KGB followed its own policy. Thus they had their own fight in which it was unclear who is stronger — Gorbachev or KGB?" (see [Bonner (1998), p. 129]). If this fight did really take place (and this is to the highest degree probable) then, at the time when Andrei Dmitrievich was suffering from forced nutrition in a Gorky hospital, it was developing very fast and in a certain direction.

As Sakharov writes in his memoirs [Sakharov (1990), p. 4] on July 11, i.e. after having suffered for three months, he terminated his hunger strike: "... having not born a torture of complete isolation from Lyusya and thoughts on her solitude and physical conditions," and was returned home from the hospital. However, on July 25 he resumed the hunger strike and after two days was again forcedly taken to the hospital. Of course he knew nothing on the development of the above-mentioned "fight" upstairs but A.D. writes [Sakharov (1990), p. 7] that already on September 5 there arrived the same Sokolov who had visited him on May 31.

However, "... previously Sokolov spoke to me in a very hard manner. His aim was probably to force me to stop the hunger strike by creating an impression of its total hopelessness... This time (on September 5, 1985) Sokolov ... was very friendly, almost soft... Sokolov said: "Mikhail Sergeevich [Gorbachev] has read your letter[13] ... M.S. Gorbachev instructed a

---

[13] Apparently the Sakharov's letter sent to Gorbachev in the last days of July in which A.D. promised "to stop, save for exceptional cases, all public activities if a trip of E.G. would be allowed" was meant. A.D. mentions in the same book that began writing it a month before.

group of comrades to ... consider a question of a positive decision on your petition. In reality I think that a decision of Lyusya's trip had at that time already been taken at the highest level but KGB, following its goals, had postponed its implementation." It was implemented after one more month when Elena Georgievna did at last get an official permission to go to the USA. There she underwent, at first, a conservative treatment and then finally open heart surgery. This did radically change her state of health, one might think — saved her life. In the above-mentioned article in *Ogonyok* Elena Georgievna writes that the protests of world community and a concern of statesmen in the West did bring freedom for her and A.D. while "new or old government is an issue of secondary importance."

One cannot agree with this at all. Due to some reason "with the old government," in 1984, such a hunger strike did not help. Also, during all the years of exile, "protests of thousands of foreign scientists," "Sakharov's day" and everything else that Elena Georgievna lists did not lead to this freedom but rather made the situation harder: E.G. was sentenced to exile, its regime was strengthened, it went up to a theft of A.D.'s manuscripts (during which A.D., who was sitting alone in the car, got some spray in the face that made him lose consciousness for some time), etc. It seems to me that a declaration "New or old government is an issue of secondary importance" is deeply unjust. The "old government," even when yielding in some cases, for example, when expelling Pluysh, Ginzburg and others (let me repeat: It is not even clear whether it did yield or expelled in the same way as with Solzhenitsyn) did this without changing its repressive policy in general. The "new government," on the contrary, did release all human rights activists. Of course, one cannot exclude, that precisely because it was Gorbachev (and not Chernenko, as it had been during the fruitless horrible hunger strike of 1984) as head of the country a pressure of foreign public opinion helped the new leader to overcome the resistance of KGB.

Andrei Dmitrievich returned from the hospital being not at all the man who had not even been old before all the hunger strikes. Once I was asked: Why did I so passionately persuade A.D. to not go on hunger strikes? the answer is very simple: I did not want him dying, experiencing again and again the torturing things he was subjected to. I knew that the authorities would not yield, that the protests of all the world did not mean anything to them. A.D. considered the failure of the hunger strike of 1984 as a defeat

The summer of 1973. Zakhadzor, Armenia. At the All-union conference on gravity.

and weakness of will. In fact he showed extraordinary courage but the result was predestined.

It is not difficult to imagine how the hunger strike of 1985 would have ended with if, say, Grishin or Sherbitsky, or Romanov, not Gorbachev, had been elected. I do not doubt that Sakharov himself thought that with these hunger strikes he won a victory over the authorities. It would have been cruel to dissuade him of this. Moreover, it is possible, that it was this conviction that gave him new belief in his strength and helped him in future struggles. Let it be so. All is well that ends well.

3.9 A Release

Another year had passed before freedom came to Andrei Dmitrievich at such a scale that nobody had been able to dream of before. During this year of 1986 I was sick and my domestic situation did not allow me to leave the house. Therefore our communication with Sakharov was restricted to letter exchange. The salutary changes in the country did however get stronger and, correspondingly, there grew our hopes on a change in his fate. We were impatiently waiting for it and sought for good signs. At last there came a memorable day when, late in the evening, a representative of "organs" brought to Sakharov's apartment two technicians that hastily installed a telephone. When leaving the person in charge of this operation said, "Tomorrow you will get an important call". This call did come. It was M.S. Gorbachev.

I learned about it two or three days after from the stories of those who listened to foreign "voices." They told something unbelievable: That Gorbachev reportedly invited Sakharov to return to Moscow and "resume his patriotic activity". I decided that this was an error, a result of double translation. It was probably said "... begin working for the benefit of the country" and in the course of translating from Russian into English and back there resulted "patriotic activity." Having got the telephone number — 2669569 — I called A.D., laughing from happiness. I asked: "When will you finally arrive?" He answered: "Elena Georgievna cannot go out if it is colder than −10. One promises that it will get warmer on Monday. If it will really be so we will come on Tuesday morning." Repeating some half-sensible words I told, as a joke: "Andrei Dmitrievich, do you remember? On Tuesday at three there is a seminar, as usual. OK, do not take it seriously, you will be tired and in general occupied by other things."

As I know from stories told by others on Tuesday, December 23, early in the morning when it was still dark, a crowd of photo-, kino-, tele- and simple journalists met Elena Georgievna and Andrei Dmitrievich at the railway station. Darkness surrendered to photo flashes. It was obviously laughable to expect that Sakharov would have come to the seminar (only afterwards did I see the news record in which, still at the railway station, probably answering a question of some journalist he tells: "First of all I will go to the institute"). Later around noon V.Ya Fainberg went to Sakharov's apartment in his car, simply to ask how he could help. Sakharov said, however, that he will go to FIAN.

Knowing nothing of this all I came to the department at 2.30 p.m. and saw a corridor filled with an improbable news: A.D. is already here, in his office, with the same carton card with his name (now having become yellow) that had been on the door before his exile. There stood the same antique carved wooden desk that had gone to him after the death of Igor Evgen'evich. We embraced and in the same state of joyful agitation

I accompanied him to the institute's conference hall where there had already gathered a lot of people. Everybody already knew everything and met Sakharov with applause. He took his usual place and I, as a chairman, started telling something disordered. I began with a phrase said by actors in mass scenes when they picture a crowd and its noise: "What could I say when there is nothing to say." And then told, for unclear reasons, the story of how Igor Evgen'evich and I gave Andrei Dmitirevich "four" at his Ph.D. exam. And asked myself: "Dear God, why am I telling all this?" Ritus exclaimed: "Out of overwhelming feelings." Everybody laughed.

Due to a completely occasional coincidence the talk was on that very baryonic asymmetry of the world that Andrei Dmitrievich had explained a quarter of a century ago. The speaker started with the words: "As shown by Andrei Dmitrievich ..." (in general, during all these years the fearful name of Sakharov and his works were openly mentioned at the seminar. Not only us but other physicists in Moscow were carefully ensuring that the necessary references to his works did appear in the papers in our main scientific journals, ZHETP and *Uspekhi Fizicheskih Nauk*. Once during the Gorky exile period Andrei Dmitrievich submitted, as an academician [this was his right] a paper by B.L. Altshuler to the journal *Doklady Akademii Nauk SSSR*. In this case there arose a disarray, the editorial board was hesitant but finally one succeeded in "pushing this paper through" and it appeared with the comment "Submitted by academician A.D. Sakharov".)

After the seminar, after the handshaking of old colleagues, we, the four "elders," again went together with Andrei Dmitrievich to his office and there began infinite conversations.[14] A.D. told in more detail about his conversation with M.S. Gorbachev. It is described in his memoirs and I will not reproduce it here. It turned out that Gorbachev had really said "Come back and resume your patriotic activity." This meant a full recognition of the rightfulness of what Sakharov had told about 18 years ago. This conversation is remarkable by the scale at which Gorbachev acted and the ever-present feeling of self-respect with which Andrei Dmitrievich met his freedom and immediately began discussing others, his comrades in the human rights movement.

---

[14]In his memoirs, see *Literaturnaya gazeta* of December 19, 1990, A.B. Migdal writes that having learned about Sakharov's arrival he and some other theorists went to this first seminar. This is not exact. He, as well as the well-known journalists Yu.M. Rost and O.P. Moroz, came to the second seminar, in a week. Yu.M. Rost took many photos.

I will not write on the three last years of his life after exile. These were already visible for everyone. And although one could tell a lot of interesting things here as well, I should stop — the story is already too long.

Sakharov became a part of the revolution that had commenced in our country (carrying a modest name of perestroika) that he had prophetically called for already in 1968. Its basic ideas coincided with his. The most impossible, unreal dreams were coming true — glasnost, free speeches at the meetings and demonstrations, liquidation of all-embracing censorship, freedom of religion and return of churches, the end of confrontation with all the "foreign" world that became our friend — everything that is necessary for democracy but not sufficient for it. There remained (and remains) a task of changing the people's mentality in the spirit of democratic responsibility of everyone for the society, of deep respect towards the law and a readiness to conscious self-restriction in the interest of society, of tolerance towards the opinion of others — the task requiring many decades to fulfill.

A.D. Sakharov acted abroad as "a messenger of perestroika," leading politicians of the West to believe in his words. He put forward new constructive ideas of immense importance. One cannot but wonder when reading the materials on his talk at the Forum for Nuclear Free World and International Security in February of 1987. Sketching the way to disarmament, discussing the nuclear strategy [Sakharov (1990), p. 51], he spoke as an expert and his ideas (rejection of the "package principle" and others) were realized in the foreign policy of our country (of course, I cannot judge to what extent these were understood without him). In the course of his intensive political activity that followed there were some concrete things with which one could disagree — nobody, even the great ones, is insured from mistakes. He symbolized however before the whole world and, first of all, before his own people, purity and justice, freedom from cheap politicizing (one cannot therefore consider him as a politician who sometimes needs this), insensitivity to attacks by people that remained in the clutches of former ideology that had been hammered into them for decades, people that had not grown up to truly democratic consciousness.

I would like to finish my memoirs referring to this period by several words on the purely personal qualities of Andrei Dmitrievich.

The late friend of A.D. from university and his Ph.D. studies, M.S. Rabinovich, (see his memoirs in [Keldysh (1996)]) tells that at this time he felt himself, in essence, lonely. The same words I have recently heard from Elena Georgievna. V.L. Ginzburg (see his article in [Keldysh (1996)]) thinks that one can characterize A.D. by the word used with respect to Einstein by his biographer A. Pais: apartness. Indeed, often, when talking to him, especially when the subject was not related to something everyday, usual, I experienced a feeling that in parallel to our conversation some internal life, that by no means diminished his interest to the subject of the discussion, was going on within him. He, inside himself, was simply continuously working on something related to the subject of the conversation but the result of this work surfaced quite seldom.

Even if he was indeed "lonely," "apart," this incomprehensibly went along with his emotionality and strength of feelings towards other people. Having been in real "solitude" he would have felt himself cold.

Indeed, he himself writes what a shock did he experience after the death of his first wife Klavdiya Alekseevna. The strength of his feelings towards Elena Georgievna, "Lyusya," can be judged by anyone from the two volumes of his memoirs: *Memoirs* and *Gorky-Moskva, Then Everywhere.*

I recall one episode in FIAN that took place in the seventies. I approached the stairs leading to the conference hall and saw Andrei Dmitirevich coming down it. With half-bent arms, awkwardly coming down the stairs in this pose and, as almost always, making intervals between the pronounced words, he almost cried to me: "Evgeni L'vovich! Terrible disaster, terrible disaster! Lyuba (his younger daughter) gave birth to a dead child, more exactly, he died right after the birth. Terrible disaster, terrible disaster." And, in two years, he was late to the seminar that had already begun and, having sat next to me, said in a low voice, shining: "Lyuba bore a child, everything is fine."

To phrase this shortly, this apparently dryish, correct, "apart" person was at the same time paradoxically deeply emotional, even passionate. He was a true friend both to the friends of his youth and to his fellow participants in the human rights movement, as is also seen in his memoirs. He could write to his friends a greeting card and sign it "With great love. Kiss. Andrei."

He had much tenderness towards people, much love and need to be heard. Lonely? Apart? No, everything was more complicated. Like in his scientific life one can hardly understand everything. Let us recall the words of an outstanding physicist Ya.B. Zeldovich cited by V.L. Ginzburg in his memoirs: "... I can understand and range other physicists, but Andrei Dmitrievich is something different, something special."

In one of the years of the Gorky exile my wife and I began sending examples of handwritten texts of some persons to an acquaintance of ours,

L.V. Gorokhova, who was convincing us that she was an expert in graphology. This was done, in part, to cheer an invalid chained to her chair. We chose the people whom he could not know and could not know their handwriting.

The texts were sent, numbered, by post and she dictated the results of the analysis over the phone. In the first mailing there were two texts, one by Andrei Dmitrievich and one by another academician who was, one could say, almost opposite to him in terms of his personal characteristics. The conclusion on this (second) person stroke us with its accuracy even in detail: "... (very) clever, cunning (or with a little admixture of it)... Kind, but mostly "for himself." Tender. Attitude to people and humanity as a whole is, generally speaking, bad (probably due to arrogance)... is not honest (will not dig into the pocket, will not kill), etc." Here, however, is the analysis of the handwriting of A.D.:

"Directness. Honesty. Kindness. Naivity, sometimes along with infantility. Undoubtedly clever. A mind not egocentric, humane. Accepts humanity with kindness. Undoubtedly gifted. Attitude towards himself even too modest. Therefore had problems. No speech about careerism. Will accomplish what he does when not under compulsion. Does his work with inclination. Must be happy personally. A generous soul. Loves people, in particular those close to him. Capable of sacrifice (not very pronounced). One could go with him on any reconnaissance mission (the usual final criterion by this expert in graphology was "could one go with this person on a reconnaissance mission"). In a dangerous situation will act so that it is better for others, not him."

Is graphology indeed an exact science?

Last time I saw him on Monday, December 11, 1989, on the day when, following his appeal, there took place a two-hours political strike. In FIAN, at 10 in the morning there was a two hours meeting at which he gave a brilliant speech. I was approaching the main building when a man in a short coat and a cap with ear-flaps went out of the car. He energetically went, almost ran, up the stairs of the main entrance and stopped waiting for me. Because of poor eyesight I could not recognize him but from the figure and the gait it seemed that it was somebody else. Only from the gesture by his hand, after having come closer, I recognized him. An intense political life of the last three years had almost made younger this man who had so much suffered from terrible hunger strikes.

And, nevertheless, in three days he fell.

3.10 Addendum

3.10.1 *Letters and telegrams of Gorky period*

Here I give, with necessary short comments, excerpts of my correspondence with Andrei Dmitrievich Sakharov in the period of his Gor'ky exile. They are possibly not as significant as those quoted in the text but characterize an everyday connection between Sakharov and the Theoretical Department on a variety of issues that are in some way important. Most often he sent letters to the institute's address but sometimes also to me at home. He was signing them somewhat more formally than before the exile because for the whole letter, including its contents, it was taken into account that it would also be read by "extra eyes."

I omit only the places related to purely domestic errands of A.D. (all of them are marked with < >). My inserted comments are denoted with (— E.F.).

It is necessary to take into account that letters arrived with a big delay — two or three weeks and more, telegrams — on the same day. I have already mentioned in the text that I was responsible for supplying Andrei Dmitrievich (and Elena Georgievna when she was locked in Gorky) with medicines. In the correspondence much space is devoted to this problem. In 1986 (see the text) I did not go to Gorky and there were few of our trips there in general therefore in the letters one also discussed scientific papers by Sakharov.

1. Telegram to FIAN of 14.01.83.

Theory Department Feinberg
Waiting your other staff members arrival inform fs[15] if possible ask get Bormotova[16] medicines sustac forte timoptic mono mack capsules complamin no-shpa Yours Sakharov.

On the telegram — my pencil note "+ nootropil" — I evidently took into account the next telegram.

2. Telegram to FIAN of 17.01.83.

Theory Department Feinberg
Dear Evgeni L'vovich additionally monomak capsules timoptik other medicines ask try get Bormotova nootropil also piracetam tablets Sakharov

---

[15] Full stop.
[16] Bormotova was a head of polyclinic of AS USSR to which Sakharov was attached.

3. Telegram of 15.02.85 probably expressing an agreement to our planned trip.

Dear Evgeni L'vovich joy waiting you Linde please bring nitromak Sakharov

However, there came A.D. Linde and D.S. Chernavski, I was probably not feeling myself well like in the whole beginning of 1985.

4. At my home address there came the tragic telegram of 17.04.85 which was clearly an answer to my letter sent on 09.04.85 in which I was, with despair, trying to dissuade A.D. from the hunger strike of 1985, wrote on the impossibility of my visit because of illness and on a possibility of sending medicines with one of his children.

Medicines no urgent need currently have everything categorically object against sending medicines my children their visit leaving academy hunger strike distressed your position absence any understanding situation responsibility for all my actions should rest only on me alone this is my right of free person attempts putting responsibility wife depriving (her. — E.F.) children health freedom unsustainable for me Sakharov

As seen from the dates this telegram was sent the next day after the beginning of the hunger strike when he was not yet forcibly taken to the hospital (which took place after several days). This horrible telegram should be compared to what is written in the corresponding part of the text on Sakharov's hunger strikes.

It is difficult to describe a comical episode related to this telegram but to characterize an atmosphere of that time it is worth describing this as well. I think this episode was related precisely with this telegram. At that day in Moscow there was a terrible shower. I got a call from the telegraph. Judging from the voice this was an elderly man who, in the fearful voice, stumbling, began saying something chaotic:

'You know ... You see ... there is a telegram to you and we cannot deliver it because of the rain.'

'Then read it to me,' I said

'I do not know ... You understand — this is ... from Gorky.'

'Oh,' I said, 'I see, and what is a signature?'

'Precisely... This is a problem ... You know, here it is written ...'

'Sakharov?' I asked

'Y-e-e-es.'

Having pronounced a horrible name he evidently felt relieved and finally read me the whole text.

5. Telegram to FIAN of 02.09.85.

Dear Evgeni L'vovich ask send again banderole medicines which Elena Georgievna impulsively sent back beg pardon respectfully Sakharov

This telegram was sent from the hospital during the third hunger strike (see below item 8) and needs explanations.

When Andrei Dmitrievich began the last hunger strike and later was put into hospital Elena Georgievna, in order to let friends know about this fact, that she was alone, gathered all the presents sent to the birthday of A.D. on 21.05.85 and mailed them back. She did the same with the medicines: Put them into a large box and mailed them to me at the FIAN address. I got it in a post office damaged, supplied with an official act composed at Gorky main post office by a commission of three people. I am keeping this act. On all the boxes it was written that the sender was Elena Georgievna. However none of us understood the meaning of this action.

6. After the end of the fruitless hunger strike of 1984 A.D. sent me a letter, half of which was devoted to a detailed exposition of questions related to his request to organize, as he wanted it at that time, the selling of his dacha near Moscow. I omit all this and give the second half of the letter:

> '... independently of all this I would be very glad if you could visit me during the next visit of physicists. Your previous visit was an eternity ago!
>
> Brief information on us. I am gradually recovering from the hospital experience (these are the very torments of forced feeding that A.D. described in his letter to Aleksandrov, cf. journal *Znamya* (Nmb. 2, 1990) — E.F.) and am even glancing from time to time at the desk. Elena Georgievna is on the whole worse than when you visited us. She does not have a possibility of checking her eyes, from the feelings — frequent pains, narrowing of the sight field (irreversible!), fog from new precipitates. With the heart — everyday pains, everyday massive doses of prolonged nitropreparates and nitroglycerine. You know that her eye medicines harm her heart and vice versa. For almost a month an exacerbation of diskogenic radiculitis, for 25 days is taking analginum. She cannot do an analysis of blood (in Gorky one can not do a paid analysis of blood or a cardiogram). (This remark is worth a comment: evidently Elena Georgievna and Andrei Dmitrievich did not trust tests in the district polyclinic thinking that these could have been falsified on KGB request. In truth, it is not clear to me whether these fears were grounded. KGB could not exert influence on paid laboratories

and polyclinics. — E.F.) One-two times a month one needs an ambulance, its help is temporary and not very big.

Evgeni L'vovich! How is your health and the health of Valentina Josephovna? Warm greeting to her from Elena Georgievna and myself. Best regards from both of us.

16/XII 84

Respectfully, A. Sakharov'

In my reply of 31.03.85 I give a report on his errands and write in detail about medicines — what is available, what is not, what can be replaced by something else, etc. I am finishing with a wish: "All the best to you, first of all health and equilibrium and, as a consequence of this — good scientific work." It is not excluded that by this wish I was not consoling him (and E.G.) but rather irritated. This was however not shown.

7. I will not give a big letter to A.D. on 13.03.85. It is entirely devoted to a list of medicines that he and Elena Georgievna needed. It contained 12 items including a discussion of all details. Now it is clear that he was worrying about providing a stock of medicines for E.G. for the time of his new hunger strike on which he had already made a firm decision. Against each of the 12 items there are my remarks in red: What is sent, how much, what has not been yet found, etc.

8. There passed another half a year, the third hunger strike and new terrible worries. At last, there arrived a new letter:

'Dear Evgeni L'vovich!

As you probably know last week Elena Georgievna got a permission to visit her mother, children and grandchildren. For us this is an event. There also ended our six-month, with a brief interruption in July, separation. In the same days there arrived a parcel with medicines from you, thank you very much for it. I have already written to you that Elena Georgievna sent medicines back to you not out of offence or "bad temper" but as an only possible sign that she is alone, without me. In turn, I sent you a telegram from the hospital, where I was parted with her, in a state of extreme anxiety with respect to her health and the lack of medicines. I had all grounds to assume that my telegram did not mean that I was with Elena Georgievna — indeed, last year I had already sent from the hospital proofs of an article. Elena Georgievna did all the more have all grounds to assume that she would be understood rightly.

After getting this permission for the trip I sent a telegram to the Presidium of USSR AS to Anatoli Petrovich Aleksandrov in which I asked to consider my request on leaving the Academy invalid from May 10, 1985. Elena Georgievna went to the West at the end of November, we wanted to spend this month together.

In December or any other time convenient for you I would be glad to receive a visit of you and other staff members of Theoretical Department, hope to hear a lot of new things about auperstrings (or about something that has replaced them at the frontier?) and on other scientific issues. Before the trip I would ask to inform about it in advance Boris Birger, friends of Elena Georgievna would like to send me something through him.

I would ask you to convey best wishes to Vitali Lazarevich and show him this letter. Greetings to all staff members of Theoreetical Department.

Best wishes to Valentina Josephovna.

Yours, A. Sakharov'

One sees that this is a letter of an almost happy person. I answered him by writing: "We are all extremely glad that you are alive, that a nightmare of these six months is over," etc. In the same letter I told, in order to cheer him up, that there appeared a paper in which there figures an idea put forward one or two years ago by Andrei Dmitrievich himself (on a variant of field theory in multidimensional space in which the signature corresponds not to one temporal dimension but to three, etc.).

9. There came the year of 1986, the last (which we did not know then) year of the exile and persecution of Andrei Dmitrievich. He began working in science. He wanted to enlarge a circle of colleagues and friends coming to Gorky. He sent the following letter (24.10.1986):

'Dear Evgeni L'vovich!

With a delay I am answering to your letter in which you ask me on my wishes with respect to visits of physicists. I would very much like to have the visits of Boris L'vovich Altshuler and Yuri Abramovich Golfand (to discuss supersymmetry, hypotheses of Kaluza-Klein type, etc.). I am asking you, Vitali Lazarevich Ginzburg and Sergei Ivanovich Nikolksi to organize this trip so that there will be no misunderstandings "at the entrance door". The fact that B.L. Altshuler is not a staff member of FIAN (he was immediately hired to the Theory Department according to Sakharov's desire after his return to Moscow. — E.F.) is not essential — it is completely sufficient that he regularly attends the seminar in FIAN and has a pass to FIAN. I would ask to display perseverance that is necessary in such cases (After that A.D. writes about a desirability of a visit of some staff members of the department — E.F.) I would like to ask you to convey my congratulations with the recent jubilee to Vitali Lazarevich (I learned about it too late) ...'

Of course a visit of the physicists mentioned by Sakharov would have been pleasant for him but, as it immediately became clear, it would have been extremely difficult (if possible) to "solve" this question: both were active dissidents. Altshuler, who had been fired from a lecturing position in a technical university, worked as a street cleaner (although was really intensively working in science) did, together with an even more well-known dissident mathematician N.N. Meiman, regularly take part in the work of our seminar. We included them into a list allowing to get a pass to the seminar using their old credentials from their former working places. Of course a representative of "organs" in the institute could not fail to know it but he pretended he did not notice anything. He probably knew that this was really about a scientific participation. However, a trip of such people to Gorky was a completely different business. However, very soon the question was no longer actual.

10. Let me finally present an exchange of letters related to the last scientific work of A.D. done in Gorky in 1986. It was reflected in the form of letters because during all that year I did not come to Gorky. Here is the letter of A.D. from 29.05.86:

> Dear Evgeni L'vovich!
> I am sending my note "Evaporation of mini black holes and high energy physics." I have big doubts whether everything written there is not trivial and in any case this is dividing up the bearskin before the bear is shot (because no black hole was so far observed — E.F.). It is also not good that many estimates are not followed down to numbers (a remark very characteristic of A.D., he liked to push everything down to concrete numbers. — E.F.) (especially with respect to the rotating hole. Possibly this is also known). I would like to ask to give my manuscript to one of the knowledgeable people, probably to V. Frolov, with a request to look at it critically and mercilessly. If at the end the note will nevertheless be good for publication I would like to ask Frolov (unfortunately I do not know his name and patronymic) to equip it with references. Here I do not have anything at hand including the book of Frolov on the existence of which I have learned recently. This lack of literature is one

of the reasons for my uncertainty. If the note will be prepared for print I would like to ask you for advice on where to send it — may be, to *Pis'ma v ZHETF* ("Journal of Experimental and Theoretical Physics" — our main physics journal issues for rapid publications of small-size articles a supplement *Pis'ma v ZHETF* — E.F.) and help with the formalities (having in mind an organization of the expertise conforming an absence of secret elements, etc. — E.F.). Very best wishes to Valentina Josephovna and you.

May 29, 1986

Yours, Sakharov'

P.S. The paper of Curir mentioned in the text has the following coordinates: Physics Letters, vol. 161B, n-b (probably one should read "number" or "numbers". — E.F.) 4,5,6.31 Oct. 1985. A. Curir "On the Energy Emission by a Kerr black hole in the superradiation range." I am writing this just in case Frolov missed it.

P.S.S. (sic! — E.F.) Instead of a reference to Okun's book it would be better to give a direct reference.

The uncertainty of Andrei Dmitrievich expressed in this letter is explained by the fact that he had not been previously working on this special topic (and therefore had not been closely following the corresponding literature. As told in the text at that time he was carried away by the theory of superstrings). Valeri Pavlovich Frolov, a staff member of the Laboratory of electrons of high energy in FIAN, is a specialist in relativistic astrophysics as a whole and in particular in black holes. Probably Sakharov ran across a paper by Curir and got an idea of his note. It is clear how harmful was his scientific isolation to his work. I am also giving my response so that one could see how the Theory Department tried to overcome this isolation.

Dear Andrei Dmitrievich!

As you understand, the manuscript on radiation of mini black holes sent by you has arrived with some delay. According to your wish it was discussed by specialists, in the first place with Frolov (by the way his book with Novikov mentioned by you is still not out of print, its sales will begin only in September and, possibly, even later). The result of the discussion was quite positive: A new idea was put forward, estimates carried out and in general it is quite actual from the point of view of ideas currently developed in cosmology. One only made two remarks: 1) In the estimates you neglect possible existence of a cloud of already emitted particles which could influence the effect but, as far as I understand, you mention this possibility in the text yourself; 2) An estimate of the change in the angular momentum was performed (evidently you did not know that) by Page but only for emission of massless particles. Therefore in this place we allowed ourselves to make an insertion — one phrase: For massless particles this question was considered by Page and gave the corresponding reference. Besides that we composed a regular reference list for the literature mentioned in the text of the article.

Sending you a copy of the finalized text we are simultaneously preparing it for publication in *Pis'ma v ZHETF* and, without waiting for your response to this letter, will send it to the journal. If you would like to make some corrections then, although they publish fast, there will still be time.

I am using this opportunity to congratulate Elena Georgievna and you with the success of so dangerous heart surgery she had.

Valentina Josephovna as well as Vitali Lazarevich asked to convey to you greetings and best wishes.

All the good (and in the hope for good)

17.06.86

                                                         Feinberg

Without any doubt Frolov discussed the work together with A.D. Linde, I do not know whether anybody else did participate.

A little remark: On rereading my letter I see how many unnecessary words did I use. For example, on "heart surgery." In a private letter the word "heart" is seemingly superfluous. However, one had to write in such a way that an "extra reader" would not develop suspicions or even uncertainties, everything had to be clear for him otherwise it could have happened that a letter would not go through.

11. In response there came a telegram of 25.06.86:

Ask send photocopies Page articles Physrev[17] D13 D14 before my reading ask delay sending my paper changes possible Sakharov

This telegram (the last one from him from Gorky) meant that the critique of "specialists" of his work of mini black holes moved A.D. to begin modifying and rewriting the paper. Possibly he had some new thoughts himself. The life in science continued.

After half a year Elena Georgievna and Andrei Dmitirievich returned to Moscow.

---

[17] *The Physical Review.*

S. I. Vavilov

Chapter 4

VAVILOV, Sergei Ivanovich (1891–1951)

4.1 Nine Scars on the Heart[1]

There opened folders with the inscription "Keep Forever" and the earth covering the mass graves of those who had become the "Gulag dust", we hear the voices of those who had to keep silent and the terrible truth about a long but recent past entered the lives of our contemporaries. One of the episodes of this past was the life and death of the world famous scientist, a biologist, academician Nikolai Ivanovich Vavilov.

There was however another academician Vavilov, his loving and loved younger brother Sergei. As recalled in recently published memoirs of a person close to Nikolai Ivanovich, 'He would often tell: It's not me, it's Sergei who is worth it!'

Alas, Sergei Ivanovich has recently become an object of ill-disposed light-weighted judgments of some journalists and cinema people who, with an incredible easiness, contrast him with his brother who perished in prison. It is difficult to refrain from exploiting such a sensational possibility when in two and a half years after the death of his brother Sergei Ivanovich became a President of the USSR Academy of Sciences, was a person respected in Stalinist times and pronounced all the ritual words that were obligatory for a high-ranked person at those times and sound so horrible nowadays.

In reality, however, everything is much more complicated. The time has come to talk without omissions about this Shakespearian, as termed by my clever interlocutor, the film director Aleksander Proshkin, situation. I feel obliged to tell everything I know about it because my whole scientific life took place in the remarkable institute organized by Sergei Ivanovich,

---

[1] Significantly revised and expanded version of the article in the magazine *Nauka i Zhizn* (v. 8, p. 34, 1990).

the P.N. Lebedev Physical Institute of USSR Academy of Sciences, FIAN. I feel that I have a right to write about him also because I have thought about this "Shakespearian situation" a lot and in the course of many years gradually learned from people closer to him the facts that always confirmed the understanding of this situation that I already had. This is what I will write about.

Sergei Ivanovich was, as one says, a "God-blessed physicist." This can be seen, for example, from *how* he did come to his perhaps highest scientific achievement, to the discovery of Vavilov–Cherenkov radiation.

Here he showed not only a feeling for physics by grasping an occasionally found weak and seemingly irrelevant phenomenon, not only an art of an experimentalist and an exceptionally fine understanding of laws of optics that lead him to a conclusion on the genuine unusualness of this phenomenon, but also a courage of a true scientist in making this conclusion public although very few believed in it at that time. There appeared prickly, almost scoffing jokes and even Joliot-Curie was noticed to secretly move one element of the experimental installation during a demonstration of the phenomenon (at that time experimental possibilities allowed to observe this radiation only in complete darkness). It was clear that in spite of all his polite words and smiles he also did not believe. Soon, however, it became absolutely clear who was right. The method of observation suggested by Sergei Ivanovich was reliable, the measurements of Cherenkov (a Ph.D. student of Vavilov) were unobjectionable and astonishingly accurate and Vavilov's conclusions on the unusualness and novelty of the phenomenon were correct.

There exists a story on how the composer Carl Philipp Emanuel Bach (who was at his time more famous than his father, the great Bach) was invited by the most enlightened Kaiser Friedrich to live and work at his Berlin court. The composer was very happy. After some time, however, he wrote in a letter: 'First I thought that Friedrich loves music, then I understood that he loves only the music for flute and now I see that he loves only his flute.'

Sergei Ivanovich loved and knew the music of the whole of physics.

In two years, after he had become head of a small department in Leningrad, the department became a separate institute, this very FIAN, and moved to Moscow.[2] Sergei Ivanovich expanded it tenfold to

---

[2] The birth of this institute and the institute itself are described in Section 4.3. The author apologizes for covering once again some points from that Chapter.

a "polyphysical" one and invited people known to everybody in the scientific community to head various laboratories. These were L.I. Mandelstam and N.D. Papaleksi, G.S. Landsberg, I.E. Tamm, D.V. Skobel'tsyn, who soon moved from Leningrad and N.N. Andreev (who, by the way, was the first to give a course on relativity theory in Moscow university).

Sergei Ivanovich himself became head of a small laboratory of luminescence and concentrated most of his own scientific research in a big optical institute in Leningrad. He had become its scientific leader earlier. After becoming a director of FIAN he shared his time between "the two capitals." It should not be a surprise that such a scientific staff was allowed to once more increase its size tenfold after the war. Sincere respect, full trust and goodwill did establish between all these people including Sergei Ivanovich himself. With them they brought the best traditions of the Russian intelligentsia which, in the limits of the possible, exerted influence on other people working in the institute. Most significant here were a devotion to work, generosity in sharing scientific ideas, honesty in evaluation of their own successes and failures and those of others, complete absence of smugness even by recognized leaders that never became bossy, a respectful support of talent in whoever showed it, a Ph.D. student or an academician, and, most importantly, a freedom of thought. It was probably due to this why during the first 20–25 years of its existence (19 of them with Sergei Ivanovich) FIAN produced so many remarkable results: Two discoveries awarded with Nobel Prizes and one more that formally was not awarded the prize due to the foolishly imposed secrecy (but recognized in the world and afterwards awarded in the USA by the presidential prize of the same rank, "Atom for Peace"). Besides that, principal foundations of thermonuclear fission in its both aspects, uncontrollable (military) and controllable (peaceful) were worked out. To this one should add many other fundamental achievements.

In the next quarter of a century, which we timidly call an 'epoch of stagnation', the institute produced many valuable results. The fate of a country did however influence the life of FIAN as well. Was it accidental that the phenomenon of Sakharov, who began his Ph.D. studies in the institute in 1945, was at that time born in FIAN?

Much of the basis laid at the time of Sergei Ivanovich was carried through many difficult years and is felt even nowadays when new generations take the fate of the institute in their hands.

The denunciations, so common in the thirties and forties, were seldom in our institute and when they happened they were not investigated. They were "shelved". In the decades to follow the traditions established in the distant past helped many staff members of different generations to keep an elementary decency in difficult political situations.

I have already mentioned that Sergei Ivanovich knew and loved "the music of the whole of physics." This was however not all. He was devoted to the culture, past and present, of the whole of humanity.

He chose physics as a dominant interest but at the same time his erudition in fine arts, literature and history was immense. He was a bibliophile, I would even say a "biblioman." Each Sunday he toured the second-hand bookstores. He did not just possess books, he was in command of everything in them. When he became president of the Academy of Sciences he was visited by historians, philologists, historians of science, historians of fine arts who met with complete understanding. The discussions were "on equal terms." There was nothing astonishing in the fact that when the academy, together with the union of writers and Ministry of Culture, was celebrating the jubilee of the *Word on Igor's Regiment* or the 150th anniversary of Pushkin in the Lycee at Tsarskoe Selo the voluminous opening word was given not by a philologist but by the president-physicist.

In 1946 Sergei Ivanovich delivered a fundamental talk, "Physics of Lucretius," at a meeting celebrating the jubilee of Lucretius Carus. He began it with the words:

> There hardly exists another poetic or scientific work of ancient times, even when recalling creations by Homer, Euripides, Euclid, Archimedes, Virgil, and Ovid that has carried to our days, through thousands of years, such freshness and topicality as the unfading poem of Lucretius. It was admired by Cicero and Virgil, it was under heavy attacks from "fathers of church" who rightly saw in Lucretius a terrible danger for them. This poem determined many features of *Weltanschauung* of Newton and Lomonosov, caused delight in Herzen, was of profound interest for young Marx and served as a banner for the mechanistic materialism of L. Buechner. Lucretius was perhaps read by Turgenev's Bazarov and the heroes of A. France did not part with the precious book in the most critical moments of their lives.
>
> Such a two-thousand-year-long ambivalence, a most rare case in the life of culture, is worth special attention. What are the origins of the strength of Lucretius? In his poetry, wonderful but, in the opinion of many, inferior to that of Virgil, Ovid and others? In his *Weltanschauung* and scientific knowledge in which he is basically a true follower of his deified teacher Epicures?

The attractiveness of Lucretius is in neither of these two things considered separately. It doubtlessly lies in the astonishing and unique in its effectiveness merger of its philosophical content, eternal in its truthfulness and breadth, with its poetic form. ("Poetry is a perception of one's truth," said Osip Mandelstam.)

Knowing Sergei Ivanovich one can confidently affirm that he had not just heard about all of the authors he mentioned but had read them, most of them in original language.

His literary style, in particular that of his numerous articles and books on history of science, was splendid. The clear, exact and capacious text is such that one wants to read it aloud. Such are his book on Newton, one of the best in the world literature, and numerous articles on scientists of the past — on Galilei and Lomonosov, on Euler and Faraday, on Petrov and Huygens, on Grimaldi and Lebedev, etc. I would dare to say that much of written by Sergei Ivanovich impels one to recall the prose of Pushkin.

All this makes one think about persons from the Renaissance epoch.

A turning point in this remarkable life came at the beginning of June 1945 when Sergei Ivanovich, who had just moved to Leningrad, got a call from Moscow either from Malenkov or from Molotov and was asked to immediately return to Moscow and come to the Kremlin. Completely perplexed, Sergei Ivanovich went to Moscow. On the way to the Kremlin he visited his acquaintance, also a passionate bibliophile, E.S. Lichtenstein, an editor in chief of the publishing house of the Academy of Sciences — a clever man who was well informed on what was going on in the academy, and shared with him his perplexity. In response he heard: "Sergei Ivanovich, you will be a president." Lichtenstein told me that these words put Sergei Ivanovich in a state of horror. He waved his hands and began pronouncing phrases like "God knows what you are telling, curse that tongue of yours" etc., and in such a state he headed for the Kremlin.

Why was he so horrified? Not because he was afraid of the immense difficulties that were indeed a part of the president's work. At that time he was a head of two very big institutes and of various scientific committees (including the stratospheric one). Like his brother Sergei Ivanovich had a phenomenal capacity for work. He was never in a hurry and never hurried an interlocutor, his speech was not fast but very exact. Everything was done in time and effectively, the organizational problems were, it seemed, dealt with ease.

These were not the difficulties related to work that could horrify him. Rather, having been a very clever man, he clearly foresaw that to occupy such a high place at such a cruel time meant that he would be obliged to tell and write some things he did not think and did not believe in, that he would have to make shameful, humiliating concessions and compromises and be a helpless witness of an ignorant oppression of science. The tragic fate of his beloved brother should have provided a Kafkian decoration for all this.

Why did he nevertheless decide to take all of this upon himself? (At the same time, had it been possible to find somebody who would have dared to say "No!" to Stalin?) It seems indubitable to me that he did this because he learned who was the alternative candidate for president.

Many years after the death of Sergei Ivanovich when Lichtenstein told me all this I asked him whether I understood it right that Sergei Ivanovich had deliberately sacrificed himself when he had agreed to become a president? The reply was prompt and resolute: "This is absolutely correct." I added: "Indeed, had he refused, Lysenko would have become a president." The reaction was unexpected: "No, not Lysenko, it would have been Vyshinski." I was shocked.

A contemporary reader needs an explanation of what kind of a figure was the academician Andrei Januar'evich Vyshinski. A lawyer with a prerevolutionary education, a dean of Moscow university in the twenties, he was an active menshevik before the revolution. In the July days of 1917, having been head of one of the district Duma's in Moscow, he signed an order to find and arrest Lenin that was posted in the streets. Several years after the revolution he joined the communist party and, praying for forgiveness of old sins (and, of course, from simple careerist considerations) put all his outstanding abilities and qualifications in the service of Stalinist terror. He became a merciless prosecutor general, took an active part in preparing the famous trials in the thirties, was, as a state prosecutor, the main figure at them, but even more bloody deeds were done "behind the stage."

He also introduced as a fundamental thesis a "principle" stating that a confession of the accused is a "queen of proof" into theory and practice of court trials. This stimulated torture as a means of beating out confessions. This "principle" is in clear contradiction to all the foundations of jurisprudence in the civilized society. One could not say about this man that only his hands were covered in blood because he was in blood of his victims up to his neck. He was a traitor by his nature, could send a friendly letter to a comrade from the pre-revolutionary underground and immediately issue an order of his arrest (one of such cases that ended up in a quick death in the camp is known to me in detail). With his "principle" he also betrayed the juridical science. In exchange he became a minister of foreign affairs, outlived Stalin and died a natural death.

To tell the truth, at that time I had doubted the rightfulness of the Lichtenstein's opinion but later learned that Segrei Ivanovich had himself described Vyshinski as his competitor to his old friend academician G.S. Landsberg. I understood that from Stalin's point of view this was quite suitable a candidate. All the "atomic problem" was in those years supervised by Beria and this was at that time perhaps the most important task set for our science. Why not "... give as a Voltaire" not a sergeant major but an "intelligent" embodiment of the diabolic trinity Stalin-Beria-Vyshinski?

Moreover, not so long ago, in 1991, during a celebration of the 100th anniversary of Sergei Ivanovich, I got an even more reliable confirmation from the academician Alexander Leonidovich Yanshin who, having heard my speech containing the above written phrases, told me the following (I quote his information with his kind permission).

In 1945 when the government understood that the then president of the academy, Vladimir Leont'evich Komarov, was so old and feeble that had to be replaced in the circles of the presidium of the academy there began a discussion of his possible successors. At that time young A.L. Yanshin was a scientific secretary of the "Iron Commission" of the presidium while its president was the first vice-president academician Ivan Petrovich Bardin who was de-facto replacing the sick president. Yanshin was on close terms with him and he told him about what was going on. Bardin told Yanshin that he (a metallurgist by profession), academician A.A. Baikov (metallurgist and chemist) and some other scientists close to the orbit of presidium's work had discussed several possible candidates and had finally chosen S.I. Vavilov. With this choice Bardin went to Stalin to report it.

Having heard Bardin, Stalin, after some thought, asked: "Would Vyshinski be more suitable?" Bardin gave the reason that at that time the president had to be a physicist, a well-known scientist with whom physicists could talk, etc. Stalin contemplated some more and agreed. Somebody not knowing those times well could wonder why Stalin did not take into account that S.I. Vavilov was a brother of the executed "enemy of the people"? It does even seem that Stalin liked such situations. The wife of Kalinin was at that time in the camp while he formally remained the "president" of the country. Soon after that the wife of Molotov, the second person in the country, was imprisoned, the brother of Kaganovich was executed, etc.

One can also understand a lot from an unpublished draft of the memoirs of I.P. Bardin. An historian of science, G.E. Gorelik, had found it in the academy archiv and presented me a copy with the kind permission of using this material.

Bardin tells what a good impression Vavilov made upon him long before his presidency. He liked his businesslike approach shown, for example, at the meetings of HAC (Highest Attestation Commission, Bardin was its chairman): "He impressed me from the very beginning by his concrete statements free of empty phrases with which many HAC members did usually fill their speeches when making some point and I liked that feature of him." Then Bardin writes "... especially memorable is one of his speeches when Sergei Ivanovich did somewhat deviate from his usual even tone ... when he had to give his opinion of the so-called priority in Raman effect on which there had worked our physicists academicians Landsberg and Mandelstam. ... I remember there spoke one academician who took a formal position. Then spoke Sergei Ivanovich and I first heard him speaking loudly and violently proving his point ... he was supported by other academicians, I also supported him. This was his victory." Here a meeting of the Committee on Stalin prizes in 1943 is described.

They also occasionally met when visiting an exhibition prepared for the anniversary of the academy. Bardin was surprised by how lovingly and interestingly Vavilov described the history of the exhibits: "So, in addition to the love to science he had a great love to the history of science."

In this draft note Bardin describes a story of electing S.I. Vavilov the president as follows:

> A number of academicians was suddenly summoned. Vyacheslav Mikhailovich [Molotov] gave a speech: Vladimir Leont'evich [Komarov, the president of the academy] is sick, his health does not allow him to work. ... At the same time much work will be required from the Academy, especially in the field of physics [this was in June 1945]. What do you think about Vavilov? At that time Sergei Ivanovich was in Leningrad. We began exchanging opinions. I would not say the opinion was unanimous but almost everybody told this candidate will pass. Somebody objected and wanted to tell something but did not tell anything. There were no other candidates and nobody told one needed them. I also told that this is a suitable person, a

> physicist, relatively young and therefore it seemed to me he could be suggested ...
>
> The nomination was unanimous [apparently already at the electing General Assembly of all the academicians. — E.F.]. He began to refuse referring to the lack of time, the lack of capabilities required in such a work, then to personal circumstances (the hard family situation he had in relation to his brother). He did not mention the family matters but one felt that this was present. His appeal was not heard and, finally, he had to agree.
>
> I do not remember his speech but at the end he told he would have to give up science that he loved.

In this draft note everything is important. It is of course clear that it was not Molotov but Stalin who decided to make Vavilov a president. The words that the president should have been a physicist confirm the words by Yanshin that this was precisely the motivation used by Bardin in proving to Stalin the necessity of electing Vavilov, not Vyshinski. The helpless disordered attempt of Sergei Ivanovich to persuade the academicians not to elect him were hopeless and he himself, of course, could not but understand this because it was certainly clear to him as well that the issue had been already decided by the "lord." The awkward reasoning of loving science could not be taken into account — all this looked completely ridiculous and did hardly ever happen at these serious and solemn meetings. Such behavior of Sergei Ivanovich in this situation reflects his deep disarray, confusion and fear with respect to things he would have to endure in this position.

All this shows in what a state he was in and agrees with Lichtenstein's description of a degree of horror with which did Sergei Ivanovich react to his prediction "You will be a president."

One could ask a question: Why did not Bardin say a single word on how he persuaded Stalin to reject the candidacy of Vyshinski and make Sergei Ivanovich a president? Who could, however, dare at those times to publicly describe how Stalin was persuaded to change his opinion and write about this in one's memoirs? This omission is completely natural for that epoch.

The fate of our science, our culture, were for Sergei Ivanovich more important than his personal emotions. Everybody who has known him personally are convinced that he made this step to protect everything in our culture that it was possible to protect. Not worrying for himself, "not fearing offence."

One should however take into account that all this happened in the year of the euphoria over the great victory. The victory parade at which the commanders-in-chief of the allied armies, Eisenhower and Montgomery, had been standing on the Mausoleum next to the Soviet commanders, had just taken place. The anniversary of the academy attended by scientists from 16 countries had just been celebrated. Scientists from the academy whose clothes had been worn out to an impossible degree (or had been sold for food in the hungry times) got some new clothes, the institute's buildings were renovated. Many expected that new times were coming.

Quite seemingly reasonable people who knew very well the horrors and treachery of Stalinshina [Red Terror], the power of immensely cruel state machine, were convinced that, by its victory in the war, not only did the people deserve a right for a bigger freedom but that they would get it. One friend of mine — M.A. Leontovich, who soon became an academician asked me in 1943: "Do you really think kolkhoses [the collective farms] will exist after the war?" When the war ended many thought so. They did not understand that the system that wins never changes itself . The armies, exhausted by war, were returning to a ruined starving country and, to save itself from the fury of the disillusioned victors, the "system" had to "tighten the screws" even further, strengthen the oppression, separate itself from the world by an "iron curtain." As is well known all this soon began happening with a customary constantly increasing cruelty.

.5Could Sergei Ivanovich deceive himself in this respect? Could he think that the president could have remained "pure"? It is possible that he did have some illusions. I remember however his soberness at the beginning of the war.

The soberness of Sergei Ivanovich that contrasted the then so popular light-weighted political analysis makes me think that when agreeing to become president he understood what he would have to face.

As I will describe in another essay below, see Section 4.4, he thought about culture at a scale of millenniums. He knew that humiliation of an artist or a scientist, exploitation of his achievements by cruel power holders combined with the complete negligence to his personality, was a common phenomenon in the history of mankind. So it was in antiquity, in Renaissance, in the times of Lomonosov.

It seemed impossible for all this to happen in the XX century but it did happen, not only in our country, and the degree of humiliation and the scale of extermination of the outstanding people of science and arts did exceed everything that had ever happened in the past.

There exists a wise prayer for every day composed in our times by a professor of a chair of "applied Christianity" of the New York Theological Seminary Niebur. It reads: "God grant me the serenity to accept the things I cannot change; courage to change the things I can; and the wisdom to know the difference." I feel that this wisdom was granted to Sergei Ivanovich.

Having become a president in this environment Sergei Ivanovich wrote and pronounced all the words required at those times. "You don't cry for hair when the head is cut!" and he could not distance himself from everything inevitably linked with him while occupying this high position. Archive newsreels show us him handing, in the name of the Academy of Sciences, a greeting to a blighter of his brother (academician Lysenko) at his jubilee and how they embrace. The newsreel is black-and-white but an academician I.M. Frank who was present there told me that the face of Sergei Ivanovich was not even white but rather green. Another reel shows him delivering a speech at a general assembly of the Academy of Sciences and concludes it with thanking Stalin "for his care about science." These reels wake bitterness and compassion in him and many other academicians participating in the assembly that gave a standing ovation to these words. In the first row the cameraman selected the face of the academician Vyshinski.

This was not all. There were papers praising the "scientific genius of Stalin," a "coryphaeus of science." It is worth noting that in contrast with the usual style of Sergei Ivanovich their style is shaggy and clumsy and boils down to an at that time usual set of newspaper cliches (there existed such a joke: What is this? People have gathered and are praising Stalin, thank him? Answer: a jubilee of Tchaikovsky). There was also presiding at special sessions of the Aacademy devoted to crushing genetics and physiology (Stalin termed this "free discussions"). Vavilov accepted this steadfastly, as inevitable.

One can of course understand those scientists who worked in the demolished areas of science and could not forgive this of Sergei Ivanovich. It was difficult for them to grasp that he was powerless, that had he not obeyed the norms of behavior obligatory for his position, this would have been occupied by the Vyshinskies and Lysenkoes who would have humiliated the scientists even more and would have completely destroyed science. It is necessary to give a due appreciation to what he did for saving our culture.

Everybody knows about the destruction of genetics, about cybernetics that was strangled in the cradle, on pogrom CC resolutions on literature

S.I. Vavilov shortly before his death. Taken without warning at the laboratory of S.I. in FIAN.

and arts. Many horrible things, however, took place in the "protected" areas as well, e.g. in physics and chemistry. For example the omnipotent philosophers did not stop hunting quantum mechanics and theory of relativity. At that time in many universities, especially provincial ones, the courses on the theory of relativity and foundations of quantum mechanics were not read because of the fear of the lecturers to be accused of idealism.

A late academician of RAS, I.M. Tsidilkovski, volunteered to the front as a student, fought in a reconnaissance unit, suffering from wounds till the end of his days, did after the war graduate from the university and lectured in the Melitopol Pedagogical Institute. He dared to raise an objection against a head of the chair of marxism-leninism who stated in a "ground-defining" talk that quantum mechanics and relativity theory were ideological diversions of the imperialism and their creators were its direct agents. At the next day Tsidilkovski was summoned to the "organs" where one hammered the fist against the table and promised to "grind him into camp dust." There immediately arose a question of his expulsion from the communist party. He saved himself, following the advice of the dean, only by leaving the institute "at his own will" on the same day and leaving the town in the evening. This was typical for those days.[3]

I will not pay particular attention to the behavior of Sergei Ivanovich with respect to people in trouble but it is worth recalling that he, together with an astronomer academician G.A. Shain, wrote two letters asking for arrested astronomers to A.Ya. Vyshinsky — first in 1939 as to the prosecutor general (he occupied this position in 1936–1939 and this was perhaps a reason for "electing" him to the Academy in 1939 — who would have dared

---

[3]See Tsidilkovski, I.M. (1997), *Half a Century with Semiconductors*. Ekaterinburg: UrB RAS.

to vote "against"?) and then, in two years, to him as the deputy head of Sovnarkom. (I have xerox copies of these letters before me. Vavilov and Shein wrote that about 20% of all actively working astronomers, the most prominent ones, had been arrested). They also asked for all arrested wives giving the names of all of them. At that time this was an act demanding great courage especially because Sergei Ivanovich himself did not work in the field of astronomy and explained that he wrote as a representative of the opinion of "physical community" and as a deputy of the RSFR Supreme Council.

In 1943 Sergei Ivanovich wrote a letter "upstairs" defending his brother (he learned that he was no longer alive only at the end of 1943). He helped those who "broke away by chance" (as written by O. Bergolz) to find a job, sometimes in a completely miraculous way like it was with professor L.S. Pollak whom he knew only from occasional meetings in the second-hand bookshop in Leningrad in the thirties. In general his help to other people was very wide. His salary was mainly used for sending money to the people in need according to the list (I knew this at that time from Anna Illarionovna Stroganova, his secretary-referent of many years in FIAN). After his death the amount of money left on his account was equal to his monthly salary.

One should specially mention the care which Sergei Ivanovich showed towards the family of his brother. The memoirs of the younger son of Nikolai Ivanovich, Yuri, who is still working in FIAN, have already been published. In them he quotes the words of gratefulness from the letter of his mother to Sergei Ivanovich: "We would not live through these times without your help."

A still stronger impression one gets from a description of numerous touching signs of attention to Yuri Nikolaevich himself at times when he was a schoolboy, then a student, then a diploma student in FIAN. For some time he stayed at Sergei Ivanovich's place, on weekends he was taken to the dacha, etc. Once, on the way from the dacha, Sergei Ivanovich told him that a position of the president was a "dog's" one and that he would have preferred to work as a plumber. The memoirs end up with the words: "I loved Sergei Ivanovich a lot. His premature death in January 1951 was, like the death of my father, a heavy grief for me and my mother."

His main occupation during the presidency was the saving of and the development of science in the places he could do it. His close coworker of many years, V.V. Antonov-Romanovski, who is still working in FIAN, recalls how once, having not noticed the especially concerned state of Sergei Ivanovich (which is nothing to wonder about, he was always fresh and restrained) complained about some insignificant organizational matter, on current work. Sergei Ivanovich lifted at him sorrowful eyes and said: "Oh, Vsevolod Vasil'evich, now I have to save our physics, and you ..." V.V. cannot forget this look up until now. This was in 1948 when another "free discussion" was being prepared, this time on physics. Leading physicists, Sergei Ivanovich among them, managed to divert this danger.

It is even more important to recall all the constructive things he contributed to the development of our culture. Having become president he launched a gigantic organizational activity. I.P. Bardin wrote that during the period of the presidency of Sergei Ivanovich there began and was partially completed construction of more than 50 new buildings of institutes and other offices of the academy including the botanic garden, etc. He took an active part in organizing several academies of sciences in Soviet republics. He completely reformed the editing activity of the academy. He founded the series 'Classics of Science' and 'Literary Monuments' and understood very well who had to be summoned to this activity, who could accomplish it at a high scientific level. As is well known the authority of these editions and their popularity are nowadays extremely high. He founded and headed a society that is nowadays called "Znanie." Having been a brilliant propagandist of science (it is sufficient to read one of his popular books, e.g. the many times reprinted book *An Eye and the Sun* or the more serious one *Experimental Foundations of Relativity Theory*) he was personally and very actively involved in this. He was an editor-in-chief of the Great Soviet Encyclopedia and simultaneously a scientific editor of our central scientific journal on physics.

When Sergei Ivanovich passed away I, as a member of the editorial board of the journal, got the task of preparing the obituary. Having written and reread the text I doubted the possibility of publishing it: Who could have believed that a single person could accomplish all this, that his speeches were not prepared by his referents? However, I was a witness of how it happened myself. In particular every issue of the journal under preparation, all articles in it, were discussed by the editorial board once a month in his office when the staff of the presidium had already left the building and nobody could disturb. There arose many difficulties characteristic for those times. For example there came papers from the physicist Yu.B. Rumer who had served his term in jail and lived in exile in the far north of Siberia in Eniseisk. He could not present some documents that were at that time absolutely mandatory: Recommendation of a scientific institute, confirmation that the paper did not contain classified information, etc. Sergei Ivanovich simply neglected all this and took the responsibility on himself. (One could add that Sergei Ivanovich ensured Rumer's move to Novosibirsk but suddenly died before he could ensure his job there). These meetings lasted no longer than one hour.

As to his speeches and articles, who could have written them for him in such a splendid style, so cleverly and with such rich content? His editing of the encyclopedia was also not formal and bossy at all. He wrote some articles and proofread many articles written by others himself. When he passed away at night, towards morning, on his desk there was a proof of such an article which he had been proofreading until midnight. The proofmarks were first made by a steady hand, then the handwriting became less clear and shaky and, finally, abruptly terminated.

Modern experimental psychology performs wide studies of an interrelation between various features of higher nervous activity, in particular between creative activity and emotions. In a number of behavioristic experiments in animals one obtained results allowing to conclude that when a specimen has a specially strong and productive research instinct this goes along with "courage (low index of fear), friendliness and non-aggressiveness." (Simonov, V.P. (1975) *Higher Nervous Activity in Humans*. Moscow: Nauka, p. 23). It was probably not accidental that these features were characteristic of Sergei Ivanovich as well. They formed an emotional basis of his talent as researcher, organizer of large-scale research and attractive intelligent person.

I have written that Sergei Ivanovich was a man from the Renaissance epoch. The Renaissance did however leave us not only the enlightened images of artists and thinkers. It also left us inquisition. In our days the fate of Nikolai Ivanovich Vavilov is, perhaps not quite correctly, but with good grounds compared with that of Giordano Bruno.

In this case when speaking about the fate of Sergei Ivanovich (and leaving aside a scale of the genius) it appears natural to recall the fate of Galilei who at several interrogations ensured the inquisition tribunal that he did not hold to the Copernican point of view and then, on his knees, publicly denounced it in a detailed horrible statement. Only after this did the tribunal agree that he was not an "incorrigible sinner" and therefore did not have to be burned to death.

According to the verdict, however, he was left "under strong suspicion" and was considered, according to an official formulation, "a prisoner of inquisition," spending the last 9 years of his life practically under house arrest (in a tiny place called Arcetri) under a strict ban of discussing the heretical issues of a structure of the solar system with anybody. In fact he did not pronounce the words "It is nevertheless rotating." This is a clear, albeit beautiful, fantasy (as a "prisoner of inquisition" he would have been immediately "automatically" executed). However he did something more — wrote the second of his great books in which the laws of mechanics were formulated. These books are commonly considered to mark a beginning of the history of new scientific physics. His abdication became not his humiliation and shame but that of the church and the whole epoch.

In analogy to this when remembering how much did Sergei Ivanovich Vavilov, completely neglecting his health, do for our culture we feel thankfulness towards him and shame for the horrible epoch in our history that gave birth to this tragedy different from that in the life of his brother.

These comparisons are of course somewhat risky. Any careerist can justify his shameful deeds explaining that only in this way could he ensure a possibility of scientific work, of making "contribution to science." However, Sergei Ivanovich, as president, did so much — infinitely much — of good, did sacrifice himself so evidently that no room is left for questioning the nobleness of his motives.

I earlier quoted the words of the modern "prayer" and said that the wisdom allowing to see a difference between the possible and impossible was granted to him. He was also granted with the courage to fight for what he could change (academician D.V. Skobeltsyn who worked in close contact with him told me how once during a discussion in the president's office Sergei Ivanovich, who was about to leave for a report to Stalin — such reports were told to take place once or twice a year, — told him: "Each time I go I do not know where I will return, to home or to Lubyanka;" the same words were heard from him by his closest disciple, academician I.M. Frank).

Was he however granted the ability to peacefully accept what he could not change? Hardly. Nobody knows what he went through. The years of his presidency from 1945 until January 1951 when he died were the years of the constantly growing horror in the life of science, culture, our entire society, all the country. His sufferings were indubitably increasing as well. Even if he had some illusions at the beginning they should have faded. At the same time the necessity of saving what could still be saved was increasing. It was already impossible to retreat. But what was hidden before his apparent restraint?

Those close to him knew how stubbornly (and unfortunately successfully) did he resist the suggestions to go to the doctor. Some think he was meaningly heading towards his death. It was not accidental that after his death one found on his heart (as then told) nine scars.

4.2 Sergei Ivanovich Vavilov and His Time

Like his brother Nikolai Ivanovich (the brothers had indeed much in common) Sergei Ivanovich was not just a remarkable person: His fate, his formation as an outstanding scientist and organizer, his immense erudition both in natural sciences and humanities, his true belonging to the intelligentsia (I would also recall the word gentleman) deserve special attention. Each period of his life, changes in his activities and behavior were astonishingly deeply related to the fundamental transformations that his country and his people were going through.

His grandfather was a serf and his father came to Moscow in the seventies of XIX century from Volokolamsk and at the beginning was a boy "on errands" to a merchant. According to the recollections of S.I. he was self-educated and in less than 20 years (by the time S.I. was born) had become a wealthy independent merchant, "read and wrote a lot" and doubtlessly belonged to the intelligentsia. He was twice elected to Moscow city Duma and played an active role there. He supervised charities, was one of the initiators and active participants in the construction and development of Moscow tramway. Besides that he was a close business partner of the management of the at that time biggest Prokhorovskaya Trehgornaya textile mill at Presnya and was developing its business relations with the East, the major consumer of its textile production.

How could this all happen?

Sergei Ivanovich was born in 1891, 30 years after the abolishment of serfdom when, during the 20 years of the truly great reforms of Alexander II, they had already exerted a profound influence on the life of the country. These reforms had been so well matched that in spite of some "counter-reforms" that came relatively soon even the stubbornly conservative politics of Alexander III and Nikolai II who could not understand the necessity of a further widening of reform-related transformation could not stop a fast development of the country they caused. The conservatism did just provoke revolutionary bursts and upheavals and in the end lead the country and the monarchy itself to a catastrophe.

In just half a century (from the beginning of the reforms in 1861 till the First World War) a retarded country with serfdom, recruit-based army, corporal punishment, a degree of lawlessness in all areas of life that was impossible in Europe, did radically change. It is sufficient to recall the pre-reform court that often delivered its judgment in the absence of plaintiff and defendant. The court reform (even after the adoption of certain new laws that somewhat weakened its progressive elements) did bring the court system closer to the international norms (jury, irremovability of judges and investigators and their administrative independence, etc.). There arose a state with a rapidly growing industry, brilliant corps of engineers, widespread railroad network, modern (although not so big) fleet that had been built after Tsushima, developed court system, rapidly expanding network of brilliant gymnasia and universities, remarkable intelligentsia and zemstvo[4] that did so much good.

Although we rightfully reproach the tsarist Russia of that period for the poverty of the uneducated people's masses one should not forget how education was spreading and how a feeling of human dignity was arising in these masses. A proletarisation of landless peasants gave birth to a sharp polarization of material and spiritual levels of the people and political confrontation. Each worker's maevka[5] was an episode of a struggle for personal rights, for the feeling of human dignity for all its participants.

Nikolai Il'ich Vavilov.

This atmosphere of general progress in society and increasing confrontation could not be quenched or at least tempered by a stupid stubbornness of tsars, frightened by the assassination of Alexander II, patronizing Black Hundreds ideas and realizing certain "counter-reforms."

New Russia was entering the world. It is characteristic that when Bulgaria was freed as a result of the Russian-Turkish war of 1877–1878 and became an independent state Russian generals gave it constitution which at that time was one of the most progressive

---
[4]Local elected authorities.
[5]Worker's meeting.

in Europe. At the same time one did not dare to do the same in Russia.

The formation of the brothers Vavilov as persons took place in that very period when this radical renovation of the country was already felt. Everything however was changing so fast that old was closely interlaced with new, sometimes in a very strange way.

The father of S.I. Vavilov was already a rich man but until the summer of 1905 the family lived in a one-story wooden house with an attic in one of the sidestreets in Presnya, near the church in the modern Zamorenkov street. The whole district consisted of such houses usually inhabited by people related to the all-dominating "Trekhgorka." There remained, however, eighteenth century nobility estates, fallen into decay, that had been built when this district was still outside of the city. Only in 1905 did the Vavilov father buy a wooden house in one of these estates with big rooms, high ceilings and even a dancing hall. It was completely rebuilt.

S.I. writes about his mother in his autobiographical notes: "Mother was from a working family [let us note that these were highly qualified workers, artists-engravers — E.F.]. For all her life until the death in 1938, never "madam-like," washed linen, cleaned floors and cooked herself.. She woke up at 5 a.m. ... It was difficult to be simpler, kinder, more industrious and democratic than my mother." Both sons loved her a lot. She taught S.I. to read using Tolstoy's ABC, then he began attending a small private school where he prepared for going to a commercial college (his father probably prepared him for commercial activity). Unlike at the gymnasiums, ancient languages were not taught here. Later, however, S.I. learned Latin himself when preparing for university. He read Ovid, Virgil and his beloved Lucretius Carus by heart (he knew several other languages, he had a talent for this).

To a large extent the childhood of S.I. took place among the children of Trekhgorka workers. They got the same impressions. The earliest things S.I. remembers were related to the coronation of Nikolas II and the famous Khodynka located not far from Presnya. He was six when, like all the Presnya kids, he was looking through the fence at an infinite stream of waggons with dead and crippled bodies coming from Khodynka to Prokhorovsky hospital (about a thousand and a half people were killed and about the same amount badly injured). Discussions on this terrible and symbolic beginning of the reign of Nikolas II, his fault, the fault of city authorities was for a long time providing food for talk. Everyone knew, of course, that the dull-witted tsar did not even have enough sense to command a dirge but in the evening not attend the planned ball given by the French ambassador. Together with the empress he opened the ball with a quadrille. There exists, however, evidence that they were strongly distressed by what had happened, went together to hospitals where the injured were. Nikolas paid with his own money for burying the dead not in a mass grave but in coffins and gave substantial money to the families. Nevertheless, for the people Khodynka did remain a terrible symbol.

It is no wonder that already during his boyhood S.I. was firmly considering himself to be a democrat and liberal but in his notes he describes this ironically ("all this was light-weighted, not ripe"). Until the age of 15, however, under the influence of his mother he believed in Christ. The elder brother, Nikolai Ivanovich, declared himself an atheist much earlier.

Of course, for a scientist in the field of natural philosophy and in general for a person with such a mentality, atheism is simpler and more natural than religiousness. During our times we also see how few scientists are religious. This is different for people with an artistic and generically humanities-related intellect. A figurative metaphoric thinking, the religion, in which the truth is provided in the form of parables, artistic circumlocution, is closer to them. There is nothing surprising, then, in the fact that academician I.P. Pavlov who was a son of a priest was a convinced atheist. Already at the time of his graduation from the military medical academy he warned his religious fiancee about this. In the recently published memoirs of his student and colleague professor, M.K. Petrova, she describes his statements on this issue. At the very end he agreed that religion is perhaps necessary for weak people.

One should say that atheism was in general a common phenomenon within Russian intelligentsia. Around the turn of the century ideological swings were a generic phenomenon. Along with passions for theosophy, disputes of religious philosophers, Tolstoyism and an infinite number of other deviations from the official church there was a growing number of simple intelligents-atheists who thought that a man works out the moral norms himself. Such a highly moral person as Chekhov, a son of a shopkeeper, who sang in the church chorus in his youth, wrote in a letter to Dyagilev a year before his death: "I have long lost my belief and can only glance with surprise at each believing intelligent."

It is natural that from the outside, the life of the family and of Sergei Ivanovich himself during his youth looked quiet. It was shaken only by external events and by incessant internal spiritual labor and the ideological swings of Sergei Ivanovich. We shall return to this later. In the sketch of his autobiography S.I. makes some laconic comments on this: "The beginning of the XX century. Discussions at home... Some not very clear for a child but distinct subterranean revolutionary shocks, student meetings, assassination of Bogolepov, revolutionary dirges in Vagan'kovskoe graveyard. And, nevertheless, peal, fist fights on the ice of the Moskva river, pancake week feast in Presnya." This working class life was also still flowing around.

Along with this there was a life in commercial college from which Sergei Ivanovich graduated in 1909. In his notes S.I. gives astonishingly detailed individual characteristics of fellow students and of numerous teachers that often changed (demonstrating in particular his remarkable memory). Of special interest is a story on how "in senior forms there appeared a scientist-theologian A.I. Artobolevski. He was a clever and tactful man and had to teach during the least suitable times — after the revolution of 1905. There arose eternal discussions on the creation of the world, on Darwinism, on proofs of the existence of God. In my group I was the main theological opponent and energetically fought with theological constructions of Ivan Alekseevich ... All "fathers" taken together did not strengthen but at the same time did not weaken the religious beliefs of students. Internal evolution in this field went its own way independent from "fathers" and school God's Law."

These words are very important. I have already said when mentioning Chekhov that the atheistic intelligentsia did work out its moral code itself although some religiously established moral norms (in Russia predominantly the Christian ones) exerted some influence on this. Ideological swings at the borderline between XIX and XX centuries were in Russia extremely strong. S.I. Vavilov did not escape them as well. When he had graduated from the college and was taking the entrance exams for the university he wrote, assessing his development, that before the age of 15, i.e. before the revolution of 1905, he was a "dreamer, mystic, deep believer. Then I tried to become a poet, philosopher, world viewer ... I went through pessimism and optimism, joy and despair, "scientific religion"." He bought and studied numerous books on philosophy, among then a book of certain Il'in [Lenin] *Materialism and Empiriocritisism* — of course, not knowing the true name of the author. Following an example of his elder brother he organized a group of friends and fellow students. They gathered at home and discussed a "vast range of issues" in philosophy, literature, arts, politics. Only a few participants were however "at the appropriate level." "I had to carry it on myself," writes S.I. in his autobiographical notes. "I wrote essays on Tolstoy,

Gogol, Tuytchev, Mach, decadents, on suicides as a social phenomenon." Eventually the group broke apart.

The passionate personality of S.I. could not be satisfied with what the college gave him. It was already mentioned that he studied Latin and other languages himself. He read Mechnikov, *Foundations of Chemistry* by Mendeleev, Timiryazev, attended the meetings of the Society of Natural Philosophy in the Polytechnic Museum. And, along with this, there was a devotion to arts, a deep knowledge and understanding of it.

Everything around was however "boiling." By that time impressions from Khodynka and Bogolepov assassination were enriched by discussions of other terrorist acts, "there was some unrest." In 1904 the foolish tsar Nikolai went for a stupid, unnecessary and shameful reckless bloody gamble by starting a Russian–Japanese war. In the words of S.I. it caused in the society "an unspeakable concern. Sorrowful war without bright spots. Black shroud over Russia. It was extremely pitiful."

Then there followed the "Bloody Sunday" of 1905 (the tsar who sanctioned it[6] again did not even command dirges for hundreds killed). A gap grew between the people and the authorities: they were on the brink of war.

This stupid and merciless way of ruling the country with the, by that time far developed economy, social movements and spritual life could only lead to heavy consequences. Revolution broke out precisely in Presnya where its own "government," a Soviet of worker's deputies and a Revolutionary tribunal channelled the accumulated fury onto police and kasaks. S.I. writes that poor and even rich sympathized the rebels. It was quite natural that brothers Vavilov did also sympathize with the workers from the environment that had been the one of their childhood. They helped (S.I. writes: "actively") in erecting barricades, helped the wounded, took some home.

The rebellion was cruelly crushed in Presnya and other places where it caused response (e.g. along the Siberian railway). There began Stolypin terror. Still the monarchy understood a necessity of concessions. There appeared a constitution (although a "curtailed" one) and a Duma (although a consultative one). Elections to the first Duma became a broad political campaign. Political meetings were gathering, sometimes even in Vavilov's

---

[6] A day before the minister of internal affairs, Svyatopolk-Mirsky, went to the tsar to Tsarskoe Selo where Nikolas was spending the winter and reported on military preparations: 40,000 soldiers had been gathered, partly from Pskov and other nearby places, cannons had been prepared. The intention was to act in a hard way. Nikolas approved this.

house. The father considered himself to be a "left Octobrist." Although it is customarily said that the revolution of 1905 was defeated, all these transformations did essentially change the political atmosphere in the society and the country.

In his notes S.I. writes about himself: "As I can remember myself (from the age of 5, "the Khodynka") I always felt myself being a "left, "democrat", "for people"... However, my left views and democratism never turned to politics, its hard ways and even cruelty. Nowadays this is called "softness." This is the source of my organic distance from political parties. The revolution of 1905 frightened me. I rushed into philosophy, into arts."

The October revolution did sharply change the life of the family. The father understood what threatened himself and his capital and in 1918 left to go abroad. Sergei Ivanovich had already in 1914 graduated from the faculty of physics and mathematics of Moscow University, but refused to stay there "for preparation for professorship" and was therefore mobilized. For four years he was in the fighting army, was in German captivity but escaped. He had a perspective of scientific work opening in front of him (Nikolai Ivanovich had already been a professor and in 1916 had undertaken the first of his travels — to the East).[7]

All the family except the father stayed in Moscow. Apparently the loss of capital did not worry them. They lived the same life as everybody — in cold and hunger. His nephew S.I. Ipat'ev recalls queues for food rations and their redistribution within the family: "The bread in the form of black flat cake is taken from the sac by Sergei Ivanovhich who apparently plays the main role in this." A rapid development of science that began soon did inspire the brothers. They could not but cherish joyful hopes when, during the very first still hungry years, research institutes of Western type previously absent in Russia began appearing. Mainly in Petrograd: Radium or X-ray-radiological, optical, Institute of Physics and Technics, etc. A pre-war development of the country had already prepared a lot of young people for scientific work. The new authorities were clearly intending to develop science with all possible means. This became obvious when NEP got stronger. With many scientists this caused a loyal attitude towards authorities that in the literary circles was termed "fellow travelers" (this meant "although not allies but in any case fellow travellers"). Nikolai Ivanovich, for example, did

---

[7] We leave aside an important part of the life of S.I., his above-mentioned very serious devotion to arts or, as he called it, an "aestheticism." Because of this, before the war, he had twice traveled to Italy and had published two essays on architecture of the cities in Northern Italy.

think in the twenties that the collective farming system provided especially good conditions for selection work, the issue most important for him.

Of course Sharikovs, as well as many Soviet officials, especially of low and middle rank, did not quite understand the difference between scientists and intelligentsia in general that were materially prosperous in tsarist times and the middle class "bourgeois." With a feeling of satisfaction they treated the intelligentsia as lower class people and saw a restoration of social justice in their humiliation. For young scientists who finally got a possibility of doing science, however, this mere possibility was more important than the hardships and even horrors brought by Soviet power.

What did people like the Vavilov brothers expect from the future? It is sufficient to give an example. In the time of the golden age of NEP they persuaded their father to return to the Motherland and in 1927 he came to Leningrad but fell ill on the way and soon died (perhaps fortunately: The Stalin terror that was widening would hardly have left him untouched).

The brothers did apparently take a loss of the former material level of life easily. The only thing they needed from Soviet power was the same what Archimedes demanded from a Roman soldier: *Noli turbare itae circuls meos*! (Do not touch my draughts!)

A subsequent transformation of this power into a totalitarian Stalinist system could not remain unnoticed or misunderstood by S.I. He was too clever and had thought over too many things already at the time of the ideological swings of his youth, "hardness and even cruelty" was too foreign for him, the distancing from party membership so characteristic of him was too organic for him to remain an unthinking observer. In the thirties and forties he helped the victims of the "red wheel" as he could. He wrote letters to highest authorities in defence of the arrested scientists, even those he did not know personally, helped them financially. Even leaving aside the tragedy of the arrest of his beloved brother his emotional feelings were quite heavy. Those who knew him sufficiently closely clearly saw that his position could be understood as follows: In our country, like in other countries in different periods, there were good times and reasonable rulers and awful times with cruel tyrants. His duty as a scientist was to live through hard times and do everything possible for the saving and the development of science and culture in general, and to help others in doing the same.

S.I. behaved himself with restraint and fulfilled this duty of his with incredible energy. He worked in science himself and organized new scientific institutes, scientific commissions and councils; he was becoming one of the leading figures of the country's rapidly developing science.

At the same time he wrote during these years several articles on philosophy in which one finds standard phrases that were ritual for conservative Soviet philosophers. Reading them nowadays is not comfortable. Nevertheless if we discuss contents of these papers, 'Dialectics of Light Phenomena' (1934), 'V.I. Lenin and Physics' (1934), etc., we can state that these were not written "to please the bosses" but rather quite candidly. We have already mentioned that in his youth he was interested in philosophy, read a lot, bought many books on philosophy including *Materialism and Empiriocritisism* by Lenin. As for the uncomfortable ritual phrases — well, they were obligatory. Only with respect to what he wrote in this field already at the time of his presidency could one experience a feeling of deep regret. At that time he was obliged, like some others as well, to call Stalin a "coryphaeus of science" and this was a humiliation he accepted in order to have a possibility of doing the immense work for our science that he accomplished. He sacrificed himself to science and did it knowingly — like Galileo who on demand of inquisition did publicly, standing on his knees in the church, renounced the heliocentric views of Copernicus (but was not therefore staked and could write the second of his great books on mechanics that marked a beginning of the physics of new age).

When in 1945 Stalin unexpectedly offered Sergei Ivanovich to become a president of the Academy of Sciences (his beloved brother had died in jail two and a half years before) he reacted to this offer with horror. He knew that in this new position he would have to pronounce horrible ritual words, take part in criminal deeds following Stalin's orders (later it turned out that whole branches of science were demolished) but to say "no" to Stalin — nobody could dare to do this at that time, the consequences could be completely unpredictable. The agreement of S.I. was in no way a manifestation of softness. He knew, in addition, that if he did not become a president Stalin would appoint one of his favourites who would terminally kill our science. Nowadays we know that initially Stalin wanted to appoint Vyshinski, not even Lysenko, a president of the Academy. However, the vice-president of the Academy I.P. Bardin who practically replaced a sick, almost in marasmus, president Komarov and represented an opinion of several leading academicians, could make Stalin change his mind so that he agreed with their choice of S.I. as a candidate. Again, at this time tragically, there interlaced with one another his fate and the time in which he lived...

S.I. did compensate this humiliation by the gigantic in its scale and incredibly fruitful work on support and development of our science. What he achieved during the five years of his presidency amazes with the scope, thoughtfulness, success, amount of what was done. However, this demanded from him such physical efforts and moral sufferings that ended in his premature death. Look at this photo. It was made by L.V. Sukhov, who worked in FIAN, when S.I. was in his laboratory and did not know that he was being photographed. It is sufficient to compare it to the earlier ones in order to see: S.I. was approaching his death. Like many others I really think that Sergei Ivanovich sacrificed himself for our science and that we should gratefully bow at his deeds.

4.3 Vavilov and Vavilov's FIAN

I was working with Sergei Ivanovich Vavilov as a director for 15 years. Officially I became a staff member of the Physical Institute only in 1938 but already in 1935, having had become a Ph.D. student of I.E. Tamm in Moscow University, I had come to FIAN and, enchanted by its atmosphere of devotion to science, mutual benevolence going together with tactfully imposed quality standards which had looked so different from what at that time one had usually met in other places I had practically moved to the institute "in Miussy." Many years, however, passed after the death of Sergei Ivanovich before I understood something, in my opinion, important in his personality.

Thus I would like to begin my story with three "constituents" of this personality on which, in my opinion, either nothing or too little was written.

I. First, I think that Sergei Ivanovich felt himself and possibly identified himself with a link in an endless history of the world and, most importantly, national culture. A historical approach to perceiving culture and thinking it over is quite rare. It was in the highest degree a characteristic of Sergei Ivanovich. I dare to express an opinion that he perceived a history of culture as a unified and continuous aspiration of human spirit for knowledge and perfection, unified despite all deviations from this main line due to historical, national and social conditions, despite backward moves and modifications sometimes leading to perversion of culture. All this shows itself in particular when reading his books, articles and talks on the issues in the history of science that are to a large extent published in the third volume of his collected works.

In his youth he wrote essays on the culture of cities in Northern Italy created by great artists serving all-powerful albeit sometimes generous rulers. In his ripe years he translated Newton's *Optics* and published his remarkable biography elucidated by an understanding of the epoch in which a scientific genius could peacefully create in a favorable atmosphere of the stable university tradition. At the same time Newton, summoned by the king to reform the coinage, was not only brilliantly solving technical problems but had to participate in exposing the counterfeiters that inevitably ended on gallows. During all his life Vavilov propagandized a research by Lomonosov who had to beg for time and money for it and for a single well received ode got as a reward a sum that was more than three times larger than his professor salary.[8]

Therefore he could also understand those horrible and senseless things that happened in our country, in particular in the field of culture. He knew well that from a broad perspective many similar things had already happened.

Sergei Ivanovich saw however that with all the strange turns and difficulties of the fate of culture it constitutes a pride of humanity and could hardly tame his admiration when writing about it. He felt himself being a successor of the past who was deeply and personally responsible for the future.

Neither by origin nor by social group or conditions in which he had been brought up was he a descendant of Pushkin and Derzhavin, Newton or Euler. Nevertheless this grandchild of a serf was astonishingly in his place in the office of the President of the Academy of Sciences in the Neskuchny palace surrounded by the antique portraits of his predecessors-presidents and the founder of the Academy Peter the First. He was there by the right given by the true continuity of culture. The fact that it was him in particular who was there can to a certain extent be considered as an occasion. There was by us, however, not so many people that had so much deserved it and shared the deep aspiration of doing everything possible to ensure a dignified continuation of the history of national culture.

Everything we know about Sergei Ivanovich tells one thing: This aspiration was dominating his life and played the main role in it. For this he was ready to sacrifice everything. Vavilov did not live up to 60 and during the last decade of his life he had, as described in Section 4.1, to endure much. His feelings were hermetically locked within him but were tragically heavy. Work during the war and especially afterwards in the office of president, immense in its scope and psychological intensity, also had its effect.

---

[8] Cf. Kapitsa, P.L. (1977). *Experiment, Theory, Practice*. 3rd edition, Moscow: Nauka, pp. 255–272.

He performed it following his sense of duty and endured more than a man could endure. And he died, died simply because the physical and nervous capabilities of his body were exhausted.

Now I will try to substantiate the above statements by facts. In essence one needs to substantiate two statements. First, that Sergei Ivanovich understood the contemporary, and in fact any other, period of development of science and culture in general as primarily being a part of a unique process of their development. Second, that he put his duty as one of the successors and developers of this culture who due to the circumstances was in a special position higher than the so-called personal interests (here I write "so-called" because fulfillment of his duty was for him the "personal interest").

To substantiate the first statement one could first recall his above-mentioned personal contribution to the history of culture. The latin of Newtonian *Optics* connected him not only with the university and monastery culture of middle-aged, renaissance and post-renaissance Europe but also with ancient Rome. He knew *De Rerum Naturae* by Lucretius Carus almost by heart. Entire physics, from atomistic of ancient Greece to theory of relativity by Einstein was open to his eyes.

A history of national science constituted a special domain of his interest. In his director's office in FIAN there stood, and are still standing, shelves with glass doors on which a caring hand had put the first specimen of electrotype invented by Jacoby that he had produced himself, a gilded wheat ear and other similar wonders. These things were a century and a half old. Near them there were miniature devices used by P.N. Lebedev in his unique experiments at the beginning of the XX century. On the wall there was a large portrait of Lomonosov. In this environment there took place discussions on the nature of radiation of relativistic electrons, on details of experiments and theory of Vavilov–Cherenkov effect. One discussed the results of research on radio geodesy, radio range finding, soundness of principles of creating new particle accelerators. In relation with gigantic dimensions (dozens and hundreds of meters) and weight (thousands of tons) of these "devices" that were soon built this had been mentioned. In the same neighborhood of Lebedev devices the first words on thermonuclear fission were pronounced. There was nothing unnatural in this coexistence of physics from different centuries. This was a visible continuity of science.

It would however be completely insufficient to talk only of the historical continuity of physics perceived by Vavilov, of the "vertical" unity of science. A "horizontal" one of different branches of culture within the given epoch was also clear to him. It can be seen, for example, from the fact that he wrote not only about the architecture and art of Italy but also about Galileo. Books on Leonardo da Vinci occupied, as his son tells, more than "one cubic meter" in his home library. This is however not all. An understanding of the unity of natural sciences and humanities showed itself with special clarity when Sergei Ivanovich Vavilov took the responsibilities of the president of the academy and had to care about all sciences.

The most important thing in this gigantic activity was that a specialist in the history of literature or art, a historian, a philologist, a philosopher could come to the president to talk about their problems with him as an expert, discuss them and find support.

Vavilov knew how fundamental is the role of publishing activity in the development of science. He surely knew the bitter words of Lomonosov from the annual report (of 1756) in which the situation with some research that had not been completed is in particular explained by the fact that "long publishing process kills a desire to comment" (as quoted by Pushkin in *A Travel from Moscow to Petersburg*).

Therefore an expansion of the publishing activity in the academy was a subject of his permanent concern. Should one tell in addition that it was not by chance that it was this encyclopaedist who supervised an edition of the *Great Soviet Encyclopedia*? As always he worked very seriously on this as well, worked as an editor, wrote articles and discussed those written by others.

All the facets of the history of culture and its relation to contemporary times were inseparably connected for him.

As almost any big scientist, Sergei Ivanovich, cherished applications of science to practical needs (Einstein had many patents on inventions; Newton reformed the coinage technique so efficiently that even a hundred years later the English government did not allow to share the production secrets with the French delegation; the examples are numerous). As a scientific leader of the state optical institute in Leningrad, Vavilov did a lot for the optical industry including the defense one. From his laboratory in FIAN there came out, in active collaboration with the Moscow factory 'Electrozavod,' fluorescent lamps that were broadly used in industry and everyday life and helped to achieve a gigantic economy of electric energy.

Such applied activity was organically characteristic of Vavilov. It was also tradition for Vavilov's FIAN as a whole. It was not accidental that during the war Sergei Ivanovich was a representative of the State Defence Committee. When he was in evacuation with SOI (State Optical Institute in Leningrad, he was its scientific leader along with having been a director of FIAN) in Yoshkar-Ola he did not stop such research and oriented the whole institute towards it.

And, suddenly, it was him who in the middle of the war wrote a book on Newton — the book in which there was nothing momentary, nothing related to the times in which it was written. Based on original documents it presents an original point of view on various aspects of Newton's life and his main works. Vavilov analyzes even his theological works. Are there many physicists knowing what does 'Arian' mean? Sergei Ivanovich points out that Newton was Arian, i.e. a follower of Arius who, in the dispute with Athanasius at the Council of Nicaea, denied the divine nature of Christ. Some could exclaim: "What an absurdity!" One could still understand that this was interesting for Vavilov but why should this have been published, especially during wartime?

An explanation could apparently lie in the fact that Sergei Ivanovich believed that it was not sufficient to just save the culture from physical extermination at the moment of mortal danger for the country. One had to save and hand over to future humanity all the diversity of culture that had been developing for millenniums, through labors and trials, achievements and failures. It would be naive to consider Newton's interest in theology as an eccentricity of a genius. It is necessary to grasp the spirit of this bygone time when science was not so separated from religion as nowadays and was not so alien towards religious *Weltanschauung*. It was important for science of the Newtonian epoch. If we want to comprehend Newton as a cultural phenomenon we have to understand this particular side as well. The culture, saved from fascism and handed over to future generations, should not have been simplified and less rich. Otherwise the humanity would have been pushed backwards even when technically enriched with radars, missiles and atomic energy necessary for the victory.

All said could probably justify the first statement on *how* Sergei Ivanovich comprehended the unity of past, modern and future culture.

Let us now turn to the second statement on the fact that Vavilov considered his duty with respect to the developing world and, especially, national culture being above the so-called private interests. In essence we are talking about his altruism in the broadest sense of this word. It was probably evident for anybody who could directly observe the deeds and behavior of Sergei Ivanovich but, nevertheless, there follow some facts. The main evidence is given, in first place, by his incredibly immense activity in the office of president.

His astonishing organizational talent (we shall tell more on this below) did certainly play an essential role in this. With all his capacity for work it was clear that he did not think about himself at all. He could easily get rid of many of these "loads," nobody would have reproached him. He also did not need this for himself, for his fame. He was already a president, his portrait would after his death hang among those of the other chosen. He had more than enough fame. Could it be that he was overwhelmed by the thirst for power, for domination? Such a hypothesis is of course absurd even from the point of view of logic: In order to dominate one does not need to honestly do everything oneself, it is sufficient to sign documents prepared by others. Even without this reasoning everyone remembering the real character of Vavilov and witnessing his deeds understands the absurdity of this assumption.

Already at the beginning of the thirties, just after he had become an academician and practically a co-director of the Institute for Physics and Mathematics of the Academy of Sciences in Leningrad, he understood, with a clarity rare for those times, the exceptional importance of research on the physics of atomic nucleus for future science and technics. Everybody in the institute expected the new director to "align" everybody onto research in his field of expertise, optics. What happened was however completely different. Although Sergei Ivanovich himself had never been engaged in the nuclear research and was probably not going to be personally involved in it in future he directed almost all the staff members and Ph.D. students, who were at that time approximately of age 25, towards research in nuclear physics.

When the physics department of the Institute for Physics and Mathematics became an independent FIAN and in 1934 moved, together with the academy, to Moscow, Vavilov, who headed it, began forming an essentially new institute. He organized many completely new laboratories and departments and invited the most prominent and already very well known Moscow physicists to head them. Leonid Isaakovich Mandelstam and Nikolai Dmitrievich Papaleksi (who had moved from Leningrad) headed the laboratory of oscillations (in essence a laboratory of radiophysics); Igor Evgen'evich Tamm headed a theoretical department; Grigori Samuilovich Landsberg, an optical laboratory; Sergei Nikolaevich Rzhevkin and Nikolai Nikolaevich Andreev from Leningrad, an acoustic one.

For all these outstanding scientists the exceptionally favorable working conditions at FIAN were created. They did not have anything similar working, for example, at Moscow University. Of particular great importance was an atmosphere of mutual respect, benevolence and care. Sergei Ivanovich left for himself only a small laboratory of luminescence. His personal scientific research was mainly concentrated in the state optical institute in Leningrad where (as has been already mentioned) he had become a scientific leader (even transferring there some research from his laboratory in FIAN).

He did however have to temporarily become a head of the laboratory of atomic nucleus — simply because in Moscow there was no real nuclear expert and there was no possibility of appointing anybody else. It was also probably necessary to encourage the young researchers who had been pointed by him into this direction and pay special attention to them. One could not allow that the researchers in this laboratory would have felt themselves abandoned. In this laboratory there continued the research of P.A. Cherenkov on studying the radiation he had discovered with Vavilov, here there worked I.M. Frank who together with I.E. Tamm had several years after this discovery given an explanation and a theory of it. Already because of this a participation of Sergei Ivanovich in the work of this laboratory was not purely administrative.[9] As soon as one managed to persuade Dmitry Vladimirovich Skobeltsyn to move to Moscow from Leningrad he became a head of this laboratory. Did this behavior look like an aspiration to dominate, capture more, elevate oneself?

It is said that academician Aleksei Nikolaevich Krylov whose attitude towards Sergei Ivanovich was very good (in any case it was him and L.I. Mandelstam who suggested his candidacy for the Academy) but liked to make jokes once said: "Sergei Ivanovich is a remarkable man. He organized an institute and was not afraid of inviting there physicists who were stronger than himself." It is hardly reasonable to give such relative grades of strength and weakness to scientists of such a scale. The fact however is that five years before that G.S. Landsberg and L.I. Mandelstam had made a world class discovery (combinational scattering of light), one of the few of this scale in the history of our physics, and were at the peak of their creative activity. I.E. Tamm did already have a solid international reputation backed by first class papers on quantum theory of radiation processes and

---

[9]In this respect see the article of I.M. Frank in the journal *Uspekhi Fizicheskih Nauk*, (v. 91, p. 11, 1967), where he describes in details the beginning of research on nuclear physics in FIAN, the obstacles and also the skepticism from all sides that Sergei Ivanovich had to overcome after deciding to entrust this new complicated project to very young people.

solid state (the "Tamm levels" had been already discovered) and theory of beta-forces.

Sergei Ivanovich still had to wait for the recognition of his highest and most remarkable achievement which at that point in time was still in its infancy — the discovery of Vavilov–Cherenkov effect that made him famous and was marked by a Nobel Prize already after his death. Nevertheless the scale of his talent and personality, the level and significance of his scientific works were already evident for anybody who was in contact with him.

Therefore in the atmosphere of FIAN nobody thought that the presence of Mandelstam and other outstanding physicists could even to a slightest extent diminish the authority of Sergei Ivanovich, deep respect to him as a physicist, a director, a person. It was equally impossible to imagine that the director Vavilov would have created obstacles for them, applied "pressure" when insisting on his own point of view which was different from theirs. It is impossible to recall even a single fact of their conflict, moreover, a single shadow of misunderstanding between them.

All this behavior of the director (appearing, alas, so astonishing if one thinks about the practice of many modern institutes) showed the same thing: When creating his institute and creating it in *this fashion* Sergei Ivanovich cared, above all and always, on what would be more useful for the science. Shallow passions were completely foreign for this great personality. Sergei Ivanovich fulfilled his above-mentioned internal duty that directed his service, that of heir to the culture of the past, responsible before the culture of future.

It is of interest to summarize the main results of the research on nuclear and high energy physics obtained still during the life of Sergei Ivanovich in the laboratory that he organized "on an empty spot" from "green" youngsters and the theoretical department headed by I.E. Tamm that worked in close contact with (and was for some time organizationally a part of) this laboratory:

(1) The Vavilov–Cherenkov effect was discovered and thoroughly investigated, both experimentally and theoretically;
(2) Completely new principles of acceleration of electrons and protons that allowed to surpass the relativistic barrier for the achievable energy were discovered and theoretically studied (see below). These principles underlay the future accelerators for the whole world. First synchrotrons were built in FIAN and a construction (in Dubna) of phasotron and synchrophasotron did begin (Veksler);

(3) Basic principles of thermonuclear fission were discovered (by Sakharov and Ginzburg for the uncontrollable, by Tamm for the controllable ones).

At that time, in the middle of the thirties, Sergei Ivanovich did however find the scale of the research on nuclear physics still insufficient. He found it necessary to concentrate in Moscow, in the Academy of Sciences, more powerful scientific forces. In particular there was a discussion, with his participation, on the possibility for several nuclear experts from Leningrad Institute of Physics and Technics which belonged to Narkomtyazhprom (people's comissariat for heavy industry) to move to Moscow because here, in the Academy of Sciences, one could have created more favorable conditions for research on nuclear physics (one should not forget that at that time one considered this topic as not having any applied significance and even not promising it in the foreseeable future).[10]

One told that some understood this as an aspiration of Sergei Ivanovich "to take everything to himself" and practically destroy the Leningrad nuclear school. The above-given description of how the Moscow FIAN was organized is probably showing the absurdity of such understanding of what was for Sergei Ivanovich a driving force in his activity, which scale, altruism and honesty were characteristic of him. At the same time the above-listed results in nuclear physics show that this organizer of science could only create, not destroy.

II. The second feature of Sergei Ivanovich which I would like to draw the reader's attention to is the scale of his thought and organizational activity.

---

[10] I recall a meeting of the FIAN scientific council specially devoted to the plans of laboratories having applied character. This was perhaps in 1938. The laboratories reported on important studies directly related to their fundamental research — spectral analysis of metals, radiogeodesy, luminescent lamps, etc. When it was the turn of the laboratory of atomic nucleus its representative began babbling something on the possibility of measuring the width of reservoir walls from scattering of gamma rays from the radioactive source that was available in the institute placed inside the reservoir. One of the council members, a well-known physicist B.M. Vul, could not contain himself and said: "Usage of physics in people's economy is a serious business and we are indeed doing many things that are really important. However, one should not make a game out of it. Physics of atomic nucleus is a very important field of fundamental scientific research and should be developed but it does not have practical significance and nobody knows whether it will ever have it. One should openly tell this and not demand applied work from the laboratory." Everybody agreed with this and the discussion continued. Nowadays it is difficult to believe in this because it took place only four years before the first chain reaction of uranium was realized in the laboratory conditions.

The scale of understanding of historic phenomena is of course clear already from how he perceived the history of science and culture in general and the scale of his organizational talent — from what he achieved as president of the Academy of Sciences. Nevertheless I would like to support this by particular facts I myself was a witness of.

Sergei Ivanovich became head of the physics section of the Institute of Physics and Mathematics in 1932. It included one or two dozen staff and Ph.D. students. In the Moscow FIAN organized by Vavilov in 1934 it had, by the end of the thirties, eight to ten times more people and the area of the institute building "in Miussy" was also eight to ten times bigger than that of several small rooms in Leningrad. However, for the central physics institute in Moscow this was not sufficient. Vavilov understood that the collective of FIAN which at that time already included twenty doctors of science, among them six or seven academicians and corresponding members, was a ready nucleus for a bigger center that the academy needed.

Therefore already before the war, i.e. five to six years after the move to Moscow, a project of a new, much bigger building was ready and a piece of land for it amidst the vast field in two kilometers from Kaluzhskaya zastava (now Gagarin Square) was chosen. There had been already built a small two-story building of the Acoustic laboratory (which in area and volume was roughly half of the Miussky FIAN but is now completely lost among the big new buildings). A new, bigger building which is now the main one, was built right after the war. Sergei Ivanovich, however, did not live to enjoy this. The institute moved in the year of Vavilov's death. Only this building exceeded the area of old FIAN again at a factor of eight to ten and the number of staff did rapidly increase in the same proportion.

Therefore in 18 years of the directorship of Sergei Ivanovich the institute grew in two gigantic steps both in the number of staff members and the area at a factor of 50–100.

One cannot exclude that Sergei Ivanovich himself, who began his scientific work in the small laboratory of P.N. Lebedev, would have felt himself better not in the giant institute but in a charming building of the type of the pre-war FIAN in Miusskaya Square. He saw however the tendencies in the development of science, a growth of the influence of physics on society. It could be that already before the war he foresaw the future gigantic development of science and, correspondingly, a future development of FIAN.

For example, I can judge this from the following fact. Once, before the war, one suggested the laboratories to express their opinion on the main building that at that time was still in the project stage. A draft laid on the giant desk in the director's office, everybody stood around. Sergei Ivanovich suddenly said: "First of all one should surround the territory with a fence." I was startled and naively exclaimed: "What do we need such immense territory for?" Sergei Ivanovich quietly answered: "We shall make use of it, *all* the area will be used."

And indeed when Dmitry Vladimirovich Skobel'tsyn replaced Sergei Ivanovich as a director he energetically continued the development of the institute. The area was rapidly filled with new bigger and bigger tall buildings. Nowadays the territory is filled so densely that the main building bears its name partly due to tradition and partly because of hosting the direction and all-institute facilities — conference hall, library, etc. It is already not the biggest one. For a natural development of the institute one had to build, already a long time ago, the new buildings outside of Moscow (in Troitsk).

Looking back one could guess that Sergei Ivanovich foresaw this from the very beginning. From the list of scientists invited to the institute back in 1934 one could judge that each laboratory did have a chance to reach at least the size of a small institute. This did indeed happen.

Another example. At the end of the thirties it was as clear as it is nowadays that nuclear physics needed particle accelerators at high energies. In 1934 a Lawrence cyclotron performed a revolution. With this cyclotron it was however principally impossible to reach relativistic energies and even in the nonrelativistic domain a growing energy of the particles demanded an increase of a solid magnet to a big size. By us a cyclotron with pole's diameters equalling 1 m — approximately the same as in the American ones was built (in Radium Institute in Leningrad) but this turned out to be such a complex problem that only in 1940, thanks to a gigantic work of the then young I.V. Kurchatov and his comrades, one could finally start working on it.

Sergei Ivanovich understood that serious nuclear physics was unimaginable without a big accelerator. In the process of its construction he could seemingly rely only on his inexperienced collective. Later, in 1940, one made a bold decision to create a "cyclotron brigade" with a task of studying a problem of constructing a cyclotron with the pole's diameter of several meters and starting its design. Up to now this decision looks almost unbelievable to me. The "cyclotron brigade" included the same "green youngsters": Veksler, Vernov, Groshev, Cherenkov and myself. An analysis of the problem was intensively progressing, there were hot discussions on possible variants resulting only in convincing ourselves again and again that the problem was incredibly difficult. At the same time there began a construction of model cyclotron for accelerating electrons to be used for carrying out experiments on checking various ideas.

All this did however abruptly change when in February 1944, V.I. Veksler who, irrespective of the current topic of his interests, had during all these years been contemplating on the problem of acceleration did literally cut this Gordian knot: He discovered that it was possible to *jump over* the relativistic barrier.[11] The possibility of constructing accelerators of a completely new class discovered by him paved a new way for the accelerator technique in the whole world.

One can say that the remarkable Veksler solution of the problem was an unexpected result of the activity that the "cyclotron brigade" organized by Sergei Ivanovich had begun before the war in order to solve a gigantic problem. Instead of a plain solution a brilliant shortcut was found and the whole activity took a scale unimaginable ten years ago at times of the organization of the "cyclotron brigade." I would like to stress once again what I began this story with: A scale of the decision of Sergei Ivanovich on constructing a giant cyclotron.[12]

When the Great Patriotic War broke out the first days were characterized by a mixed atmosphere of deep worry and hope that overwhelmed everybody, in the atmosphere of intensive activity on mobilizing the forces, in comprehension of disaster that had fallen upon the country. Along with this there was an atmosphere of facile optimism: Before the war in the mass consciousness there had been planted a conviction that in the future war "we shall destroy the enemy on its territory," that "there is no question of our victory but the task is to minimize the losses," etc.

Improbable as it seems nowadays after all we have learned and endured there was quite a few people who did not comprehend the scale and importance of the events. On one of those days we were preparing an institute's wall newspaper for which Sergei Ivanovich wrote an article. I remember very well one phrase from it: "Our Motherland is threatened by a greatest danger since Mamaj invasion." Then he told about a necessity of maximal intensification of efforts, on willingness to bring sacrifices for saving the

---

[11] S.N. Vernov recalled (in a private discussion with myself) that at the meetings of the "cyclotron brigade" with Vavilov, to whom the incredible complexity of the problem was described, Sergei Ivanovich used to reiterate: "It is impossible that one could not jump over the relativistic limit." Myself, I do not remember this.

[12] Here it is appropriate to recall how, in the words of I.M. Frank, Vavilov said about discoveries that it is impossible to plan a discovery, that it is always unexpected but arises from a diligently and cleverly undertaken rationally planned research. It was precisely this style that characterized the best works at Vavilov's FIAN.

House in Presnya in Moscow in which Vavilov's lived.

Motherland, on everybody's duty, on tasks of working for defense, etc. The general tone was stern but not at all panicked, it was rather inspiring and mobilizing. According to the rules existing at that time the wall newspaper should have been checked by a secretary of the institute's party organization before it could be hanged. At that time the secretary was Veksler. After having read the article by Vavilov he became indignant: "What Mamaj! What is this panic! This article should be sent directly to NKVD!" This was of course not done but the article was removed from the wall newspaper. Vavilov, on the contrary, understood everything from the very first moment.

We left Moscow one month after the beginning of the war and because of this our work in the new place could start without delay. For example, very soon the experts in acoustics constructed an acoustic sweep-net for blowing up German floating mines which at the beginning of the war brought heavy losses to our fleet. Some staff members of the acoustic laboratory were spending much time at the front. The optical laboratory continued to work on the methods of spectral analysis of metals aimed at detecting new elements. This was of exceptional importance, first of all, for the express sorting of metal from damaged home and foreign weaponry so that, for example, valuable quality steel would not sent to general remelting but directly reused. One could organize a production of corresponding devices — steeloscopes. People from plants in the Urals and other places as well as those from the front were coming over for devices and instructions. I was myself a witness of how a representative of a Stalingrad plant came over either in late spring or in summer of 1942 when Hitler's troops were already marching towards the city. All this was possible because Vavilov, who had understood the scale of the future struggle, had insisted on a well thought over thorough evacuation of the institute's equipment.

III. Finally, on the personal features of Sergei Ivanovich. Other memoirs tell much on the style of his behavior, his friendliness, attention to everyone he was working with, readiness to help and what they say is good. I can only witness once again that this is not a memorial gloss or exaggeration but the truth. It has been justly and adequately written on his incredible capability for work and readiness to accept new loads however difficult for him this was.

I would however like to add something. First of all, try to understand how kindness and apparently time-consuming attention to people could go along with an astonishingly productive activity demanding exceptionally pure work-related qualities. One would see from what follows that these "kind" features of the character of Sergei Ivanovich did not hinder the work but rather helped it. Sergei Ivanovich did not just work a lot, he successfully completed many of his undertakings. How did he manage to do this?

The simplest answer to this is that he possessed the talent of an organizer and, in the first place, of that of his own work. One could recall an American aphorism: "If you have a problem go for help to a busy person, one who is not busy would never find time." This went along with a particular feature of the behavior of Sergei Ivanovich that was stressed by many — absence of rush. He seemed to have never hurried and never been late. He was not hurrying his interlocutor but, as well marked by G.P. Faerman "if you turn back when leaving his office while turning the doorknob you would see that Sergei Ivanovich is already writing something. His ability to switch... was remarkable."

I recall how in the beginning of 1944, soon after the re-evacuation to Moscow, he went to the laboratory of atomic nucleus, called I.M. Frank, L.V. Groshev and myself and said in his usual slow quiet mellow bass: "Comrades, look, we have to start working on the nuclear problem. This project is very significant and important but needs more physicists working on it. We cannot stay aside. You should talk to Kurchatov, spend some time there, examine the situation and choose a problem to work on." I think this was everything that was said. I recall that we were not even

Sergei and Nikolai Vavilov with their mother A.M. Vavilov (December 25, 1916), old style. The photo made during S.I. Vavilov's visit to Moscow from the front.

sitting, the conversation took place when we were standing at a small table in front of the window. A feeling of an atmosphere of quiet discussion of an important problem did however form right away. We were, of course, to a certain extent prepared for this by the general atmosphere of that time.

After that Sergei Ivanovich made a joke: "Earlier there were two methods of establishing truth — deduction and induction. Nowadays we have three — deduction, induction and information." After that he left. It is funny that at that time we did not understand his joke on information. This was for sure a clear hint on getting information through espionage. I understood that such information did exist only later when after one of my reports on neutron multiplication in uranium-carbon medium made at one of the I.V. Kurchatov special seminars, Igor Vasil'evich gave me this advice: "You assume the diameter of a uranium rod is equal to this. Take half a centimeter more."

It is worth paying attention to one detail: When talking to us Vavilov did not say that the work would be interesting or promised some profit for us personally or to FIAN. There was only one argument — *this is necessary*. Several years after Sergei Ivanovich gave the same argument when he summoned me to his office and offered me a position on the editorial board of the *Journal of Experimental and Theoretical Physics* which he headed. He said that he understood how burdensome this additional work was but motivated its necessity by just saying: "This is what is needed. Now, after Tamm has left "to the object", there is only one theoretician left in the editorial board — Ya.P. Terletski. You understand how difficult for us it would be to deal with him". A gifted physicist, Terletski had by that time marked himself with "ideological" attacks. In particular, in an article in *Issues in Philosophy*, he furiously attacked the famous course by Landau and Lifshitz for the idealism allegedly filling the course and in general fought for "ideological purity." We know now that in 1945–1950 he was head of Department C of the Ministry of Interior and, on a spy mission, visited Bohr in Copenhagen.

Sergei Ivanovich apparently did not understand that I perceived his offer as a special honor. These words, said by somebody who took an unprecedented load upon his shoulders only because "this was needed," sounded like an irrefutably convincing argument.

A second remarkable purely human feature of his behavior that supported the business-like manner rather than hindered it was trust with which Sergei Ivanovich endowed his colleagues (and other people he met) who, in turn, felt the same towards him. One could say that here, if one rephrases a well-known Latin proverb, there acted a principle: "I trust for you to trust me."

Due to this support of his colleagues Vavilov could do, as a scientist and a director of FIAN, things in which he was irreplaceable. Here it is worth mentioning one of his important features — he believed that a "prophet in his Motherland" *could exist*.

One editor of a provincial newspaper who always rejected verses brought by the authors was caught in a preset trap — he rejected among other verses brought to him those by Blok. Later he justified this saying "I could not expect new Blok entering my office." Sergei Ivanovich was mocking a hunt after discoveries but was always ready that his colleague would bring him something new and valuable. Unlike that editor he was of course capable of distinguishing discovery from nonsense and this was no less important than him having had his own experience that a big discovery *could* be made.

His Ph.D. student P.A. Cherenkov, who was working on a topic, suggested to him by Vavilov, did accidentally find a weak luminescence of *pure* liquid under the influence of γ-rays from radioactive source — so weak that to notice and study it one had to stay (to adopt the vision) for two–three hours in the darkness and use a special protocol suggested by Vavilov. This luminescence *hindered* his research which was studying the luminescence of a substance dissolved in the liquid. Cherenkov was terribly depressed and told his comrades: "This is it with my thesis." Sergei Ivanovich however did not neglect the unexpectedly seen luminescence but began studying it together with Cherenkov and, in particular, did some measurements himself.

He convinced himself that the observations were reliable, various measurements thought of and realized together with Cherenkov were accurate enough and, after deep consideration, he came to the conclusion that this was an unusual completely new kind of radiation. To arrive at such a conclusion one needed not only a clear understanding of the laws of radiation of light, not only a trust towards experimental data obtained, but also a great scientific courage. This courage, based on objectivity and conviction, helped to withstand a volley of mockery ("in FIAN they study ghosts in the darkness").

An atmosphere of expectation of discovery going along with a kind but uncompromising critique was characteristic of FIAN. All the above-described features constituted elements of the system, of the spirit created by Sergei Ivanovich and his closest colleagues at FIAN. The fruits of this system are evident. In the above-described examples from the life of the institute I mentioned some names but how many equally brilliant names were not mentioned! Many dozens of outstanding physicists well known both at home and far abroad did form as mature scientists in Vavilov's time. There is no doubt that each one of them gratefully remembers Vavilov, as an organizer of FIAN.

All that has been said illustrates the remarkably effective scientific and organizational activity of Sergei Ivanovich and his excellent relations with his colleagues in the institute. This might result in a picture of a resplendent life of a scientist, a multi-talented person, whose outstanding personal qualities could be realized without obstacles in full sincerity and of an equally resplendent life of the institute he headed.[13]

---

[13] Here I am not describing the tragic side of the life of Sergei Ivanovich after 1940, the year of the arrest of his brother Nikolai Ivanovich, which in essence led to the premature death of Sergei Ivanovich himself. This is described in Section 4.1 'Nine Scars on the Heart' while here I am writing about the institute, about FIAN.

This impression could have however arisen only by myself, a young Ph.D. student who joined the institute after a heavy atmosphere of student's milieu of that time (when the attitude towards students from the intelligentsia was extremely unfriendly) and, as told in the first lines of this essay, enchanted by the remarkable scientific atmosphere of FIAN and relations between the staff members.

In reality no such paradise could freely exist at that terrible time of the "large purge" of the second half of the thirties. The institute was but an island of decency in the sea of evil and horror but waves of this sea did sometimes sweep over the island as well. It could not exist independently from what was going on in the country in which fear reigned above everything.

When in the twenties and at the beginning of the thirties Sergei Ivanovich wrote articles on, for example, Lenin and his role in philosophy, we can have no doubt, remembering his interest towards Marxist philosophy in his youth and also that his brother, a biologist-academician Nikolai Ivanovich, was after the revolution also captured by an enthusiasm of creating a national world scale science and that both of them (at least during those years) were conscious "fellow travelers" of the authorities (a terminology that at that time was applied to a certain group of writers), that Sergei Ivanovich wrote these articles candidly and seriously. This was not camouflage or mimicry. After all, it was not accidental that the brothers persuaded their father to return to the country from emigration in 1928.

Time was however going on and there began a horrible period of collectivization. It was already impossible to shut one's eyes on an ignorant ideological pressure upon science, on all the horror of Stalinism. Could the brothers Vavilov fail to see it and keep their beliefs? Many decades ago I once discussed this with my elder friend, the cleverest Solomon Mendelevich Raisky. A student and a close collaborator of G.S. Landsberg for many years, a member of the vast clan around the family of L.I. Mandelstam to whom he was close as well, he understood people well and explained to me: "Sergei Ivanovich is a deeply feeling Russian patriot. He knows that in the long history of Russia there were good and evil tsars, quiet periods of country's grandeur and those of chaos and devastation. Russia withstood this. One should do everything possible for Russia to also withstand the evil now and to come to a new bright period — at any personal price."

I would not have recalled this conversation were it not for an appearance of the publication of an interview given in 1933 (long before his arrest in 1940) to a Paris newspaper (in the Russian translation made in NKVD) that was found in his investigatory file. It was Yuri Nikolaevich Vavilov who drew my attention to it. One part of it is remarkable. Here it is.

A reporter's question: "In 1916, in tsar times, were you already an employee of the emperor's government?" Answer: "Why this interest? In Europe one always speaks about government. In Russia, even in tsar time, we always spoke of state. In 1916 I was already in the state service. This is right. I was an assistant professor. I remained in the service of the state, the Russian state, the state of my Motherland. This is quite natural." (see *Vestnik Rossijskoj Akademii Nauk*, N11, 1997).

When taking into account that the brothers were very close to each other, were in close contact and when living in Leningrad at the same time spoke with each other every day, in the worst case over phone — does it not confirm the above-cited words of S.M. Raisky?

Let us however return to FIAN. Sergei Ivanovich organized the institute's administration in a very particular way. All the above-mentioned outstanding scientists were not, like himself, members of the party. When organizing the Moscow FIAN Vavilov offered a position of deputy director on scientific work to Boris Mikhailovich Gessen who at the same time remained a dean of the physics faculty of the university and a director of the research institute at this faculty. Gessen was a close friend of I.E. Tamm from their gymnasium years in Elisavetgrad. However, while the

S.I. Vavilov in his office in FIAN with the instruments of P.N. Lebedev, 1947.

social-democrat Tamm did after the October revolution break his political activity and fully devoted himself to science (although keeping the socialist ideals of his youth, see the corresponding essays in this book), Gessen became a convinced communist, worked in party structures, graduated from the institute of red professorship that prepared party activists educated in Marxism, became a professor of MSU on philosophy of natural sciences (I attended his lectures, these were the ones of a very cultured and educated person) and a corresponding member of USSR AS, Gessen had infinite respect towards Mandelstam and stressed it by his behavior in the university and was a friend not only of Tamm but also of Landsberg. It seemed obvious that by this clever action Vavilov strengthened a "good" party leadership in FIAN.

Besides that in FIAN there formed a group of younger party members. Vavilov relied upon them in organizational matters. These were B.M. Vul and D.I. Mash, volunteers of the civil war. There was a passionately convinced party member M.A. Divil'kovski, a son of a close collaborator of Lenin, who had grown up in Switzerland. He was 17 when he, as a secretary, accompanied V.V. Vorovski at the Lausanne conference and was wounded when Vorovski was killed. There was also Vladimir Iosifovich Veksler who had had, at his previous place of work, certain problems with his Trotskyist past. Because of this he behaved himself strictly "in a party way." There was, finally, a sympathetic Mikhail Ivanovich Filippov, in appearance a simple country lad.

What unified them besides the party membership was also that they really loved science, were in varying degree talented, really worked in the laboratories and (with en exception of events that will be described below) seemed to have deep respect towards big scientists gathered in this institute. They occupied, in succession, important positions of a secretary of party organization and deputy director of general issues (except maybe for the general questions of scientific activities).

It happened so that right after the creation of the institute in Moscow, on December 1, 1934, Kirov was assassinated. Shortly before this, in the same year, he got in the CC elections at the 17th party congress more votes than Stalin. A group of influential party functionaries "from the second row" (secretaries of the biggest regional committees that had shown themselves during the civil war) offered Kirov to become a secretary general but he refused. It was clear what should have followed.

Stalin demonstrated his grief in unprecedented forms (for example he walked with funerary procession from October railway station at which the coffin with Kirov's body arrived to the Red Square) and the country was swept by an equally unprecedented terror. Almost every day the newspapers published the lists (with many dozens of names) of "conspirators" executed in one or another part of the country.

In August 1936 Gessen was arrested and soon executed. As usual, the reason was not given. It was also usual as well that in the university there began meetings searching for the facts of sabotage of an "enemy of the people," sometimes ridiculous ones. For example, one accused Gessen of compilation of a sabotage university program in physics. The people that had been more or less close to him or simply had close working contact with Gessen were condemned for "a loss of watchfulness," one demanded them to look for new facts of "sabotage," to repent for a loss of vigilance, to condemn Gessen and demand a heavy sentence for him. Such was a reign of fear that only very few could withstand it.

Of course, at the meeting of the physics faculty one insisted, first of all, that everything from the above list was done by people very close to Gessen. However, when one demanded an explanation from Landsberg he just said, quietly and with dignity, that he was not aware of any kind of sabotage activity. And added: "The physics program that was discussed was compiled by me, not by him." The watchful and spiteful conveners of the meeting were outraged: "Do you understand what you are saying! Think it over and come out again." (The meeting lasted for two days). Nevertheless when at the following day he was forced to speak out again he only said with the same dignity: "I have nothing to add to what I said last time."

I am wondering myself that up to now I have no evidence that something like this was happening at that time in FIAN. It seemed that then the storm had passed.

The terror was however developing. There were arrests and executions at the Kharkov Institute of Physics and Technics and among Leningrad physicists and astronomers. The atmosphere was darkening more and more. In this situation there took place a famous March meeting of CC VKP(b) [Central Committee of Bolshevist Communist Party] at which an issue of "watchfulness" was raised to a new height. Torture was practically allowed.

Now one could not hide from it even in FIAN. In April 1937 there took place a so-called general meeting of the institute's active, i.e. in essence that of all more or less significant staff members, to discuss the Stalin speech at the CC meeting and its decisions. Not only was I not present, I was not aware neither of what was going on at the institute's meeting nor of the very fact of it (I was formally a Ph.D. student at MSU, besides, as usual, in spring I was in bed with another exacerbation of tuberculosis). However, "manuscripts do not burn" — there remained a shorthand report in FIAN's archive. It was found by a historian of science, G.E. Gorelik, who showed it to me.

What was happening at this meeting does not conform to a picture of a paradise island in a storming sea of evil at all. The introductory talk was made by Sergei Ivanovich. I can present no arguments supporting the understanding that I have developed: That his attitude towards what was going on in the country was no longer that of an ally or "fellow traveler." Nevertheless I am sure that such a clever person could not fail to understand that there came horrible times when he had to save people, culture and his institute in particular. Fooling themselves out of fear, many intelligents were even then trying to find some reasons for what was going on. Vavilov, however, was a soberly thinking man. For him it was doubtlessly clear that for saving the institute he had to assume a part of the shame of the time upon himself. Romans said: "Doing something — do it." He had to give an introductory speech of "due" character. This is what he did. In truth, Vavilov used some tricks.

The talk was composed entirely from citations from that of Stalin "sandwiched" with phrases of the type: "The following statement of comrade Stalin is very important," "We have to pay special attention to the instruction of comrade Stalin on ... ," "The meeting has stressed that in our work we will have to follow ... ," etc. Of course, this did not change things a lot. Nevertheless formally everything was correct. There began a "discussion," i.e. the same as had previously happened in the university. Furious attacks concentrated on the fact that "saboteur Gessen" was brought to FIAN by the group interest of his friends Tamm and Landsberg. Accusation of "gruppovshina" was at the same time very dangerous. This groupism was almost a conspiracy, the most terrible of which authorities responded to with awful punishments.

Vavilov did immediately stress that Gessen was invited to FIAN by him and that he assumed complete responsibility for this. Nevertheless Divil'kovski, one of the most arduous of attackers, without paying attention to what Vavilov had said, told that "... in the university I.E. Tamm and G.S. Landsberg did not demonstrate a willingness to help the activists to fully uncover the roots of this problem." A future academician B.M. Vul gave an equally sharp speech: "A responsibility for Gessen lies on the group that led him (one should read: pushed him — E.F.), on directorate, on everyone who was under the influence of this group."

Reading the speeches of Tamm and a staff member of the Theoretical Department, a well-known physicist Rumer, is incredibly difficult. Tamm's situation was psychologically terrible. Within one year he had lost a close friend Gessen, beloved brother and beloved talented disciple S.P. Shubin — both had been executed. His men'shevist past which was in itself sufficient for an arrest was hanging over him. He knew that he was under constant undercover surveillance. Already before the "big purge" he lived through a terrible tragedy in the family (see p. 54).

Beyond this there was also a grief of disillusionment in the dreams of his youth. He saw in what the socialism which he had desired so passionately was turning into. One should also recall that philosophers and ideologists in general were hunting him as a bourgeois idealist in physics.

Tamm, of course, did not do anything indecent with respect to Gessen and others, denied knowing anything that could have had been considered as sabotage on their part and anti-Soviet actions in general, etc., but could not refrain from recalling that he had been close to bolsheviks, how in June 1917, as a delegate of the First Congress of Soviets from Elisavetgrad, he had been the only non-bolshevik who voted together with bolsheviks condemning the new offensive at the front begun by Kerenski (and how Lenin who had noticed it had pointed him out as the "only honest person" belonging to other parties).

The speech of Rumer was almost hysterical. He was under a constant threat of arrest as well. He had been working in Germany for 6 years and at those times this was sufficient in order to be accused of being a recruited spy.

It is not only difficult to read all this but also to give the names of the furious inquisitors. Here is Divil'kovski. When there began the Great Patriotic War and all of us were intensively preparing for evacuation I was once a witness of how he entered the dining room where we all dined, fresh, in a half-military *yungsturm*[14] uniform. Somebody shouted: "Maksim, are you intending to go to evacuation?". He proudly answered: "What? The world revolution is beginning and I will go to evacuation?" He went to the front as a volunteer leaving a beautiful wife and three small children and did not come back. In order to get to the front in spite of the September

---

[14]Originally the uniform of the "Rot Jungsturm", a youth communist organization in Germany in the twenties. This uniform was very popular in the USSR in the twenties and the thirties.

decree of the state committee for defense that forbade mobilization of scientists, M.I. Filippov did stay in Moscow as a representative of the evacuated presidium of the academy, then found a moment, managed to get into the army and also perished.

One could of course say that convinced followers of Hitler showed heroism and readily gave their lives for Furer and his vile aims. "And nevertheless, nevertheless ..."

What was however the result of all this, was FIAN saved? Yes, it was saved. One had to certain things to satisfy the watchful bosses with something though. The theoretical Department of I.E. Tamm was dissolved and its staff members were enlisted to other laboratories according to the topic on which each researcher worked. After reevacuation to Moscow from Kazan in 1943 this department did somehow quietly and naturally reappear. Theoretical seminars under supervision of I.E. Tamm had been working during all these years as usual. The institute itself remained to stay one of the "islands of decency" on which the waves of the surrounding sea of evil were felt only seldom. The spirit that had reigned in it before continued to exist.

Here are the facts: I know for sure that political information coming to partkom was not sent anywhere and did not lead to anything bad. During all pre-war years, in the institute with such a gathering of non-party intelligents that were suspicious for the "organy," only one person was arrested — Yuli Borisovich Rumer. This was only because he was a close friend of L.D. Landau (who worked in the institute of Kapitsa) who had been arrested for preparing an anti-Stalin leaflet. Rumer was also accused in this (see the Section 'Landau and others'). They were arrested simultaneously (afterwards he was freed from this accusation. Nevertheless he did pay his "term." Fortunately for him — in the privileged "sharashka," an aviation construction bureau in jail).

It is worth noting a small detail: L.I. Mandelstam never participated in the meetings of this sort neither in the university, nor in FIAN. Even in the university where he experienced a lot of problems he was persecuted for this. What could one say about FIAN then! The respect to him as a scientist was unusually high.

Dmitry Vladimirovich Skobel'tsyn told me that once before the war, when he had been in the office of Sergei Ivanovich, one vigilant member of the above-described "ruling" party group had come there in outrage: "Sergei Ivanovich, this is not right. Leonid Isaakovich Mandelstam has a full position in our institute and comes here only once a week." In the words of Skobel'tsyn, Vavilov did somehow darken and told very harshly: "Remember, the whole FIAN rests upon Mandelstam." The watchful one became silent and went away.

I think that this episode tells much both on Vavilov himself and on Vavilov's FIAN.

4.4 What Gave Birth to the Vavilovs?

There unwittingly arises a question: From where did this miracle of the Vavilov brothers, physicist Sergei and one of the leading scholars in botany in the world Nikolai, arise? The first was a president of the Academy of Science (perhaps one of the best in its history), the second, a president of the Academy of Agricultural Sciences (which he himself founded). Both were people of incredible talent, energy, initiative, with the widest scope of knowledge. In recollections of the people that were often in contact with them, they were people of immense charm, always ready to help. How did these grandchildren of serfs form?

Everyone can of course recall Lomonosov and say that there is no miracle in this (by the way, he was not a serf). Russian peasantry did often give birth to remarkable people. This required however either general conditions favoring finding talents hidden in the dark people's mass or special outstanding strong-willed personalities. Lomonosov embodies the second case, the brothers Vavilov the first one because at that time this case was by no means exceptional. A miracle was an epoch of great reforms of Alexander II and its fruits that in a few decades showed to the world a new Russia.

For the third time in three centuries Russia made an attempt of "entering Europe." For the first time it was punched there by the merciless truncheon of Peter the Great. Then, in the words of Pushkin, "Russia entered Europe as a ship put on the sea, in the thunder of guns and the knock of an axe." However, having declared to the world its greatness and its hidden potential, Russia gave European education only to a thin upper layer of its society. The main mass of the population continued to remain slavishly deprived of human rights, dark and even not knowing of the possibility of a different life.

For the second time Russia just "visited" Europe when chasing the defeated Napoleon. It established its imperial grandeur even more firmly but the people inhabiting the boundless space did continue to remain in the same feudal absence of rights. However, those who went until Paris did learn something on the possibility of a different life. This showed itself in particular in a rapid growth of intermediate layers, first of all of raznochin intelligentsia.

Now, for the third time, there however happened something much more essential. There began a Europization of the country based on the introduction of many democratic principles that did in a half of the century change the country.

Personal freedom instead of serfdom. An incorruptible jury, new court code, a code of laws, brilliant lawyers instead of the pre-reform court that carried out decisions even in the absence of the plaintiff or accused (and having a right not only to acquit or sentence but also to "acquit but leave under suspicion"). Universal military service instead of rekrutchina,[15] with the abolishment of corporal punishment. Abolition of preliminary censorship that gave considerable freedom to a rapidly developing press. Changes in the economy that turned "Tit Tityches"[16] of A.N. Ostrovsky to Chekhov's Lopakhin's — active, initiative and cultured industrialists and bankers. An intensive development of a network of schools and universities. Finally, a growing development of modern industry and network of railways covering the immense country.

Although at the same time there took place a proletarization of peasants and there remained a poverty of the people and a sharp division of society into a growing educated layer and still dark masses. Although one needed the revolution of 1905 to tear from the incapable tsar new reforms and creation of parliament — albeit only with consultative rights. Although there remained many remnants of the feudal past, a shame of pogroms and many other things but nevertheless all this did create a new country.

One of the remarkable particular results of this transformation was craving of the "new Russians" that had become rich to high culture.

This is a special movement characteristic of our country. It gave Mamontov who made millions on construction of a northern railroad and went bankrupt because of the generous support of many talented artists, actors and even opera. It gave Alekseev (Stanislavsky) who created Hudozhestvennyj theater on his own money, the money of one of the directors of a large factory. It gave brothers Tretyakov, connoisseurs of painting, who presented Moscow (and forbade taking entrance fees!) the famous gallery of painting.

---

[15] The system of conscription when only a certain percent of potentially eligible conscripts is drafted but those drafted serve for 25 years.
[16] The despotic merchant Tit Titych is a character from the play by A.N. Ostrovsky *Hangover at Somebody Else's Feast*.

It gave Morozov and Shyukin, creators of richest and most modern collections of impressionist painting, the first in the world, and so on, without end. A wide competent patronage of arts and science became a characteristic feature of the epoch.

Striving for culture did however reveal itself not only in patronage. Famous producers of broadcloth brothers Chetverikov, Sergei and Dmitri,[17] had many children. Not a single one from the eight children of Dmitry did follow the father's steps. One son, Sergei Dmitrievich, became a distinguished scientist-petrographer, a professor of Moscow University, one daughter, Ekaterina Dmitrievna, became an art critic. From four children of Sergei only one did follow the father. Another one, also Sergei, became an outstanding originally thinking professor of genetics (hunted down in the years of lysenkovshina). The third became a mathematician.

The father of brothers Vavilov came to Moscow from a village by foot, became an errand boy at the biggest Prokhorovskaya (Trekhgornaya) in Presnya manufactory and ended up being one of its influential bosses. His cleverness and later great capabilities showed themselves very soon. After revolution he emigrated but due to the efforts of the sons returned home and soon, in 1928, died.

The brothers Vavilov origin is in this group of the newly rich "new Russians" striving for culture.[18] A talent of the people that was glowing and not dying for centuries did enflame in all these people.

Saying that they were becoming highly cultured persons is not enough. Very many of them were becoming true Russian intelligents, accepted moral norms worked out by working intelligentsia originating from poorer, even poor raznochin layers.

In another place (see the beginning of the Section 2.1) I have already written on how diverse was the intelligentsia in Russia — on talented and worthy wealthy engineers and lawyers, on religious philosophers, etc. I have also written that its medium-wealthy working layer can be considered as the basic and most important one. One could say that it was most brightly personified by Chekhov, a highly moral tireless worker. A hardly explainable but nevertheless easily understandable word "decency," a feeling of duty, modest level of existence, honest labor, feeling of responsibility for others, an unconditional primacy of soul over material profit — these were perhaps

---

[17] "You do not even know what Chetverikov cloth is," a talkative elderly taxi driver told me at the end of the sixties. "And I am still wearing a coat made of this cloth".
[18] Let me add that the son of their sister Aleksandra became a professor of botany.

the "key words" for the majority of these (actually, other as well) Russian intelligents. For most of them this moral code lead to materialistic and atheistic *Weltanschauung*.

Apparently, the brothers Vavilov, who had helped in constructing barricades in Presnya where they had lived, did calmly face a loss of property resulting from the October revolution, lived "as everybody," did enthusiastically take part in creating the big national science of the XX century. Characteristically they joined the above-discussed intelligentsia even in their *Weltanschauung*.

One reads in the autobiographical notes that Sergei Ivanovich began writing during the last years of his life: "Nikolai Ivanovich became atheist and materialist very early." On himself he writes that at the age of 15–16, in the course of discussions with a theologian, a teacher of "God's Law" I.A. Artobolevski, he "... was the main theological opponent in the class and energetically dismantled theological constructions of Ivan Alekseevich." In general in his youth he experienced ideological swings. He read marxist philosophical literature, Marx, Engels, Lunacharski, etc., did even buy *Materialism and Empiriocriticism* by Il'in (Lenin) — certainly without having any idea on who the author was. "I tried to become a poet, a philosopher observing the world;" "went through pessimism and optimism, joy and despair, scientific religion," etc.[19]

Similar swings were characteristic of the whole intelligentsia. From worshiping antiquity to theosophy, from religious Slavophilism to devotion to revolution which would wipe away the world of injustice and oppression. However, in the center there remained the main thing — the above-described working medium-wealth intelligentsia. The brothers Vavilov joined this environment and shared its fate. Due to this mutual understanding and respect between Sergei Ivanovich and other leading scientists of his institute, not at all intelligents in the first generation, were established so easily.

This was how we knew him in FIAN, in an atmosphere of the beloved scientific work so dear to him.

His somewhat different unexpected features were revealed when he became president of the Academy of Sciences. These are colorfully described in the already mentioned (see Section 4.1) unpublished memoirs of the at

---

[19]S.I. Vavilov. 'Beginning of an Autobiography.' In Vavilov, S.I. (1991). *Essays and Memoirs.* Third enlarged edition. Moscow: Nauka, pp. 104–106.

Vavilov's time, first vice-president I.P. Bardin (who played such an important role in electing Vavilov as president). Bardin writes that after election:

> There began common work. In general one should say directly that the character of Sergei Ivanovich was autocratic. He did not hinder others but nevertheless inherited something from his father. My collisions with him were mainly on construction. He was in this sense completely uncompromising: physics and nothing else, we should build a physical institute, everything else is second sort. This was in fact necessary and he kept this line firmly. Thanks to him the physical institute was built on such a scale, that many things were bought, other things done. All this did boost physics a lot ... Nevertheless with Sergei Ivanovich one was building an Institute of organic chemistry, began building an Institute of metallurgy, started reconstruction if the Radium institute, building of Pulkovo observatory, Mangush, an observatory in Alma-Ata ... I had (albeit under his protection) to take all the blows on myself. He could really talk to the Ministry of Finance — this is true, the merchant was seen in this as well. He could, completely imperceptibly, convince Zverev.[20]

Hereditary features of this astonishingly diverse personality revealed themselves quite differently. In such a way the liberation from serfdom of a third of the country's population, liberation from retrograde "norms" of life of all the people did lead, through stormy development at an epoch of truly great reforms, not only to an appearance of many rich capitalists but, through them, to a new blossoming of arts and science.

One involuntarily asks: Would an appearance of "new Russians" help to increase a general level of culture nowadays when there took place an equally powerful transformation of the country characterized by many features that, it seems, are similar to those described above? The time is of course different, an economic growth of those times is in the distant future and only numerous nightclubs, casinos and villas all over the world are representing the "signs" of their culture. Let us hope that this is still the first generation.

---

[20] Here one describes a narkom-minister of finance, A.G. Zverev, who during 23 years, 1938–1960, both in Stalin's time and afterwards, used his immense power in an extremely hard way.

M. A. Leontovich

Chapter 5

LEONTOVICH, Mikhail Aleksandrovich (1903–1981)

5.1 Mosaic*

Writing these recollections was long delayed. I could not find the basic pivot, the main thought around which it would have been possible to streamline what I can tell when recalling Mikhail Aleksandrovich. At the same time a necessity of describing him was becoming more and more acute: He meant too much for me, for my scientific work, too often did I admire him but ... also experienced distress.

The pivot I have just mentioned was however not found. Let us then think of what is written below as elements of a mosaic from which, after adding other elements, a reader could himself reconstruct his own portrait of this original, outstanding person.

5.1.1 The first meeting

The beginning of the thirties in Moscow University (and probably in all our higher education institutions) was marked by dismantling old forms of studying process and, for a modern reader improbable, experimentation. A number of lectures was drastically reduced, that of seminars did increase and these were usually even not seminars but something totally incomprehensible: A brigade-laboratory method. The group of 25–30 students was divided into 5–6 brigades, each of them working over the problem, using books and discussing the problem together. The process of passing the test at a lecturer was collective, any member of a brigade could answer questions (sometimes it was only the brigadier) and if the answers were correct all

---

*Published in the book *Memoirs on M.A. Leontovich*, 1st edn. (1990), 2nd edn. (1996). Moscow: Nauka.

the members of a brigade got good marks. There were many other things. For example, students collectively controlled and changed the marks given by lecturers, etc. It is not difficult to imagine the ensuing chaos.

By 1934 the senselessness of this became evident and there began a gradual return to the old forms. One problem was that big lecture audiences that became obsolete with the brigade-laboratory method were already reconstructed for the seminar-laboratory classes. For that an auditorium was simply cut into narrow and long pencil-box-like rooms. At one end there was a door, at another one, a window.

One of the first measures aimed at the liquidation of follies at the physics department of Moscow University was a creation (for the students of fourth and fifth years) of a specialty and a group of "theoreticians and opticians" (before that the specialization in theory did not exist "so that one does not diverge from practice"). The strange combination of theorists and opticians was probably explained by the fact that the group was created by co-workers of L.I. Mandelstam and under his patronage.

Foolish ideas occupied not only heads of administrators but those of students. For example, I knew that I wanted to be a theorist but thought that before one becomes a theorist one should grow into a good experimentalist. Therefore only two weeks after the beginning of classes in the new specialty, I rushed, with a humble appeal of joining these, to the dean Boris Mikhailovich Gessen. He did not let me finish and said "Run fast to the lecture, Leontovich is right now lecturing on electron theory".

I found the pencil-box room I needed in the "new building" in Mokhovaya and opened the door. A very long table stretched from the entrance to the window. At one side in a row were sitting all the students of the group. In a narrow passage between the table and the opposite long wall there stood an extremely high blackboard (the height of the former lecture auditorium was immense, this was left intact). At the blackboard there was an old "Vienna" chair on which there stood a very thin and very tall and in some way one-dimensional man in his thirties. Stretching his right hand with a chalk held still higher he was finishing writing a formula there, at the boundless height. Having finished he turned his face to the students, not leaving the chair — just slightly bending his arm, and pronounced several short phrases separated by pauses. These stumbled a little when leaving his mouth and due to some reason some phrases or some words were said much louder than the others.

At this moment I saw at the elbow of his black jacket, which was thickly covered by chalk, a big hole from which there hung a half torn away patch sewn from inside. I apologized, took a place at the table and began writing. The lecture was going on. The grotesque shape of the auditorium, the grotesque position of the lecturer and his very figure, the unevenness of his voice — all that was composing a strange harmony of disharmony.

Astonishingly it turned out, like it also happened later when Mikhail Aleksandrovich was lecturing us in statistical physics and optics in more normal auditoriums but in the same manner of speech, that his phrases were exact, laconic, easily turned into notes and summed up to a very consistent and very convincing total. When he was lecturing there arose a feeling that you were learning only the above-water part of an iceberg. A student reached the underwater one later, when thinking over the lecture that, as it turned out, led precisely to this. These lectures did improve much in the mess that had formed in the student's heads during previous years.

When Mikhail Aleksandrovich was simply talking, at some moment, the thorny tension suddenly left him and the speech began flowing smoothly and quietly and his face became beautiful. When he walked along, quietly moving his long legs, holding a cigarette between third and fourth fingers of a dropped and somewhat put aside arm, in his thoughtful tall figure there was some quiet sense of rhythm, non-invented significance.

One often says about good scientists that they can extract the heart of the matter, separate it from everything unimportant around it. A real scientist does really need this quality. At a risk of sounding routine I would still say that Mikhail Aleksandrovich possessed it to the highest degree. During a scientific conversation he was usually listening, listening and then would suddenly say: "Wait." He would take a piece of paper and write, strongly pushing a soft pencil, one–two drastically simplified formulae: "This is what you want to say?" With the same laconicism the essential was being stressed in his lectures. Because of this his books are so thin. His speech, as an opponent at my doctor dissertation comprised, as far as I remember, several phrases: "First, the candidate was able to reduce an integral equation in the plane to that depending on one variable. Second, ... Third, ... Due to this he could do that and that. These are essential results and the candidate is worth ...". The whole text, I think, was no longer than one page (in 1944, this was still acceptable).

Let us however return to the first meeting with him.

A patch on the elbow of a professor or "almost professor," especially a torn away one, should not be particularly surprising. The reason lies not only in the fact that these were hungry, uncomfortable, barefoot years when a hat or a tie was for a young man an improper symbol of a desire to stand out or a sign of a putrid bourgeois influence. A total disregard to appearance, clothes, everyday details was characteristic not only of the son of a distinguished Kiev professor, Mikhail Aleksandrovich Leontovich, but of his closest friends distinguished, like himself, by true intelligence, education and high spiritual qualities. The source of this striving to simplification, this despising of everyday details was the same pertinacious desire to select the main, really significant in life as well. This main was for them the spiritual world and value of a person. Everything of secondary importance, was external and was pushed aside.

They were four closest friends who carried their friendship from student years to the ends of their lives: Mikhail Aleksandrovich Leontovich ('Min'ka'); Aleksandr Aleksandrovich Andronov ('Shurka',) one of the inventors of theories of control and self-excited oscillations, a man of inexpressible charm, eagerness and strong mind, and unbounded humanity; Petr Sergeevich Novikov (a first-rate scientist in the field of mathematical logics "Petr who is even more clever than Shurka, as once explained by Ekaterina Aleksandrovna, a sister of Leontovich and a wife of Andronov, the man that seemed to understand everything in people, the man of immense benevolence whose behavior was always very modest) and Nikolai Nikolaevich Parijski, a strong and authoritative astronomer who outlived them all.

Together with another student of Mandelstam, Gabriel Semenovich Gorelik, Maria Tikhonovna Grekhova and others, in 1931 Andronov moved to Gorky where they decided to organize a new scientific center. All their efforts and for some, all their lives, were devoted to this. This center was organized. Nowadays in the city there are several big scientific institutes headed and directed by their disciples and disciples of disciples. After the move Andronov was given a, at that time large, three-room apartment. I was lucky to visit it during two years, 1944 and 1945, when I regularly came to Gorky to lecture at the university. A spirit of hospitality and benevolence reigned there. An entrance door did not have a lock, it was "locked" from inside by a broom put through the door handle. This continued until the time when academician Andronov was elected deputy of the Supreme Counsil of RSFR. A stream of visitors who, unlike students and colleagues, were not always capable of grasping the possibility of using such technique forced to switch to conventional methods.

Petr Sergeevich Novikov possessed a fine aesthetic feeling, liked painting. In 1951 I spent three ravishing weeks wandering with him and his sympathetic colleague in mountainous Crimea from Bahchisarai to Sudak. In a happy mood we were in time for our flight in Simferopol airport and even went for lunch. However, I carelessly pressed a big filled tomato and its liquid content poured over my whole breast. Having finished laughing we finished the lunch and I said, "We still have a couple of minutes, I will go and change my shirt." In extreme amazement Petr Sergeevich waved both hands and stared at me with the smiling slits of his eyes: "How could you pay attention to such things?"

Conversations with these people, even just their questions and remarks (each of them was more eager to listen to an interlocutor rather than to speak himself) on any of an infinite diversity of topics was a true delight. In these conversations there was one remarkable property — it was impossible to tell to any of them something that was not your true thought, true point of view. Even stupidity was pardoned provided it was not a thoughtless repetition of that of somebody else. To be insincere or repeat some popular platitude was impossible, a tongue would not have turned to do this.

This domination of the true, spiritual, sincere was what determined communication with Leontovich and all of them.

A scientist should "by definition" think independently and by himself. He exists in order to find something new that was not seen, was not understood by others, his predecessors, even his teachers, even those whose authority for him is extremely high. Phrasing it in a more vulgar way — he is paid for this. Therefore an absence of blind reverence of established authorities, dogmas, dominating points of view is mandatory for him. Of course, if one possesses this quality it cannot show itself only in the domain of science but becomes characteristic of the pattern of his behavior in life as a whole. Mikhail Aleksandrovich was a good scientist and, correspondingly, this quality was to a highest degree characteristic of him. He defended a clearly formulated position of his own with conviction, perseverance and courage.

5.1.2 The war

On Monday June 23, 1941 in a small room in the Theoretical Department of FIAN, there gathered its deafened, perplexed staff members: head of department I.E. Tamm, M.A. Leontovich, D.I. Blokhintsev, M.A. Markov, V.L. Ginzburg and myself. Everybody was squeezed not only by the grief fallen upon our country but by our suddenly uncovered uselessness. Who needed then our nuclear physics, principal questions of theory of elementary particles and theoretical physics in general to which all the passion and all the forces of each of us had been devoted to?

In a special position was only Mikhail Aleksandrovich. He had long been working on a theory of propagation of radio waves in relation to the works of L.I. Mandelstam, N.D. Papaleksi and their collaborators on a radio-interference method of measuring distances, radiogeodesy, etc. It was clear that his knowledge would in some form be required. There had already been invented "Leontovich approximate boundary conditions" that would later turn out to be very useful. He just had to leave optics and thermodynamics and work on important applications of radiophysics. Other members of the staff began feverishly fantasizing and inventing possible applied topics (let us recall that at that time I.V. Kurchatov stopped his research in nuclear physics and began working on demagnetization of ships). Mikhail Aleksandrovich mentioned however that in the theory of propagation of radio waves there were difficult unsolved problems which were of practical importance. Before that time one had considered an ideally flat and uniform surface while a reliability of sea and air radio navigation was strongly dependent on inhomogeneities and roughness of this surface. At his advice shortly afterwards I began working on this problem and became fully occupied by it.

After a month FIAN was evacuated to Kazan. Mikhail Aleksandrovich also went there but only for a short time. In Moscow, under the leadership of Semen Emmanuilovich Hajkin, also a disciple of Mandelstam and a remarkable radiophysicist, there began a research on one direction in radiolocation and Leontovich joined Khajkin to work on theoretical problems.

In February 1943 I came to Moscow and went to Mikhail Aleksandrovich to show the results (on the problem he had proposed) obtained during the year that had passed. I went to the old FIAN building that at that time was used for production of elements of radiolocation technique — to the same small room in Theoretical Department in which we had gathered on the second day of the war. Mikhail Aleksandrovich was visibly content with what I had done. Several times he interrupted my report with his "wait" and formulated in his condensed way its essence (in my work I was helped by an effrontery of the youth and an ignorance in the field that was new for me — due to this I was not subjected to a hypnosis of traditional methods). As usual, he was however somewhat dry, did not say any laudatory word stronger than "... yes, correct." Then we went to his place where I spent the night (his house was — a rare thing then — albeit weakly, but heated) and talked till midnight about everything that was tortuous then. When we were going along a corridor of FIAN I heard a continuous roar coming from our previously so comfortable conference hall. Now there stood vacuum pumps that pumped away air from special radio valves produced there.

Mikhail Aleksandrovich was very proud of his work and told with somewhat naive arrogance: "Do you really think that we will return this building to FIAN? In Kazan they are doing something useless and we here are busy with real thing." In reality this evaluation was of course unjust. During the war in FIAN one did much for the front, did something that only highly qualified physicists could do. In half a year after our conversation the building was nevertheless returned to FIAN. The institute moved there in September.

The naive arrogance I have just mentioned that was so contradictory to a soberness of his self-evaluation was in general frequent. I thought many times that one of his ancestors had added to his blood a good portion of blood of some proud Polish nobleman.

Mikhail Aleksandrovich continued to work on applied radiophysics in one leading institute. He did however reestablish his link to FIAN in 1946 when at the Moscow Mechanical Institute (that later became Moscow Engineering-Physics Institute) there began a teaching program of physicists on defence-related atomic and nuclear problems. He joined a chair of theoretical physics organized by Tamm. Initially it also included Issak Yakovlevich Pomeranchuk and myself.

The project was growing fast and there appeared new professors (A.B. Migdal, A.S. Kompaneets and many others) and also young assistants (mainly graduates of the same institute). In 1948 I.E. Tamm was summoned to work on special topics. (Under these nondescriptive words hid the work on constructing a hydrogen bomb based on ideas of A.D. Sakharov and V.L. Ginzburg.) Mikhail Aleksandrovich became a head of the chair that was already large. Applied radiophysics did however remain his main domain of work.

5.1.3 The 'character'

Difficult as it is, now I have to tell about one feature of the behavior of Mikhail Aleksandrovich. Here I speak about his, known to everyone, irascibility that sometimes transformed into a real rage. Everybody knew about this and forgave this often unjustified rage not only because of loving him. The thing is that it was directed to a large extent "isotropically," not only onto people whose position in the official hierarchy was lower or the same as his one but also on those who were "higher."

Once, at the end of the forties, he was in the office of a KGB general attached to FIAN (officially he was called a representative of the Council of Ministers of USSR). Such generals were also in other institutes working on secret projects. They helped in communication with central authorities and, to some extent, certainly controlled the work of the institutes. Of course, they paid special attention on keeping an eye on the institute's staff, to a possibility of their involvement in classified projects, etc. This was called "working with staff."

Having been in a complacent mood he gave Mikhail Aleksandrovich advice which he himself did perhaps see as a sign of special esteem and trust. The general said: "Why don't you, Mikhail Aleksandrovich, join the party? I am ready to give you a recommendation myself." As Mikhail Aleksandrovich would tell me with pleasure two days later he exploded like a mine that had carelessly been stepped on by someone. "Wha-a-at! he cried, 'You want me to join a party that plants anti-Semitism, holds peasants in kolkhoses, ...", etc. He shouted so loudly that the terrified general began running across the office and shutting the doors more tightly so that the seditious words would not have been heard and tried to calm Mikhail Aleksandrovich down. Nothing happened although at that time one could pay dearly for such speeches.

In another episode he shouted at a deputy director of the institute. I would however like to tell about an unjustified anger that I was a witness myself.

Once, in early spring of either 1950 or 1951, after a meeting of the MEPHI[1] chair, a group of its members decided to walk from MEPHI, which was at that time in Kirov (Myasnitskaya) street, to accompany Mikhail Aleksandrovich on his way home, to Prospekt Mira where he lived at that time. We were walking along wonderful spring streets chatting on various things. Suddenly, when we were already close to a metro station where we were supposed to part, Mikhail Aleksandrovich began talking about somebody who had, in his opinion, agreed to a shameful compromise and had not defended his colleague that had been a subject of anti-Semitic baiting.

In the course of the speech he was becoming more and more agitated and began cursing this person in such a way that one of us who was considered to be very amenable began to object insisting that one should not

---

[1] At that time it was still MMI — Moscow Mechanical Institute. (Ed.)

cross a person who had done much good for science and for many people, often through sacrifice, and that it was not clear whether he had been able to do anything for our colleague at those difficult times. This put Mikhail Aleksandrovich totally out of equilibrium. He literally shouted at the one who objected to him — let us call him N (the memoirs are about Leontovich, not him): "Scoundrels multiply because of people like you! People like you are worse than these scoundrels," etc. In the words of the third of us, N stopped, white as a chalk, and said quietly: "If you do not immediately apologize I will never again shake hands with you or say a single word to you." Mikhail Aleksandrovich, struck dumb, turned and started walking away. This third in our group, horrified, began running in between the quarreling crying: "Comrades, you have gone mad, what are you doing!" Mikhail Aleksandrovich stopped, returned back, shoved forward a shovel-shaped hand and angrily said "Sorry." After handshaking he told, with the same angry voice, something completely improbable: "Don't you understand that I have told all this only because I love you dearly."

The friendship was restored. It would be nice to stop at this end of a Christmas fairy tale but, alas, in 10 or 12 years I saw a new even more horrible explosion. In the presence of still larger numbers of esteemed colleagues Mikhail Aleksandrovich, in a state of extreme irritation, did pounce upon the same N. This time he reviled some staff members of FIAN who had admittedly offended him (although in reality that was him who had got angered there in FIAN, exploded, had been unjust and, I think, it was understanding of this that was a cause of his irritated state).

Knowing Mikhail Aleksandrovich one should not be specially clever to understand the main thing: It might have been (might have!) possible to avoid the worse only by stopping the conversation at any price. N, however, did not realize this. He began insistingly and discontentedly refuting Mikhail Aleksandrovich. For sure, there followed a shockingly furious explosion (with usage of some words that I do not dare to cite here). Knowing the previous story it is not difficult to guess what did follow. N said: "Once I forgave you when you apologized for your behavior. Now, even if you would apologize, I would not forgive." And he left.

Both of them had of course read Gogol, had laughed at a quarrel of Ivan Ivanovich and Ivan Nikiforovich and sympathized with their tragedy. However, having had been Moscow theoretical physicists and not small landlords in Mirgorod a hundred and fifty years before, they repeated the story of Gogol's personages but did not understand this. There ended a many-years friendly relation between two families, joint celebrations of New Year, long conversations in narrow circles. They were suffering because of the break-up very heavily during many years but neither of them could overcome himself. Two weeks after, Mikhail Aleksandrovich was coming down the stairs in his house, and met a neighbor (a cousin of N) who knew both of them well. Having passed him and turned around the corner Mikhail Aleksandrovich stopped and asked: "Do you know that I broke up with N?" "No, and who was guilty?" "Guilty is my character" said Mikhail Aleksandrovich darkly and continued walking.

Friends were able to ensure that Ivan Ivanovich and Ivan Nikiforovich met in the same hall and, pushing them on their backs, made them join their hands but, again, there jumped out an ill-fated "goose" and everything failed. With our heroes a success was nevertheless bigger. Under pressure of friends (I.E. Tamm, E.M. Lifshitz) they restored, first reserved and then even relatively friendly, relations but could not finally step over "the goose." "Boring is living in this world, gentlemen."[2]

What were then these outbursts of uncontrollable fury by Leontovich? Laxity? That heritage of unknown proud Polish nobleman? It could be. The essence is however explained by an old wisdom: "Our drawbacks are continuation of our virtues." For Mikhail Aleksandrovich these virtues were directness, carelessness to consequences, furious attitude towards falsity and injustice, readiness to stand for the hunted. Unfortunately this behavior did too easily go beyond reasonable borders and turned into an unjust fury — uncontrollable by reason.

However, it was precisely because these drawbacks of the character appeared in direct relation with its noble features that caused respect towards Mikhail Aleksandrovich, love of friends, coworkers and disciples to him, they were easily forgiven and do not overshadow his image in their memory.

5.1.4 Second scientific life

Once in 1951 Mikhail Aleksandrovich attacked me: "Listen, what is your Igor Evgen'evich doing? Is drowning himself and calls me to join! This is

---

[2] Years have passed after the death of Mikhail Aleksandrovich and my years already exceed those of his life — so why tease a reader with mentioning some N? N is myself, the author of these memoirs. The one who, agitated, tried to reconcile us then, at first time, in Prospekt Mira, was the late V.G. Levich. We really loved each other and what has been told here means much for me. It is because of this that I remember each word so distinctly.

like, you know, at the bottom of a distant pond there sit drowned people, almost completely rotten, covered with green driftweed, terrible and suddenly see that someone new is floundering on the surface and is drowning. And then they beckon him with their bony hands and cry: to u-u-s, to u-u-s, he-e-re, he-e-re!" Here he graphically stretched his hands up, with fingers crooked, and began pulling air towards him.

The thing was the following. In Kurchatovsky Institute there broadened research on controlled thermonuclear fission. There naturally arose a question on who would head a theoretical part. At one of the scientifically-organizational meetings summoned by L.P. Beria who was in charge of all the "atomic" things I.E. Tamm did strongly recommend Mikhail Aleksandrovich. An almighty and fearsome administrator, the head of all the work, was extremely surprised that there still was a big theorist who was not used in his immense system. Surprized, he asked: "Who is he?" There followed agitation, running of assistants-generals on tiptoes, whispering in Beria's ear of some words, very dangerous and uncomplimentary, but Beria declared aloud "Does not matter, you take care of and ensure this, you watch him, he will work" — and the problem was solved. Mikhail Aleksandrovich did not want to leave his radiophysics at all but clever people — Igor Vasil'evich Kurchatov and a close friend of Leontovich Igor Evgen'evich Tamm — persuaded him that the problem is interesting and was the right one for him.

Indeed, his vast knowledge, an experience of working in radiophysics and electromagnetism in general, thermodynamics, quantum mechanics[3], theory of nonlinear phenomena — all this was gathered here and was used. So there started and was developing his "second scientific life" lasting until his end and of longer duration than the first one. Its success is probably best witnessed by a bright group of physicists that grew up under his influence and in his entourage.

During this period all his activity was concentrated at Kurchatovskij Institute, he also lived there. He developed a dislike to FIAN while I worked at FIAN and lived in the south-west of Moscow. Our contacts practically stopped. We met, better said — occasionally met — very seldomly at scientific conferences, in the polyclinic, etc.

Shortly before his death I visited him in the hospital. We had a good conversation, as usual in such sad circumstances, on something foreign.

---

[3] Let us recall that a theoretical prediction of the tunneling effect and its main properties was made by M.A. Leontovich and L.I. Mandelstam and was just ably used by G.A. Gamov to explain the α-decay (unfortunately, without a reference to them).

Externally it looked dry enough, as it almost always was in conversations with Mikhail Aleksandrovich, but in essence it was very warm and very sorrowful. Several days after I went to see him again but met his relatives and close collaborators coming from the hospital. There was no need to ask. I was too late.

A. L. Mintz

Chapter 6
MINTZ, Aleksandr L'vovich (1895–1974)

6.1 Aleksandr L'vovich Tells ...

It is necessary to tell about a remarkable man, Aleksandr L'vovich Mintz. He was an outstanding scientist and engineer, grand master of radio engineering who seemed to be capable of doing everything. He was successful in accomplishing most difficult tasks. Once I was a witness of his conversation with a no less remarkable, super-talented, albeit in a very different way, man — a physicist, also an academician, G.I. Budker. They esteemed each other highly and relations between them were very good. This time, however, they began a joyful exchange of caustic remarks: Mintz addressed me, smiling, 'Here is the author of many brilliant ideas that are, however, impossible to realize.' 'Of course', replied Budker, also smiling, 'You are working only with something that you know well to be doable.'

Indeed, a realization of some remarkable ideas of Budker required a long scientific research and faced our usual organizational and financial difficulties. With all his inventiveness and unexpected brilliant ideas, also organizational ones,[1] they were often realized by his disciples only after his tragically premature death. Sometimes it was fully accomplished only abroad. Still his works got recognition all over the world. Mintz, in turn, possessed a remarkable gift of understanding of what was possible and what was not. A special feature of this gift was, however, that something that was possible for him was on the verge of being possible or even turned out to be completely impossible for others — in science and technics as well

---

[1] Here is one example. For protection from harmful radiation from an accelerator he needed much lead. This is a very expensive thing. He went to the committee for state reserves and said: "Listen, I will not charge you anything for storing 2000 tons of lead. Why fill your storage facilities with it? With us it will be safe." He got it.

as in personal behavior when he faced difficult, often mortally dangerous situations. This will become clear from the stories described below.

Being 17 years younger than Aleksandr L'vovich I met him — and this was as a nodding acquaintance — only when he was well past 50. He carried his large (not at all fat) wide-shouldered body with ease. Up until the last years of his life he usually walked fast (if, of course, this was not a lazy stroll). He radiated some kind of certitude. A seriousness of his regularly shaped face was easily replaced by an attractive smile (not at all that unnaturally sweet an abstract, perhaps covering confusion, as in one parade photo in a collection of his selected works where he is depicted with a full set of his numerous orders and medals. There he did probably pose and was unnatural). He did not use words particularly characteristic of him, spoke in a relatively even voice, neither low nor loud. Nevertheless he apparently possessed some kind of charm if, for a long time, I came home and told my wife with joy: "Today I have seen Mintz." Later when our relations became closer Aleksandr L'vovich told that he would like to get acquainted with Sakharov. This acquaintance took place (although did not become a close one).

Our rapprochement — of my wife, Valentina Dzhozefovna Konen, and myself with Aleksandr L'vovich and his wife Evgeniya Il'inichna — happened suddenly three years before his death. This immediately turned into a close friendship. What kind of a man was the one we met? I will not write about it but append to this essay what V.D. Konen wrote in her memoirs.[2]

It is finally a time to start what has been promised at the beginning — retelling the stories of Aleksandr L'vovich. These beautiful, tranquil stories enlightened by some portion of irony are deeply engraved in my memory.

One of my first questions to him was: "Is it true that there exists a Mintz cycle? There is a widespread opinion that you were arrested three times, each time made a brilliant work on radio engineering in captivity, were released and awarded by an order. Then you were arrested again and everything repeated."

He quietly laughed (it seems to me that Aleksandr L'vovich never laughed loudly and often smiled with a somewhat playful smile)[3]: "No, not quite so. First time I was arrested in the period of civil war in Rostov

---

[2] V.D. Konen. Personality of a scientist. In Mintz, A.L. (1987) *Selected Works. Articles, Speeches, Memoirs.* Moscow: Nauka. This book also contains other memoirs of his colleagues and friends.

[3] When citing in quotations marks the words of Aleksandr L'vovich I certainly do not pretend on an absolute accuracy.

where I lived with my parents and graduated from the university. Red were advancing, White were fleeing. My parents fled as well (his father, an engineer, owned a small fabric producing laboratory instruments — E.F.). I did not leave. Why? White, when leaving, hanged on each lamppost along the main street a captured Red army soldier. This made very bad impression on me. I decided that my way is not with the White and stayed living in our house. When Reds came it was necessary to find some work. I wrote an application for a position in teaching at courses of military radio engineers. Before I got an answer a Red army officer was billeted to my house. He walked through all the rooms, entered mine and said: "I will live here." I told him that this was impossible, that I lived there. He got furious and I was arrested.

'In the morning I was taken to an investigator. He said that I was accused of espionage, that I had stayed in Rostov to spy for the White. I replied to him: "Do you really think that when going to spy I would have stayed in my house, applied for working at the courses, practically for joining the army, and had an argument with a high-ranked red officer?" The investigator thought for a while and said: "Yes, you are right. Well, you are free." At that time such investigators could still exist. This was my first arrest'.

This was, let us add, the first case when self-assuredness, cleverness and self-control helped Aleksandr L'vovich to save his wife while keeping his dignity.

'Second time I was arrested was in 1931 when I had already accomplished something serious. I did not sit for a long time. I was released, probably with the help of Ordjonikidze.'[4]

'As for the third arrest — this was shortly before the war, everything was much more serious. I was arrested and waited for an interrogation for a long time. There came a day when I was taken to an investigator. I was going along a wide corridor with doors along it, cries of torture were heard through them. Finally I was led into one of the rooms. At the desk, with his back to the window, there sat an investigator. I approached, grabbed a heavy inkstand from the desk and said: "If you touch me I will hit you with this until either you kill me or I kill you." Suddenly there happens a miracle. "Not at all, Aleksandr L'vovich, I have summoned you here not at

---

[4]Unfortunately I did not memorize the story of this arrest that had been told by Aleksandr L'vovich only briefly. I even do not remember exactly whether the above-given dates are correct, this could also be in 1930. In any case, this was approximately at that time.

all for this. Comrade narkom wants to see you." Apparently he did already know what for...'

'So, they led me through stairs and corridors. Finally we enter a big room and they lead me to Beria. Near him there stands some NKVD colonel. Beria says: "This has to be done in three months. If you do it — you are free." I looked through a description of the task, thought for a while and said: "Well, I can do it, but not in three months but in six." After these words the colonel exploded, jumped to me from the side, shaked fists at my face and shouted: "How dare you! Comrade narkom extends such trust and honor to you and you are saying that you need twice more time for this!" I turned to him and said: "Do you think that I like it here so much that want to stay longer?" Beria laughed and said: "Ok, let it be your way."

'Did you do it?' I asked.

'Yes, of course. Our group that worked on this was kept in special conditions, excellent lunches were brought.'

'What was this task?'

'Well, one interesting problem.'

Even 30 and something years afterwards did Aleksandr L'vovich strictly obey the secrecy requirements because at due time he had signed a corresponding obligation.

'What was after that?'

'The project grew and I had to stay in NKVD.'

(At this moment there interferes Evgeniya Il'inichna: 'Up to now I cannot forget what a horror embraced me when, after entering a hall of our apartment, I saw his colonel cap with a blue cap-band on a table at the mirror.")

'But how did you get involved in radio engineering when you had graduated from the university as a physicist?'

There follows a story: 'Yes, and I always dreamt about working in physics. In Moscow, being a student, I even began working on a topic from physics. Life turned things differently. Then, in Rostov, when the interrogator released me, there took place a formation of regiments of Budenny's Konarmiya and I was offered to head a radio division. I was given 20 field radio stations, 200 carriages with horses, a corresponding number of soldiers and ordered to organize everything by myself. Since that I was not afraid of organizational work of any kind. When I later organized big research, designer and construction collectives, institutes — I was not afraid of anything. How was it in Konarmia you to some extent know yourself from Babel. There were various difficulties. At the very first day I was told that

I had to learn to ride well. Those who could not were pulled from their horses by soldiers who said: "Why are you torturing the animal?" In any case, in two weeks I learned it. Did I ride well? In any case, I was never subjected to such a shame. With Konarmia I went all the way through Ukraine to Poland. Of course, there was much happening there.

'By the way, here is what happened when we entered Elisavetgrad. One led me to an apartment in which I was billeted. It was dark, the room was enlightened by a candle. I asked an attendant: "Why there is no light?" He answered: "An owner of this apartment, a White, a saboteur, damaged the machine. Well, he is with us and in the morning we shall execute him." I demanded to see him. I asked: "Why is the power station not working?" He answered: "No fuel." "Is equipment in order?" "You could look it yourself." We went to the power station, I examined it. Yes, everything was in order. I ordered them to bring a barrel of fuel. It was brought, we turned the station on — there was the light. This engineer was the father of Igor Evgen'evich Tamm.'

'Did you later tell Igor Evgen'evich about this?'

'I got acquainted with him much later and told him this story not a long time ago, shortly before his death.'

How did it happen that they did not know each other? They were coevals, before the revolution they were simultaneously studying at the same faculty of physics and mathematics at Moscow University. This is described by a difference in their lifestyles. Since gymnasium years Tamm took a great interest in social-democratic ideas and in 1915 went to the front as a volunteer as a hospital attendant (or "nurse"?). After the February revolution he plunged into revolutionary activity, spoke at anti-war meetings, was a deputy in a Soviet, a delegate of the First congress of Soviets, and graduated from the university in 1918. Mintz, on the contrary, was greatly interested in science and ... operetta, even thougt of becoming an actor. As already mentioned, he graduated from university in Rostov in 1920.

Let us however return to Konarmia (the story continues):

'There were also very dangerous situations. Once, with a small group, I was forgotten by our retreating regiments. The Poles encircled us and everything would have ended very badly if Budenny himself had not have come to our rescue with a headquarters detachment. They saved us.'

'Why did not you turn to physics afterwards'

'I was an officer. After fighting was over I was sent to Moscow to a military radio laboratory that reported to a narkom of defence. There I got involved in an interesting radio engineering work. This is how it started.'

However, Aleksandr L'vovich would not have been the almost legendary Mintz if he had not have shown his character here as well.

'Once I got a difficult and urgent task from narkom and began working on it not paying attention to anything. For all the officers in the laboratory one organized obligatory political classes. Long lists of various works of "classics of marxism" and new leaders that one had to study were hanged out. I totally neglected all this. Finally the bosses got angry, complained to narkom and he summoned me. Obviously I was up to something unpleasant.

'When I entered the office of Frunze he sternly demanded explanations. I answered: "I was too busy with working on your task, comrade narkom." "And, certainly, did not accomplish it," said Frunze menacingly. "No, on the contrary, I did." I opened the door and, on my sign, several soldiers brought in apparatus and put it in front of Frunze. A test that was carried out right away was a full success. The theatrical effect prepared in advance had its effect. Frunze was very much content and, of course, I was not punished.'

This was in the beginning or in the middle of twenties and I had once read in *Novy Mir* memoirs of a well-known radio operator and polar explorer E. Krenkel, in which he wrote, in particular, on his military service at that time. According to his words soldier — radio operators were driven to torment by an ALM device for field radio stations. I asked Mintz: "Is this ALM related to you?"

Aleksandr L'vovich smiled: "Yes, we had to deal with such things as well. The army was still poor and to ensure power supply for radio equipment in all situations I introduced a simple device — a normal bicycle with dismounted wheels was put in a fixed position, a soldier sat in a saddle and began rotating pedals. A transmission triggered an "engine" that provided current for a radio station. Soldiers cursed this dumb job because sometimes they had to rotate pedals for hours

'How did a transition to powerful stations take place?' I asked Mintz. "I understood that I was lacking an engineer education and in two years finished a course in a corresponding technical institute," he replied. We had to build big radio stations for broad transmission. Within 15 years I built several of them. Each one, when launched, was for that time the most powerful in the world. The first was a Komintern radio station with a wavelength of 1450 m, that at those times was known to everybody. Then there followed a no less memorable VTSPS radio station of 1080 m. (How could I not remember them! These first words in every broadcast: 'This is Moscow. There works a Komintern radio station on the wave 1450 m,' and the same for the other station. They were working for many years, up to a television epoch, and played the same role in our life as is nowadays played by television. — E.F.) By the way, there was an interesting episode related to this station.

'You surely understand that construction of such stations did each time require new ideas. These were not always easily accepted. When I proposed a scheme for the future VTSPS radio station all leading experts, even Mikhail Vasil'evich Shuleikin and others, did unanimously declare that this scheme would not work. Nevertheless I started to build it. At some point all the radio engineering part was ready, all the elements were checked and tested, everything looked reliable (everything was done in a very short time but we were in time to the promised date). We began testing the whole assembled scheme. We turned the voltage and suddenly the main kenotron goes off.[5] We replaced it by another one, switched on — it goes off again. We put in the third one — the same. Can you imagine this situation? It is already Thursday (or Friday) and on Monday a commission of these very experts were coming to accept the work!'

'We thought about it and understood that after the complicated scheme was put together some of its parts did evidently form an unforseen oscillation contour which at the point where kenotron was located created a maximum of strong alternating voltage. We began searching for the components of this contour. We searched, searched and finally found. We removed this parasite oscillation. We put in a new kenotron — it worked! We were lucky that it was still Saturday, before the arrival of the commission, a free day. We were sitting and looking at the kenotron; one hour — it worked; two — worked; a day — worked. We turned it off, turned on again — it worked! We felt relieved. The commission arrived. We switched on the scheme, as if nothing had happened, and everybody was forced to recognize — it worked!'

'I remember,' I said, dreamfully and nostalgically, 'in the twenties I was a schoolboy. I assembled a small detector receiver and sat in the evening sticking a sharp end of a springy wire into a crystal and, suddenly, I heard *Traviata* from the Bolshoy Theater in my earphones. What a miracle was it!'

'Yes,' continued Aleksandr L'vovich with his even voice, 'it was not easy to organize those broadcasts. Special problems were with location of microphones — we had no experience. We looked where to place them.

---

[5] A radio engineering slang meaning that a large special "valve" (a complicated vacuum electric device) burns.

I even came down from the ceiling into a giant lamp of the Bolshoy Theater and tried to put microphones there as well.'

In general it did sometimes seem that in the life of our radio engineering there was no episode that had no relation to Mintz. Once we were together spending a vacation in a sanatorium near Moscow. My friend visited me and told, in particular, that there appeared a collection of memoirs of our military that took part, as advisers of Sun Yat-Sen and Chan Kai Shek, in the Chinese revolution in the twenties. This was the first time this was openly discussed in the press. Agitated I came to lunch, told Aleksandr L'vovich about this and asked, laughing, 'Were you also there?' He replied, 'No, I was not there, but my group got an order to urgently establish a radio connection between Moscow and Wuhan. At that time this was not a simple thing. For seven days we did not leave the laboratory and decided: We will not shave until we reach a success. Well, the connection was established.'

In one conversation I returned to the topic of powerful broadly broadcasting stations, 'What happened after the VTSPS radio station?'

'Most important was, of course, a construction of a radio station in middle waves in Kuibyshev, where government was evacuated during the war,' answered Mintz. 'At that time it was the most powerful in the world. When the war began I got an order to build it in the shortest possible time.'[6]

'Here was, for example, a construction of carrying columns in the main hall that were made of reinforced concrete at awful frost when a liquid concrete solution froze before it could "catch up." I suggested to warm the columns by passing strong electric current through ferro-concrete reinforcement and not build giant "warmers" to keep a necessary temperature

---

[6] One could imagine which problems faced the builders of such a station, an intricate complex of powerful electric and radio engineering constructions with eight hundred meter long antennae, in the first war winter when numerous plants producing its elements were in the process of evacuation or had just been evacuated, how much cleverness and how many ideas did one need to nevertheless build such a giant and complex construction in these conditions. Nevertheless the station began broadcasting for the whole world in November 1942. In an article about it L.A. Mintz mentions that with a planned budget of 81 millions of roubles only 81 million rouble were spent (see an article in journal *Radio Engineering* (N 11, 1974) or Mintz, A.L. (1976). *Selected Works. Radio Engineering and Large Scale Radio Constructions*, Nauka, Moscow, pp. 286–294). 'For construction I was provided with a camp of 10,000 prisoners. Among them there were good specialists as well. We were building in the hard frost of the winter of 1941–1942. I think that my greatest achievement was that in the camp there were no epidemics of typhoid. We built saunas and the workers did twice a day (I think, this is precisely what was said — E.F.) "pass through them" and their clothes was steamed. We also had, however, to overcome technical difficulties.'

around the columns as was customary at that time. (As far as I know such heating by electric current was widely spread in the construction practice in the fifties — E.F.) In such a way we managed to shorten the construction period, save time. This led, however, to an unpleasant incident.'

'When the construction was finished there arrived a governmental acceptance commission headed by a well-known builder professor (at that time also a general) Vsevolod Mikhailovich Keldysh (a father of a future president of the Academy of Sciences). He outturned to be an awful formalist. The columns, he said, had been built with violations of BNR (BNR are 'Building Norms and Regulations,' a bible for builders). I said: "But a durability of the columns has been tested, everything is in order." "No,", he said, "BNR are violated." We brought an anti-tank rifle (or gun, I do not remember exactly — E.L.). We shot at a column, in the empty hall there roars thunder but the columns stoodd still. "No," said Keldysh, "the BNR were nevertheless violated." Because of this the commission accepted the construction as "good," not "perfect." This was very unpleasant. If it would have been "perfect" we could have freed many more prisoners as with "good." An awful formalist.'

Aleksandr L'vovich was angered by this even after 30 years. Probably, here also suffered his habitual feeling of a master of situation. However, he could not hide his satisfaction that the station was built quickly and ensured communication with the world during all the war.

After the war A.L. Mintz organized a powerful radio engineering institute that formally belonged to the Academy of Sciences of USSR but in fact was supervised and was run by some ministry — either of electronic industry or of radio industry. Here he expanded a range of his topics to a new domain — a technics of particle accelerators to high energies that were necessary for scientific research in physics of nuclear particles. Mintz took a great interest in the research on this topic and was very successful in it.

This new domain connected him to nuclear physicists. Construction of accelerators was supervised by the same almighty committee that headed all the work on nuclear armament (although factually accelerators had only quite a distant relation to this problem). In this committee Aleksandr L'vovich did again have to deal with its chairman — Beria. In particular, Mintz became a member of a commission that had to choose, near Moscow, a place for the construction of a brand new gigantic accelerator.

'We went through Moscow region, studied geological conditions and many other essential factors. Finally we chose two possible places: near Kryukovo, 40 km from Moscow and near the place where the present Dubna is — 130 km away. At a meeting of the committee Beria spoke in favor of the more distand place. "From Kryukovo," said Beria, "scientists will be all the time going to Moscow and not working." I insistently opposed him stressing, in particular, that in that place there was neither railroad (and that will hinder construction) nor sufficient power supply. "That is fine," said Beria, "we will build a railroad and a power station." Of course everything was decided like Beria wanted it.'

After building the synchrophasotron, now known to everybody, they were for a long time not able to launch it. A tremendous role in its design was played by complex, specific and powerful radio technics. This is precisely where the artfulness of Aleksandr L'vovich showed itself. Not only did he participate in the design but also interfered in the lengthy process of accelerator launching. 'Engineering order' was enforced and after that the accelerator started running.

Accelerator themes found a solid place in the life of the radio engineering institute. Nevertheless Aleksandr L'vovich did not leave old topics.

Once at the beginning of the seventies he told me: "I am very happy to finally complete a large-scale task. It took 14 (if I am not mistaken — E.F.) years". I understood that I should not have asked what this task was — he would not have answered anyway.

When after the famous Khruschev speech at the twentieth party congress there began a rehabilitation of those that had been earlier sentenced, Mintz showed himself in a characteristic way. 'Once,' tells A.L., 'a head of the first department came to me (let me explain: a first department is the one in charge of secrecy issues — E.F.) and said: "Aleksandr L'vovich, we should prepare your rehabilitation on all episodes in which you were sentenced. You have to write an application on it." "Me?" I said to him, "No, I will not write. You locked me, you should rehabilitate me." He continued to persuade: "This is," he said, "pure formality that does not mean anything to you. Without this application one can not start the whole procedure." I said: "This is out of question, you manage it yourself." Well, they managed. Later I was shown my investigation files.'

A sarcastic smile appeared at the face of A.L.: 'Imagine, to corroborate an accusation in sabotage they called, as experts, two well-known scientists and they gave a conclusion aimed *against* me.'

A.L. does not give the names of these specialists but from his transparent hints I guess that one of them is a well-known and serious scientist who had occupied high posts in the management of research works of military-industrial complexes, a very capable, even talented person — but acrimonious and envious. For dozens of years, up to the end of his days, A.L. contacted him within his main job. What did both of them feel? A picturesque episode of a surprising and terrible life of our society.

I have already told that A.L. was used to being an independent "master." He acted in a way that he found necessary for accomplishing the task he was working on. An invariable success of this, important for the state, work ensured him a tolerance of authorities, albeit on the verge of the possible. This showed itself, for example, with respect to after-war time (and even wartime) anti-Semitism, stimulated and directed by party and state. A.L. himself was deeply enrooted into the Russian culture and, more generally, European one. *Never* did I hear from him not only manifestations of Jewish nationalism but even any feeling of Jewish special features. *Not a single time* did I hear from him any typical Jewish words, expressions, jokes, anecdotes. He was a true Russian European.

With his feeling of responsibility for the work he was doing, an independence of his character and high moral values it was natural that he did not pay any attention to the nationality[7] of staff he recruited to his institute. As a result he took many capable talented Jews who had not been hired by institutes directed by more law-abiding (or, better said, party-state-abiding) directors. Among these there were also the ones that were meaningly supporting this policy. Such a behavior of A.L. did certainly irritate party and state officials but up to some time they tamed themselves and tolerated the self-willed Mintz.

A.L. said, smiling: 'Once a head of the human resources division of our institute came to me and said: "Aleksandr L'vovich, I know that in your institute there is no way for non-Jews. Here is, however, my son. He is a capable young man, try and hire him." I hired him, he really turned out to be a very capable young man and I was satisfied with his work.' Of course the words of the ministry official were a wild exaggeration but they showed how were the authorities watching this, what an anger and ensuing terrible exaggeration of real facts did arise.

They remembered everything and just waited. This showed itself when the life of Aleksandr L'vovich was approaching its end. In 1970 he was 75. The state had got many remarkable results out of his outstanding activity in science and technics. In the process of his activity he had grown many valuable scientists and scientists-engineers, new "resources" as one used to

---

[7]In the USSR nationality was a question of one's "blood", not religion. One was considered Jewish if one had Jewish parents.

say. He had created a powerful institute and it was already possible to stop being tolerant with respect to this man whose independent position did so poorly fit into our system in those days.

His birthday was joyfully, cheerfully and lovingly celebrated in the institute. After the congratulating speeches at the celebration he spoke himself with short "addendum" which he finished with the words: 'Here are several additional details from the life of the academician, Hero of socialist labor, Lenin prize laureate, thrice arrested, twice rehabilitated, Aleksandr L'vovich Mintz."

In our post-perestroika time it is difficult to understand how these independent and provoking words, pronounced in the face of a first secretary of rajkom sitting next to him, sounded. Well, it was seemingly tolerated again. However, when A.L. soon had another argument on some issue with the authorities and, protesting their decision, wrote a declaration of retirement this was, to his surprise, immediately accepted. He was considered to be unbearable and no longer needed.

"'You understand,' Mintz told me, 'when a director of an institute comes to a ministry he should first go to the heads of financial, planning and human resources divisions and only then go to the minister. This was customary. I, however, went with all my problems directly to the minister. Many did not like this." (He was unable to please a yard-keeper to protect oneself from problems, his dog so that it is sweet. He was not Molchalin.) One appointed a director of the institute that was not his first choice.

So it happened that the legendary Aleskandr L'vovich was put out of action and became "simply an academician."

The president of the Academy, M.V. Keldysh, invited him and, with full respect and with care, asked which kind of activity he would have liked to choose for himself. "I said that I wanted to narrow the range of my research and concentrate on physics and technics of accelerators."

He had already been an authority in this field for a long time. He organized and chaired an international conference on accelerators, worked out and constructed a model of a principally new "stochastic" accelerator in the institute, etc. He said,

'I told Keldysh that I would have liked to organize a scientific council on accelerators but not at all of the kind of numerous academy councils on other specially selected problems. They just coordinate a work of institutes in the corresponding field, organize conferences, etc. Usually these include people that are busy with scientific work in their institutes and only one staff member — a secretary of a council. I would have liked to have, in addition, a place and four highly qualified scientific researchers with whom I would have carried out scientific research. Keldysh did promise at once that all this would be done. And, really, after some time I got two offices in the building in Vavilov Street and a secretary and a position of president for myself. However, everything got stuck at this stage — for a characteristic reason.'

The thing was that the four scientific staff members chosen by Aleskandr L'vovich turned out to be Jews! How could he make such a mistake? At that time any official (from that in the human resources department to a politbureau member) understood such choice as an attempt to create "a hornet's nest of zionism." Everybody understood this, nobody except for A.L. would not have made such a crazy step that was doomed to fail. At least one half should have been non-Jewish. How did sober, cleverest Aleksandr L'vovich fail to understand this? Probably he simply did not notice what was a result of the selection. He was so foreign to national superstitions that he neglected them, being interested only in the work itself and in how this or that person carries it out. Even if he had noticed, he thought that he had a right of choosing whoever he wanted. Indeed, with all his life he proved that he chose staff guided by the interests of work and, in this way, always achieved successes that science and state needed. He did not understand however that he was no longer that Mintz whose self-will the authorities had been tolerating for so long a time.

The beginning of the seventies was in general a time of a crawling restoration of Stalinist habits. The Thaw was long over and there began attempts of the rehabilitation of Stalinism. The scientists Mintz had chosen were of course not hired. Certainly the true reason was not announced. One killed time and invented various pretexts. It was however clear to everybody what was the reason. A.L. did insistently try to resolve this situation. He got some mendacious promises but it was already clear that everything was hopeless, in spite of the fact that highest academy officials were involved. For A.L. this was a heavy blow to his self-esteem.

Every day he always worked in his scientific council but complained that he no longer possessed his previous working ability. His health was, as one says, "swinging," sometimes quite strongly. Then there followed two terrible blows in his private life in a row: Evegnia Il'inichna with whom he had been living for many decades passed away and eight days after his only son, a scientist, a geographer that was well known to specialists around the whole world died in an airliner crash.

These were the hard days when we became closer as friends. In his everyday behavior one observed a remarkable firmness and self-control. He was sometimes even joyful, always interesting and charming in communication. It was during these years that he told the above-described stories.

He began preparing for publication his *Selected Works*. He diligently selected his papers for them, carefully read their texts. He was doing this with a carefulness that was usual of him. Once he told me: 'This is nevertheless not science but technics.' I was amazed. 'But,' I told him, 'this is simply applied physics plus remarkable engineering. Do you really cherish what you have done less than a pure scientific work? Do you consider an engineering embodiment of deep science an activity of a second sort? I cannot agree with you.' 'No, this is not what I dreamt about in my youth.' I really adhered to a different evaluation but he was pronouncing his words with a sad conviction in his rightfulness.

In general he seemed to be rethinking his life. He recalled even things of minor importance. Once he told me: 'I cannot forgive myself a weakness that I displayed when I was not elected an academician. At that time I was a corresponding member in a section of technical sciences and, when I was blackballed at the election to full membership I was very disappointed by this. I was cured by Boris L'vovich Vannikov (a clever engineer who occupied high positions, during the war — that of a minister for ammunition; after the war — that of a deputy of Beria on atomic and nuclear projects — E.F.). We were traveling in his railroad car and I complained to him on this injustice (and this was really unjust — E.F.). He replied me with a question: "Tell me, Aleksandr L'vovich, do you have a high respect towards those academicians that did not elect you?" "No, of course not." "Why do you feel offended by them then?" I am very ashamed of this weakness. Later I was nevertheless elected'.[8]

Perhaps he was telling so much about himself that he was reconsidering his past. Once when he described one of the mortally dangerous situations that he passed through honorably and I reacted in some way to his story he said, quietly, in an even voice — as if this was something usual, a phrase that I remember very well: 'A feeling of fear is unfamiliar to me.' He had a right of telling this. Danger did not paralyze him, did not throw him into

---

[8] He told the same to V.L. Ginzburg, see his memoirs on A.L. Mintz in the above-mentioned volume of *Selected Works* of Mintz and a book by Ginzburg V.L. (1992). *On Physics and Astrophysics*. Moscow: Nauka, pp. 450–456.

panic, did not take away an ability of acting cleverly and not losing dignity. We have seen enough examples of this.

Now when a translation of a book by an American "practical philosopher" Deil Carnegie *How to Stop Worrying and Start Living* appeared here everybody can read a prayer 'for every day' created by a professor of the 'chair of applied christianity' of a New York catholic seminary (I have already told about it in Section 4.1): 'God, grant me the serenity to accept the things I cannot change, courage to change the things I can and wisdom to know the difference.'

It seems that this was granted to Aleksandr L'vovich. Dignity and serenity with which he faced the inevitable did specially reveal themselves during the last years of his life. He had enough courage to fight during all his life. He seemed capable of always being able to make a difference between the possible and impossible. Only the story with choosing scientific staff for the Council on accelerators shows that he could also err. A remarkable person. A remarkable life.

6.2   Addendum. Memoirs of V.D. Konen, 'Personality of a Scientist'

My contact with Aleksandr L'vovich Mintz began in February 1972 when both of us recovered after a long medical treatment in a neurological department of a hospital of the Academy of Sciences. Aleskandr L'vovich felt very uncomfortable because of an idleness that was not habitual for him and looked for possibilities of adding color to an unbearable monotony of hospital existence. He had an idea of recalling and telling me some episodes from his numerous travels to countries in Europe, America, Asia Minor. Phenomenal memory, matchless keenness of observation, perfect understanding of the psychology of a listener — all this made A.L.'s stories exceptionally attractive. Soon, however, he felt a necessity for a change from entertaining stories of 'tourist' kind to events in his own life — a life that was long, unusual and rich with dramatic contrasts.

Diving into recollections he gradually unrolled before my eyes a canvas of years gone by with the fate of many people interwoven in it. I saw an infinite line of people of different professions and generations, nationalities and positions in society, with varying levels of mental development and moral perfection — from operetta actors (where A.L. used to have fun in his young years) to big state officials with whom he resolved important problems; from Don Cossacks that taught him riding to American magnates that showed great appreciation towards his scientific achievements. Each person he met in the course of his life was interesting and unrepeatable for him. In general his thirst towards life, his striving to cognize its different sides seemed to be insatiable. The same kaleidoscopic motley was characteristic of a "scenic background" of the reminiscences of A.L. — the diversity of colors that reflected powerful social shifts characterizing an epoch between "the end of the century" and our modern contemporaneity.

An idyllic childhood and boyhood in a highly educated and prospering family in Rostov; beginning of the First World War that caught him in Germany; university milieu in Moscow right before revolution; cruelties and carnages of white bands at Don that spurred young A.L. to join Budenny army; long wanderings with the army across the whole country with a young wife; the twenties and a beginning of a grandiose creative work in Moscow and Leningrad; frequent long trips abroad; new work of immense scale inseparable from the wartime tasks, etc.

These autobiographical stories did in turn grow into a new phase of "contemplating aloud" where there dominated a motive of "settling accounts." Openly and quietly did Aleksandr L'vovich analyze an importance of his contribution to science and strive to understand the most important things in his non-professional life. Neither before an acquaintance with A.L. nor later did I face such a high degree of soberness and fearlessness in the evaluation of one's actions and perspectives. Not a trace of showing off, not the slightest inclination towards self-deception when discussing a behavior that, after time, seemed improper to him, no attempts of "hiding a head under the wing" when discussing possibilities for work in the years to come. A.L. was clearly aware of the approaching end which, it seemed, did not frighten him. The spirit of these thoughts lives in prophetic lines in a posthumously printed book of A.L. in which he supposed that his creative activity would come to an end in 1974.

With the same disarming objectivity and directness did he tell about his personal life — not masking or softening its heavy sides. It was characteristic that A.L. was also not afraid of asking me questions that with anybody else would have seemed tactless. With him, however, everything sounded so simple, natural, benevolent that, in my turn, I opened for him facts from my biography that I would not have been able to reveal for any other stranger.

Long candid conversations in the hospital did lead to a big spiritual closeness between us. Providence allowed us to keep and develop this friendship. Not only did the days of our discharge occasionally turn out the same but, also occasionally, in several days we found ourselves in the sanatorium 'Uzkoe' where we faced a surprise. It turned out that a wife of A.L., Evgeniya Il'inichna, was my relative and despite the fact that these relations had been lost more than half a century ago we clearly remembered these young years and close relations between our families which naturally brought us to a state of deep excitement. As my husband E.L. Feinberg and A.L. had for a long time been feeling a great sympathy to each other there arose an atmosphere of true kinship. A warm and tender feeling towards both of them was strengthened by an understanding that our friendship was not destined to last long. Indeed, in a year there passed away Evgeniya Il'inichna and in three years — Aleksandr L'vovich. However, during this short period of time the mutual sympathy and need remained intense.

Because of my education and profession I lacked the possibility of paying tribute to the main side of spiritual life of A.L. — to his scientific and organizational talent. However, a look from the side has its merits. Having not been blinded by the outstanding scientific accomplishments of A.L. and his broad recognition I judged him "from aside" and without prejudice. It is all the more interesting that outside his profession he was felt as a very big and original personality.

His every word showed an immense cleverness. A conversation with A.L. captured his interlocutors not at all by the gift of storyteller or habitual charm. He always spoke quietly, the voice was dull, intonations were even, mimics and gestures extremely reserved. His speech conquered by constantly pulsating thought, by richness of content and exactness of expression.

If I would be allowed a free analogy I would compare Mintz's speech with such musical pieces in which their dynamics do not develop above *pianissimo* or *piano*, where external timbre effects are muffled and leveled but a powerful artistic effect is achieved through a concentration of melodic thought and its intense internal development. His mind was extremely multi-faceted, covering not only high scientific spheres but also the smallest very earthy and mundane phenomena and always stroked with a freshness of approach. It seemed that A.L. put away all strata of views of others, even those that were solid and firmly incorporated and penetrated anew

into the deepest essence of a phenomenon. He was a pioneer by his nature.
This gave special weight and interest to his every word.

Up to the last years an unquenchable curiosity and receptivity towards
new things lived in him.[9] At the same time he was not afraid of refraining
from expressing his opinions if the issue had not been thoroughly thought
over by him in advance. I remember how once the talk was on telepathy and
although A.L., according to his own confession, was many times witnessing
facts that could have been interpreted as clairvoyance, he did not want to
tell anything on this issue. 'I do not understand this,' he said, simply and
at the same time marking an end to the discussion of this topic.

I think that the originality of A.L. was, in addition to everything else,
predetermined by an immense cultural horizon. Both temporal and geographical ranges of his thoughts seemed unusually broad.

Unlike many of his contemporaries A.L., who was in full sense of these
words a man of our days and our society, was nevertheless in some way
belonging to the early twentieth century. In his appearance he was completely unlike that actor's stereotype of pre-revolutionary scientist that for
so many years reigned on our theater stages and in cinema. Nevertheless he
combined a thinking of our epoch with a live feeling of spiritual searchings
of the beginning of the century.

This often showed itself when we discussed literature. A.L. knew all our
newest publications, tried not to miss anything that was of some interest,
read modern works edited abroad. Nevertheless his conversation did often
contain references to authors that excited him in his younger years and to
which modern time was indifferent. For example Strindberg and Rostand
(especially the latter) were constantly mentioned in his reasonings. In the
same way his musical tastes reflected those of far separated generations.
He knew Shostakovich, Stravinsky of the Western period, Hindemith, etc.

---

[9]Here is a bright example of this. Before meeting me A.L. had no idea on what musicologists do. In the hospital I presented him my new book on Monteverdi that had just appeared — not counting on him reading it. However, he did read very attentively the part not related to analysis of music. He said: "Now I understand that musicology can be an occupation for a life" and asked me to show him my other works. On each of them he made clever and sound remarks, some of them critical.

A.L.'s attitude to films was also characteristic. He saw some of them many times not only for artistic satisfaction but rather because of their informational interest. His attention was drawn by some details in the actor's play that truthfully depicted some definite psychological situation, some of their intonations, gestures and types of movements, special features of costumes, city sights and interiors. One could say he studied them. I remember exactly that he watched the movie *Born Free*, where the main "personages" are African lions, five or six times.

and simultaneously loved composers whose plays had success in the concert programs of the times of his youth — the plays that in our time have practically disappeared from the repertoire. Sometimes A.L. hummed melodies from pre-revolution times and these old-fashioned motives did instantly bring him from our contemporaneity to the spiritual atmosphere of the beginning of the century.

In exactly the same way this typical Russian intelligent, so deeply related to Russia and everything Russian, was in some way a representative of a European educated class. Not many know, perhaps, that a pupil of Rostov gymnasium A. Mintz was in parallel educated in Germany. For many years during summer months the parents of A.L. put him into German pensions. He knew German from his baby years[10] and in Germany he had excelled in it to an extent that in each region he had been taken for a local despite of sharp distinctions in regional pronunciation.

Later this saved him from big problems. When in 1914, after Germany declared war to Russia, he found himself on the enemy's territory, he managed, due to his masterly command of German, to cross all of Germany without having been recognized as a subject of Russian Empire and found his way back to the Motherland. However, he himself liked French, which he also had an absolutely free command of from his early youth, much more than German. In the hospital he sometimes recited to me by heart, for an hour and sometimes more, verses in French, most often from his favorite *Cyrano de Bergerac* by Rostand. He also traveled all over France.

I would like to stress that his command of languages was not a formal "academic" one that is characteristic of many educated people. In this "enlightening" way A.L. knew other languages (from modern ones — English and Italian, from ancient ones — Latin, Greek and Hebrew). Elements of European culture were in his flesh and blood. They showed themselves not openly, pronouncedly, but rather in psychological nuances, in a characteristic construction of associations, in knowing many details of European everyday life that were not accessible to those familiar with Western Europe only through literature or tourist trips.

Let me also note that a manner of A.L.'s behavior was dominated by an automatism of good manners that distinguished him (in my opinion, favorably) from many with whom one has to deal with nowadays, also in the academic milieu. The Western European way was very close to A.L. It

---

[10]This was his first tongue. He said that when he had begun to speak Russian he was at first saying numbers in a German way, i.e. instead of "twenty-one," "twenty-two," he had said "one and twenty," "two and twenty," etc.

could be that because of this he could not, with all his breadth, understand and accept an "American way of life" (he often talked about it).

Aleksandr L'vovich accumulated in himself such an enormous life experience, he personified such a rich alloy of spiritual qualities and talents that presenting his integral portrait is beyond my abilities. With all our close contact during the last three years of his life, the years when Aleksandr L'vovich felt a great internal necessity of speaking on oneself openly and in detail, I could not escape a feeling of knowing very little about him, that beyond each layer of his personality that was opening for me there remained many more untouched ones.

Here I would also like to tell about one more side of his personality. I mean his ability of attracting people to himself. When I got acquainted with A.L. he did no longer possess his former attractive appearance (only his hands that were beautiful in a rare, striking way when telling about strongly pronounced aesthetic inclinations were special[11]). He also did not possess this apparent charm that instantly conquers everyone.

I have no reasons to think that A.L. was more altruistic than a majority of people and its immense self-confidence and absolute lack of pliability were seen with a naked eye. Nevertheless in the hospital I witnessed how did he attract everybody: patients from nearby rooms, doctors, nurses, hospital attendants, waiters, colleagues coming to visit him, disciples, coworkers — not speaking of the uncountable number of friends, with some of whom he was keeping close relations from the time of his youth. Later I met cold people who "flared up" with a tenderness, unusual for them, towards A.L. I knew painfully shy people who lost their shyness when contacting him. Children adored him. Probably this reaction was caused by his own interest in people (he attentively observed everyone he met).

He was also characterized by a very developed and rarely leaving him sense of humor that made others joyful as well. At his lips there usually played a reserved very characteristic smile — all-understanding, somewhat teasing, infinitely benevolent. A.D. Sakharov once told me: "After several

---

[11] A.L. did very much like theater, poetry and music. In his youth, when choosing a profession, he was even hesitant with respect to physics versus actor activity. He could listen to piano for hours. Until the age of 75 he could dance in a modern way. It seems to me however that his aesthetic inclinations did especially reveal themselves in a great necessity in beauty and elegance in the everyday life. For example, he suffered when there were no flowers in his room, when he had to wear clothes made of rough cloth or write letters on bad paper. Such necessity looks all the more surprising if one recalls that he spent many years in the army and was for a long time living in small apartments. For all his life he was a cherisher of feminine beauty.

minutes of conversation with Mintz a joyful mood would not leave me for several hours." None of the photographs of A.L. known to me do not show spiritual warmth coming from him.

The death of A.L. did heavily echo not only in people of science. For many who knew him outside his professional sphere his passing away left a sorrow and irreplaceable void in the soul.

N. H. D. Bohr

Chapter 7

BOHR, Niels Hendrick David (1885–1962)

7.1 Bohr. Moscow. 1961

In May 1961, 16 months before his death, the great physicist Niels Bohr made a two-week trip to the USSR. In Moscow he three times visited FIAN and I was lucky to see him closely for a prolonged time, to be a witness (and even a participant) of his conversations with small groups of the institute's staff during each of these visits. I also saw Bohr earlier when he visited Moscow in 1934 and 1938 and during the last visit also observed him in other institutes but that was a view from the depth of overcrowded auditoriums.

Due to this reason or perhaps due to some other ones an impression made on me by Bohr this time was so strong that I did what I had never done and what one always has to do, not relying on one's memory, in such situations: In the evenings, after conversations, I wrote down what seemed most interesting to me. These were notes for myself only (thus one meets the personal pronoun "I" so often in the text). They laid untouched for quarter of a century until the 100th anniversary of Bohr when I dared to make them public at a symposium dedicated to this anniversary.

Much of what Bohr said repeated his well-known thoughts that were many times published afterwards (also by myself). Still I did not omit these places from the text. In some cases, however, his reflections of that time are new and unexpected nowadays as well. Of course, notes of this kind possibly contain inaccuracies and even errors because of a misunderstanding of what Bohr did say — his pronunciation English was often unclear. However, even with this taken into account one can not deny that notes "in fresh steps" have advantage with respect to later reminiscences, so often did I learn how deceptive is memory after just 10–15 years — and this for people with seemingly good memory.

Nevertheless before publication I tried to cross-check all suspicious places with all possible means. I asked people who had known Bohr well. I am very grateful to all of them. Some insignificant inaccuracies were indeed found and removed.[1] Here I present a corrected version, shortened names of some of my friends are replaced by their surnames, two phrases of too personal a character and too emotional are taken out and foreign words are translated. I also provide explanatory comments. Those that were already included in the edition of 1961 are correspondingly marked (1961).

The text of the notes contains many well-known physicists. Nevertheless some of them could be unknown for readers that are far from physics. It is worth mentioning them. So, Lifshitz is the academician Evgeni Mikhailovich Lifshitz (not his brother, also an academician Il'a Mikhailovich Lifshitz); Academician Nikolai Nikolaevich Semenov — organizer and director of the Institute of Chemical Physics, a Nobel laureate; Academician Dmitry Vladimirovich Skobel'tsyn — at that time a director of FIAN and Nikolai Alekseevich Dobrotin — his deputy. Academicians Aleksandr Mikhailovich Prokhorov and Nikolai Gennad'evich Basov — Nobel laureates who invented in FIAN laser and maser; Vladimir Aleksandrovich Fock, academician, an outstanding theoretical physicist; Academician Abram Fedorovich Ioffe — organizer and for many years director of the Leningrad Institute of Physics and Technics, one of the main organizers of Soviet science in general; and academician Moisej Aleksandrovich Markov — a member of staff of FIAN.

One further meets the names of Ivan Dmitrievich Rozhanski, initially a physicist, but above all an outstanding specialist in philosophy and science of antiquity (under instructions of the presidium of USSR AS he accompanied Bohr on his trip in the USSR, see his memoirs on this in [Rozhansky (1987)]); and professor Valentin Afanasjevich Petukhov (he supervised a work of the laboratory of electrons of high energies).

Among foreign scientists mentioned without explanations are Aage Bohr, a son of Niels who after the death of his father became head of Niels Bohr Institute in Copenhagen; Victor Weisskopf — an Austrian, then American famous theoretical physicist who in the thirties was working with Bohr for several years; Djerd (George) Hevesi — a physicist, radiochemist and radiobiologist who worked in England, Germany, Denmark and Sweden (a Hungarian by origin); a Nobel laureate Max von Laue — a great

---

[1]The original text with a detailed list of all corrections can be found in Pollak, L.S. (ed.) (1988). *Proceedings of the Symposium 'Niels Bohr and the science of XX century'*, Kiev: Naukova Dumka.

physicist who stayed in Hitler's Germany where he behaved himself in a very noble way, saved people and earned universal respect from physicists; Bruno Peters — an American physicist who in his youth, at Hitler's time, emigrated from Germany and at that time was living in Denmark.

These conversations with Bohr were first of all felt as a return to a lucid world of science as it had been formed at the beginning of the XX century and had kept this spirit for a short time only. At that time dramatic or even tragic events in a life of a scientist mainly occurred in a self-determining spiritual atmosphere that was not experiencing a significant interference of a state, its official ideology and material might. Einstein had all reasons to say that science is a drama of ideas. This was augmented by a drama of people.

Suddenly Bohr's conversation is intruded by a "century-wolfhound" with its iron all-covering grip that forces a scientist to face small and big Minotaura of our epoch so that his personality, his inner world can not remained hedged from a bloody neanderthal ignorance of state power.

In 1961 these were already echoes of the past but they did remain sorrowfully meaningful. Therefore even after a quarter of a century it turned out impossible to publish old notes without trying to rethink and comment what Bohr had said — which at least in one case had sounded sensational. There arose many mutually related questions and an attempt to answer them lead to a necessity of discussing at least some aspects of problems that are so actual for our century. This has been done, for one of the issues, in an addendum to the present article (see Section 7.2). A discussion of another question grew into a separate essay, 'Tragedy of Heisenberg', that follows in Section 8.5.

Let us at last see what was written in 1961:

Three days in a row, on May 10, 11 and 12, I saw Bohr. Today, on the 12th, I was sitting next to him at the same corner of a table for two hours. I had enough time to observe him.

On the 10th I was in Dubna at a meeting on weak interactions when Bohr came to see Dubna together with his son Aage. He entered the hall, stooping as usual, and stopped at a podium patiently facing the applause of the audience that raised from its places. Cameras were clicking, everybody (including intriguers known to everyone) were looking at him and happily smiling and he was standing and waiting.

His thick reddish color brows were hanging from exterior borders of the eyes in such a way that if one looked from the side they covered half of an eye, like curtains. His giant pendent forehead, quietly resting on a steady strong horizontal superciliary line, shaded the eyes. A big nose. Relatively fleshy, not flabby cheeks. Everything createsd an image which seemed to be a mixture of two images — of an old fisherman who worked a lot and a pastor.[2] He suddenly smiles, showing big, darkened jaws of an old man, pillows under his eyes became pink and lifted and the eyes joyfully shone and shyness formed around the lips. He smiled with all his heavy face and quickly got back to a concentrated slightly shady quietness.

This time his speech was not long — usual words of politeness and admiration towards "this remarkable institute". He spoke on an importance of international cooperation in science, on how glad he was to welcome in Copenhagen young people from the Soviet Union. He told how primitive had both instruments and methods of theoretical physics been at the times of his youth, 50 years ago, in comparison with what he saw now.

He was not disturbed by two microphones that were from both sides held by stretched arms of radio reporters. He was not waving away neither these nor photographers or cinema people. He had no need of coquetting by being modest, he was really modest and accepted fame quietly.

A very kind person.

On the next day there was a reception and a "standing" banquet in Kapitsa institute. The auditorium was of course full — these were theoretical physicists. A translation was made, wonderfully, by Lifshitz. The meeting was presided by Kapitsa. Bohr spoke relaxedly, sometimes not very consequently. These were reminiscences, very interesting ones. Of special interest were two topics:

1. *Relations with Einstein*. Bohr learned about the first reaction of Einstein on an old Bohr atomic theory from Hevesi who at that time worked, like Bohr himself, with Rutherford and had broad scientific interests. In Hevesi's words Einstein said: 'This is very much clear to me, this is very close to what I could have done. However, if this is right, this is an end of physics as a science'. He was repelled by an idea of jumps from one orbit to another.[3]

---

[2]It is interesting that D.S. Danin in the book *Niels Bohr* in the LFP series (M., Molodaya Gvardia, 1970), of course knowing nothing on my note, also tells that Bohr's appearance seem to mix a pastor and a fisherman.

[3]This was not written accurately. Einstein was of course repelled not by the idea of jumps as such but rather by the fact that according to the Bohr's model an electron, when starting a jump, seems unaware of an orbit it will stop (it emits a quantum of light with an energy dependent only on electron energies at initial and final orbits), see below on reaction of Rutherford.

The first personal meeting took place in 1920. Einstein still did not want to accept a dualistic concept of light. At last Bohr said: "Well, you should address the German government so that it either forbids diffraction grids and orders to consider photon a particle or forbids photoelectric cells and orders to consider photon a wave". (It is interesting that by that time Einstein had already introduced radiation coefficients and, using them together with his idea of photons, had derived the Plank formula.)

During the second personal meeting Einstein said: 'Well, to start with let us firmly fix those positions in your theory that I can accept from my point of view and, relying on this foundation, we would logically proceed further'. Bohr answered to this: 'I would betray science if I would agree to firmly fix anything in this new field in which everything is still too unclear'.

All of us, said Bohr, should be immensely grateful for Einstein for his constant critique of quantum mechanics, attacks on it, that forced us to search, again and again, for solution of deep problems and thereby erect a harmonious scheme of modern theory;

2. *On disciples.* One asked Bohr on which secret did he have to be able to attract so many brilliant disciples, to create such a school. Bohr said: 'There is no secret here. There were two fundamental principles that we always followed. First, we were never afraid of looking dumb.[4] Second, we always tried, when having a seemingly final result, to point out new questions and new uncertainties.

'When Weisskopf once came to me and told about one physicist who was working with him that he was showing disrespect to everyone I told him that here nobody takes disrespect seriously.

'In general, there was a lot of humor. Humor is necessary because makes passing through difficulties of life easier.

We told then that there exist two variants of truth. A truth of the first kind is such that an inverse statement is clearly false. Such truth is relatively trivial. There also exist, however, such truths, we are calling them *spiritual truths*, for which, with deeper understanding, inverse statements turn out to be valid as well. When such truths are found science is living through a specially interesting and important period. This is its best time'.

It was interesting to observe how, when speaking, Bohr was gradually getting agitated, eyes were beginning to smile and wrinkles to play. However, at this moment Lifshitz touched him by the sleeve and asked him to

---

[4] A wonderful error of Lifshitz who was translating: Precisely in this place he made the only mistake and, under laughter of the audience, translated: "We are never afraid of telling a disciple that he is dumb" (1961).

pause to give him a possibility to translate. Bohr, perplexed, turned around, instantly faded, darkened and obediently sat down to begin slowly flaring up again when invited to continue.

Bohr spoke a lot about Rutherford, very enthusiastically, and invited Kapitza to do the same. One showed a movie shot by Kapitza during the previous visit of Bohr when they had gone to the Moscow Sea for a picnic. Then everybody moved to banquet tables and began to eat. Bohr, together with Tamm, Semenov, Lifshitz sat at a small table at the wall and were having a conversation under the objectives of dozens of amateur photographers. In Tamm's words, when Semenov proposed a toast for the creator of modern theory of atom, Bohr said: 'I am also just a man, it is difficult for me to endure too many complementary statements'.

12 May, 1961

At last, today Bohr was in FIAN.

He came here alone and was walking up the wide stairs of a porch lit by sun in a short summer coat and a blue velour hat with a low trampled crown. He was visibly leaning forward (Jean Gabin could play him in this way). One led him to the office of Skobel'tsyn where he was met by Igor Evgen'evich Tamm and several other people: V.L. Ginzburg, I.M. Frank, N.A. Dobrotin, A.M. Prokhorov (when I.E. introduced Prokhorov, Bohr could not recognize what maser was), I.D. Rozhansky and myself. Bohr sat on an antique leather divan, took out two pipes, a leather wallet-shaped tobacco poach, a giant matchbox and immediately started an uninterrupted work with all this. There began awkward conversations, the questions.

Answering my question on whether he saw by Bohm any reasons beyond the standard ones (Tamm added: by de Broglie, Vigier and others[5]). Bohr, putting forward his upper jaw, smiled with all the might of his immense smile. Wrinkles, running from corners of the eyes, covered the cheeks, blue eyes were enflamed by a kind smile of a god who *knows* and he answered: 'O, no,..., not a slightest new argument!" He added: 'And, in general, backward motion is already impossible. Quantum mechanics, like relativity theory, marked a beginning of a new stage and it is already impossible to

---

[5] D. Bohm is an American (later English) theoretical physicist, L. de Broglie, one of the creators of wave theory of matter and his coworker A. Vigier are French theorists. All of them were for many years trying to create a theory that would have freed quantum mechanics from such incomprehensible, from the point of view of classical physics, principally important notions as a reduction of wave packets, etc.

come back'. Ginzburg asked him on what did he think of a new theory.[6] Bohr's reply was withered. When speaking he was not always 'warming up'.

After that Bohr went to the conference hall which was full — there came about 700 people. There were also philosophers which of course came to observe an idealist.[7]

After usual problems with a microphone Bohr began his lecture. Tamm translated brilliantly. It was amazing how Bohr himself and Tamm could endure such pressure. In the hall it was stifling, especially with curtained windows, albeit all three doors in the wall opposite to the podium were opened. One saw that people stood also in the foyer.

Bohr spoke about the same, almost the same as in 1934 in the great physical auditorium of old physics faculty when Tamm had also been translating. He was still in the same epoch of paradoxes and arguments with Einstein, but each time when he unraveled the essence of a paradox and explained why Einstein was wrong his face was again enlightened by a joyful and somewhat triumphant smile of cognition. Explaining the first paradox, with two slits, and speaking on the suggestion of Einstein to determine through which of them did electron pass Bohr added, seriously and even with a shade of worry: 'This was a very awesome moment. If Einstein's hypothesis had been true this would have meant a collapse of all the beautiful system that had been constructed'. Saying this he looked somewhat upwards, eyes rolled up, so one could see the whites of his big eyes. His big straight nose was provokingly lifted.

Also afterwards, when discussing other paradoxes (for example, that with a watch measuring a moment of photon emission in a cabin suspended on springs simultaneously with waiting), he added: 'This was a dramatic moment'. And in the third case, in 1927: 'This was a difficult question'.

One saw in which torments, both mental and spiritual, was his understanding of the quantum nature of matter born. He said: 'Without quantum laws the surrounding world would not exist'... 'One speaks about some Copenhagen treatment of quantum mechanics. In fact there is no such

---

[6]He meant numerous attempts of creating new theories of particles and fields that began in the middle of the fifties.

[7]I beg you pardon for giving way to accumulated irritation with respect to our vulgarizing philosophers that condemned Bohr (for "bourgeois idealism") and, together with him, our physicists that shared his views and accepted his theory. In fact after 1953 a situation in philosophy was experiencing a gradual but strong change, strong, albeit gradual change to better (see Section 7.2).

treatment. One has quantum mechanics which treats experimental information'; 'In atomic phenomena one cannot neglect an impact of the measuring device. By taking this into account we enrich our information'. He spoke about a duality principle related to the impact of observation which is absent in classical physics. 'Quantum mechanics is a simplest of sciences in which this impact does take place. One also knows, for example, how strong is this impact in psychology'; 'Quantum mechanics answers a question on results of experiments. Its conclusions are quite objective and do not experiencing subjective influences'.

Of course, he did not say anything new but this was a charming and touching *show*. Understanding him was almost impossible although when delivering a lecture he pronounced English words with much more emphasis and clarity than in a private conversation.

After the lecture we moved to the Skobeltsyn's office where we sat in pairs on four sides of a small (card) table covered for eight. I got (due to Dobrotin who pushed me) a place next to Bohr (more exactly at the same corner of the table where Bohr sat) so that I could closely observe him from the side all the time.

Bohr took out two pipes (black, smoked over), matches and his hands began an incessant motion. The same incessant slow work was also begun by his tongue. He was lighting matches all the time and was holding a burning match in the hand for a long time (the pipe, on contrary, was burning quite seldom). At the end of the conversation a tea saucer specially put before him was filled with a hill of matches. I was not always able to understand the words he was pronouncing. His ability to talk amazed me: after the two-hour lecture when he rested only when Tamm translated he was for two hours speaking at the table and stopped only when Rozhansky insistently reminded that there remained three and a half hours until his flight to Tbilisi.

In Bohr's pronunciation "titich" meant physics, "susicion" meant superposition. His fingers were not those of an old man, they were neither thin nor thick. His joints were not knotty. A narrow gold ring on the right hand. Age-related skin changes were in fact noticeable only in a small intumescence under the left eye. He spoke and lit matches slowly but somehow it turned out impossible to put in a word. He mainly addressed Tamm and Ginzburg who sat in front of him across the table — apparently, he liked Tamm.

What was amazing in discussions of scientific topics was a lack of scale in estimating what should have been broadly known and what should have not. Answering a question on whether he found it necessary to introduce a fundamental length he answered quite positively and in explanation began presenting detailed arguments that were clear without explanations — like an impossibility of constructing a quantity having a dimension of length from c and $\hbar$, pair creation at small distances, changes in notions of space and time at small distances, etc.

When Ginzburg asked about his conflict with J.J. Thompson he said that this was too short and simplified a formulation, that J.J. was a genial man who first spoke about an electromagnetic mass of an electron. J.J., however, wrote a paper that was totally erroneous and contained, in particular, a statement on diamagnetism of free electrons. Here Bohr began to explain why free electrons cannot give rise to diamagnetism in classical physics. He was explaining until Ginzburg lost patience and said that all of us knew these arguments very well. Apparently Bohr did not understand that his argumentation of 1911 was known to everyone. Bohr added: 'After that it turned out difficult to keep the former relations with J.J.'[8]

Bohr wrote on the paper with the same pen with which I am writing now. Circles and arrows were gnarled, lines shaking — it was painful to see it. However, when he was asked to sign on the back of photos that had just been made, he slowly wrote in an exact schoolboy writing 'Niels Bohr' — as 'still active old people' always do. He also signed a piece of paper given by Ginzburg. Ginzburg took it for himself (I declined his noble suggestion of sharing this piece with me by tearing it in two parts).

---

[8] In a letter to me written in 1972 a many-years coworker of Bohr, Leon Rosenfeld (1904–1974), whom I asked to confirm a correctness of his note wrote that I put too much emphasis on this episode. Rosenfeld writes: 'It is true that Bohr did not find a common language with J.J. Thompson but the question on diamagnetism of metals did not play an essential role in this. Everything was a result of an unlucky superposition of Bohr's shyness, his pronunciation and Thompson's "island" thinking (a "foreigner" who speaks English badly can not be too good).

'In reality Bohr was at that time working on his dissertation, studied Thompson's works very diligently and found several serious errors in them. He naively thought that Thompson would have been happy if one had pointed these errors out to him. Therefore when he saw him for the first time he opened a volume at one of the pages of Thompson's article containing such an error and, putting a finger to the formula, told with a sweet smile everything he was able to say: "This is wrong". He was lost when he saw that Thompson was not happy and, on the contrary, was running away from Bohr when he approached him with an open book in his hands. Nevertheless this disappointment did not affect an admiration that Bohr felt towards Thompson. He only concluded that he had to learn English better. This was what he accomplished by reading *The Posthumous Papers of the Pickwick Club* and looking for each unknown word in a dictionary'.

We spoke again about Einstein, on why did he work alone. Bohr said: 'Einstein was not only a genius, he was also a remarkable, very kind person. His smile is still before my eyes. However, he was used to doing everything by himself and he did it perfectly. Gradually this led to a development of a feeling of infallibility in him. He even began speaking in a language of orders of a day. It became possible that he answered a journalist's question on quantum mechanics: "This is nonsense". And, in fact, he did not know quantum mechanics. In order to understand it one needed mutual discussion. For him this was out of question'.

I asked how Rutherford reacted at the old quantum atomic model. Bohr replied: 'He did not say this was foolish. However, he could not put up with the fact that an electron, when beginning a transition, does not know on which particular orbit it will go. I told him that this is absolutely the same as in beta decay. However, this did not convince him'.

Ginzburg asked what did he think about statements of Fock, that after their discussions Bohr had somewhat changed his point of view. Bohr answered: 'Fock is a very kind person. When I first saw him I thought that he looks like Pierre Bezoukhov.[9] We were all very glad that he visited Copenhagen. However, I changed nothing in my views. Since 1934 I did not change a single syllable. Fock was simply thinking that our points of view were wrong, but he came, understood them and understood they were right'.

I wanted to defend Fock and I said: 'You should know that Fock vigorously defended quantum mechanics in hard times when ignorant philosophers were attacking it', Bohr smiled and said: 'Philosophers as such are very good people. However, people feel pity towards them not only in your country but everywhere — in America, in Denmark. In Denmark one says that a philosopher is an unhappy person — he will never learn anything".[10]

Bohr liked to tell anecdotes. On Dirac, of course, on how untalkative he was. On how he delivered a lecture in America and when a listener said in response to a suggestion of presenting questions: 'I do not understand this and this', Dirac answered: 'This is not a question, this is a statement'. In connection with Dirac's non-talkativeness he recalled an anecdote on an Englishman who gave a silent parrot (Bohr said in Russian, *Popogaj*) and answered to a complaint of a new master: 'Sorry, I was confused: I had two of them. I thought I had presented a talker but it turned out to be a

---

[9]Indeed, V.A. Fock was tall with stout pale face, wore spectacles with round glasses, spoke thoughtfully, etc. — in agreement with how all of us imagine Pierre Bezoukhov.
[10]See Section 7.2.

thinker'. On how after Bohr's lecture in America a student came forward and asked: "Were there indeed such donkeys who thought that electrons really move in orbits?"[11]

At last Bohr yielded to the schedule with the same readiness as usual and left. Before leaving he came to Ginzburg and told him something he did not understand. Smiling, he said goodbye and his face was again covered by a shadow of either concern or old man concentration and, carefully stepping, he went down the stairs.

17 May, 1961

Bohr returned from a three-day trip to Tbilisi and visited our institute again. I was again observing him for three hours.

Before that Bohr had been in the university where he received an honorary diploma and at 3 p.m. he went from there, with his son Aage and Rozhansky, to FIAN. We were waiting for him. Skobel'tsyn who had returned from Romania was also there.

In my understanding Bohr had to be very tired. It was perhaps due to this that it seemed to me that he was stooping somewhat stronger, watched more attentively where to put his feet than during previous visits. Nevertheless he endured all the visit (without lunch!), a tour over several laboratories, after that a conversation in Skobel'tsyn's office and the conversation was again interrupted with difficulty. This evening he still had to meet writers. In a day he had to fly back at 6 a.m. to preside a session of Danish academy of sciences in the evening.[12]

After a short conversation in the office (Skobel'tsyn, sitting vis-a-vis, described a structure of FIAN while Ginzburg, Tamm and myself were talking with Aage.[13] We talked about Peters. Bohr, in reply to the words of D.V., was gloomily nodding.) we went to see some laboratories. One showed Bohr an automatized device for reading photoemulsion tracks, the Petukhov synchrotron at 700 MeV, Basov's laser, Prokhorov's nuclei orientation device. Bohr preferred to see installations, not talk to theorists.

Under Tamm's request I accompanied Bohr. For a long time we were walking from laboratory to laboratory, up and down stairs, through the

---

[11] This refers to a so-called old Bohr model of atoms (1913). In a more consistent quantum mechanical theory atomic orbits do not exist.
[12] Rozhansky told me that in Tbilisi at some banquet Bohr consumed a whole bottle of khvanchkara [Georgian dry red wine] (1961).
[13] Aage spoke Russian well. To my surprise he recognized me (we did not see each other since his visit in 1956). He came to me and asked *"Nu, chto noven'kogo?"* (1961).

yard. He did not ask questions and it seemed he only pretended to understand. When, however, one of our big physicists, with all the undue familiarity of his behavior (he was almost pushing Bohr and habitually laughing), could not explain the physics of the process he studied and I, relying on his fragmentary remarks, tried to give an explanation (although I did not understand it myself) and failed, Bohr, who was sitting on the chair, lifted his eyes at him, looked with surprise, then looked at me and asked for printed material. This was however also lacking.

Bohr walked through the institute's yard relatively quickly, slightly bending at each step, strongly stooping, hands freely swinging, half-bent hands looking backwards — in a pose of a tall old man that Ioffe had, simply saying, in the pose of an oran-gutan. (Somebody said about Bohr: '... genial neanderthal'. Apparently from neanderthal to Bohr is also "only one step".) Each staircase was a bit of a problem. I translated to him explanations of Petukhov — he was interested. He listened, sitting on a chair with his face readily expanding into an all-embracing smile, lifted to us who were standing. A spontaneity of transition from the full seriousness to the full and joyful vivacity of all the heavy face was somehow childish.

He obediently walked and, albeit there was no time to smoke and he knew it, only from time to time he took out tobacco pouch, pipe and matches. He simply wanted to touch them, these were his talismans.

Everybody sighed with relief when we were back in Skobel'tsyn's office (he was no longer there) and sat on chairs and divan. I was again sitting face-to-face to Bohr. We had to wait for Aage who had gone to Frank's laboratory and Tamm used this possibility to invite in our young theorists. He introduced all of them and Bohr, sitting, shook hands with everyone. When Tamm began describing a work of his theoretical department (he said a couple of words) I reminded Bohr that before the war Markov wrote to him and got a detailed answer on restrictions imposed on measurability of fields by finiteness of existing charges ($Z < 137$). Bohr, perhaps out of politeness, tried hard to recall this and pretended that he did remember something related. When I however explained what was the issue he changed immediately and in the same slow and quiet way as before but in a businesslike manner said: 'There could be no absolute restriction on measurability here'. The mobilization was immediate. I refrained from discussion

(I did not remember well this work of Markov who was absent and I just aimed at reminding Bohr his name[14]).

After that young people began asking questions and Bohr, working with pipes, matches and smiles, did with pleasure begin his recollections. He often turned to me and I was again directly observing him. The eyes were blue again but the blue was pale. Apparently he was very tired.

The first story was on a Nobel medal dissolved in acid. It turned out everything was very concrete and this romantic story had purely rational reasons. Because of this it looked even better. First, this was not the medal of Bohr but of Laue. When the Nazis began confiscating gold Laue sent Bohr his Nobel and Plank medals to keep. However, when Nazis came to Denmark there arose a danger that they would find medals and because the name of Laue was on them he could have suffered. A chemist, Hevesi, suggested to dissolve them in aqua regia and this was what was done. The Plank medal that had been made in Germany contained a lot of admixtures and dissolved poorly. On the contrary the solution of the Nobel one was standing in a green bottle for a long time. After the war gold was extracted, sent to the Nobel committee in Stockholm and a new medal was made of it (he spoke about Laue with a sympathy in voice).

The second story was about fleeing from Denmark:

'We had difficult times. Of course not as difficult as you did. We learned that the Nazis had put me and my brother (a mathematician Harald) as number ones on their list of enemies (although I am not Jewish, only my mother was). One had to flee'. There followed a poorly comprehensible story on secrecy conditions that led to the fact that something went wrong with a boat. All of them did board and, having snuck past German patrols, in two and a half hours reached Sweden.[15]

The English sent for Bohr a Mosquito plane which had room only for two pilots (these were Norwegians). Bohr was placed into a bomb hatch, he was given a parachute, taught how to inflate a boat in the case of landing on water. He could not do anything of this and, as far as I understand, was going to fly on a plane for the first time. One pulled a helmet with phones for communication with pilots (who should have warned Bohr of dropping him off in the case of German attack) but the helmet turned out to be

---

[14] It turned out that the work of Markov I described to Bohr was not the one discussed in their correspondence (see [Markov (1986)]).

[15] Almost all Danish Jews whom, as it had become known, there threatened a deportation to concentration camps, were within several days brought by Danish Resistance to Sweden where refuge had been already prepared for them on the shore.

small, it was pulled on from both sides with difficulty and the phones did not function. Therefore he did not learn that he had to use an oxygen mask and in England he was taken out of the plane unconscious.

'This is horrible'. said Tamm. Bohr continued, 'No, it was more horrible later'. As a personal secretary of Bohr, Aage should have come with the next such plane (a wife and three sons remained in Sweden) but Bohr learned that Germans shot this plane sown. It turned out however that at the very last moment the English took Aage out and put in an English military who was on his way from Russia. Aage came with the third plane.

During the war Bohr and Aage traveled from the USA to England, on one way with a ship, on another on a plane. A mission of Bohr was in keeping consent among allies.[16]

In the third story he freely and readily answered a question that seemed impudent to me — on whether Jungk had described a behavior of Heisenberg correctly.[17] Bohr even got agitated: "Heisenberg is a very honest man. It is astonishing, however, how one is capable of forgetting one's views if he is gradually changing them. In the Jungk story there is not a single word of truth. (Bohr often used this expression that perhaps should not be taken literally.) Heisenberg came in the autumn of 1941 when Hitler had occupied France and was rapidly progressing in Russia. Heisenberg persuaded me that a victory of the Nazis was imminent and that it was foolish to doubt it. Nazis did not respect science and therefore treated scientists badly. It was necessary to unite and help Hitler and then, when Hitler would win the attitude towards scientists would also change. One had to collaborate with institutions created by Nazis".[18]

Bohr lit the pipe and, keeping it in his mouth and puffing, looked at me with surprise. Because of this his face was strongly stretched, his brows that hung on corners of his eyes do not hide immense somewhat yellowish whites

---

[16] At that time I did not understand what this was about. In reality the story was on how Bohr tried to persuade Roosevelt and Churchill to inform the Soviet government on works on atomic project so that allied relations could have been kept after the war. Nothing came out of this. Experienced politicians played with Bohr like with a ball, throwing him to each other.

[17] Here I mean the well-known book by R. Jungk, *Brighter Than a Thousand Suns* in which he tells that in 1941 Heisenberg came to Copenhagen to inform that Germany would not be able to produce an atomic bomb and that it was necessary to persuade English and American physicists to stop working on it. However, he could not speak directly and his cautious speech did only frighten Bohr who, after the first words on the bomb, could not understand anything so nothing came out of this conversation. The same version is given in other books (see Section 8.1).

[18] Apparently Bohr said "institutions" which meant "establishments".

and blue pupils. He was still surprised. "He thought that Hitler's victory was imminent! I could not say a direct "no" to him. I said that I could not decide on such an issue by oneself so I had to consult my colleagues". (So, Bohr did not trust him. Jungk was right that they were secretive with each other.) Bohr continued: "From what Heisenberg said we came to a conclusion that Hitler would possess atomic weapon.[19] Otherwise why was the victory imminent? I knew that atomic weapons were possible already before the war and published a note that this was probably more possible with ^{235}U rather than with ^{238}U" — (and Bohr again began to explain elementary things — why 235, not 238). "But then, with an electromagnetic method of isotope separation, Lawrence would have needed that all American power stations had been working for him for many years".

'However, later neither Heisenberg nor Weizsaecker who came with him did return to this issue[20] (either understood that I was outraged or the first defeats in Russia, near Moscow, influenced them). Gradually their views changed. I wrote about it to Jungk but this had no effect. It is astonishing, however, how people forget their words if their views have been gradually changing'. Here Ginzburg put in: 'People are inclined to forget those views that they would like to be forgotten'. At that moment Bohr was firing his pipe but understanding shined in his half-smile.

We began getting up and leaving. I approached Aage who had arrived shortly before this and asked him: 'Do you think that one should not have relations with Heisenberg?' He denied it and referred to the good attitude of Heisenberg towards his father. He said that although Heisenberg was a nationalist he 'did not like Nazis and anti-Semitism', etc. But when I said that the book of Jungk had here been more useful than harmful he repeated very decisively: 'We do not like this book'. The decisiveness was strongly stressed.[21]

Everybody was already leaving. Of course, we went to the car. Bohr bowed and sat in the car first but when he saw that we came close to the

---

[19] As evidenced by people close to Bohr he never admitted that Hitler did have a such weapon. Of course, he, like all Western and our physicists, was afraid that Hitler could have got it.

[20] A younger friend and student of Heisenberg, a theoretical physicist Carl Friedrich von Weizsaecker came with him. He waited for results of the talk of Heisenberg and Bohr in a hotel.

[21] What Bohr, according to my notes, did tell about the famous meeting with Heisenberg does radically differ from its descriptions in the literature, also in Heisenberg's memoirs. The words of Bohr provoke numerous questions. These are addressed below in Section 8.1.

car and Aage shook hands with everybody he came out again, bowing under a roof of a limousine that was too low for him, and presented everyone a handshake and a smile of his giant face.

7.2 Addendum: Physicists and Philosophers

I dare to publish this exchange of replicas about philosophers that may seem strange to the readers from younger generations and may even seem insulting to philosophers. However, this is nowadays when during the last decades the situation in philosophy did in many aspects change. New people who understand and cherish achievements in natural sciences have come to philosophy and many qualified mathematicians, physicists and biologists are professionally working in philosophy and numerous discussions on methodological problems are usually serious, without exerting any kind of pressure.

The thing is that beginning from the end of the twenties the, at that time, new physics (theory of relativity and quantum mechanics) became an object of furious attacks from the side of philosophers with high titles that found support from party organs and several professors of physics from old generations who were extremely conservative in scientific questions. This new physics was declared to contradict materialism, to be an idealistic bourgeois figment while its supporters were meant to be servants of the bourgeoisie. Later, beginning in the thirties, the same happened to genetics and other most important branches of biology, and in the forties, with cybernetics. The damage to our science and economy is hard to overestimate.

For physicists this began earlier than in other sciences and did not go to such extremes as in biology. Some leading physicists were, in vain, trying to explain the truth to the accusers (one such episode, a discussion between I.E. Tamm and conservators is described in Section 2.1. There it is described how it happened). At the end of the forties it came to the situation when in some institutes frightened professors refrained from lecturing on theory of relativity and quantum mechanics.

This "ideological critique" reached its peak in 1949–1952, in spite of the fact that by that time the new physics had in fact become stronger, had many authoritative and active defenders even when not all of them wrote articles or participated in public discussions (it is sufficient to mention the names of S.I. Vavilov, I.V. Kurchatov, A.F. Ioffe, I.E. Tamm, L.D. Landau, Ya.I. Frenkel and V.A. Fock).

For a reader to feel an atmosphere of that time let me give an excerpt from a solidly looking collection of articles [Kuznetsov (1952)].

Processes that freed a way for contemporary science in agrobiology, physiology, microbiology and cellular biology[22] are still very slow among Soviet physicists. A presence of vicious views in physics was uncovered, for example, in a discussion of the so-called duality principle of Bohr and Heisenberg. Idealistic views are nowadays penetrating physics also in relation to anti-scientific interpretations of theory of relativity'. (See Introduction in [Kuznetsov (1952)].)

One of the leading philosophers that specially dealt with physics (I.V. Kuznetsov) wrote in the same book: "What Einstein and Einsteinians set up as a physical theory cannot be recognized as a physical theory" (p. 52). He insistently follows an idea that there is no relativity theory as such — there is "a physics of fast motions" (p. 73). He writes about "reactionary bourgeois scientists that lift Einstein on a shield as an alleged creator of a new physical theory on space and time", (p. 50) and opposes to the Einstein views the earlier attempts by Lorentz who allegedly was a bigger materialist because he derived a shrinkage of moving bodies from concrete physical laws of electromagnetic field (p. 71).

At the same time the main importance, main beauty of the relativity theory of Einstein are precisely in the fact that his new concept of space and time provides new universal laws, for example for the same shrinkage of bodies held together by *arbitrary* forces, electromagnetic, nuclear, neutrino. This is because all processes take place in space and time linked together (it is of course possible to calculate the shrinkage process for a given force but thanks to Einstein we know the result beforehand). Also a slogan given in italics: 'Denunciation of reactionary Einsteinianism in the domain of physical science is one of the most actual problems for Soviet physicists and philosophers" (p. 47). These general considerations are accompanied by illiterate 'refutations' of some concrete consequences of relativity theory, for example of 'reversibility of slowing down of time' (pp. 55–56).

Let us however return to a situation with V.A. Fock described in the main text.

I would like to recall one discussion on this very collection of philosophical papers in FIAN that took place in the difficult time of 1952.

---

[22] Cybernetics had been strangled by philosophers in its cradle, it was not even mentioned.

The conference hall was full. In presidium on the podium there sat philosophers (among them there really were, in Bohr's words, 'nice' people that wanted to make physics more 'cozy' and understandable for them, brought up on classical concepts, these people did not let themselves go to a level of politically dangerous rude statements. Based on their critique such political conclusions were however made for them by party and other organs). Their main opponent, who insistently and patiently tried to clarify the truth, was V.A. Fock. However, at long last he also could no longer sustain absurd statements and quibbles. Not changing his usual manner of speaking with pauses he indignantly said: 'We are arguing with you for 25 years. During this time we, physicists, learned dialectic materialism and know it but you, philosophers, have not learned anything in physics and do not understand it'. He put his hearing-aid into his case (Fock was somewhat deaf), he clicked, under dead silence of the audience, with the case's lock, in firm steps went down the podium, walked along a long aisle to the exit and, not quite silently, closed the door behind himself.

Nowadays this story looks almost improbable. But this took place, it really did. There was also a temptation of agreeing on the compromising and vulgarization of science — just to avoid a dangerous fury of philosophers. For example, one in general talented theorist, an author of a good university textbook, decided to please them and get rid of a claim of relativity theory on a complete equivalence of all inertial systems of reference (this means, in particular, an impossibility of an existence of classical all-penetrating ether of the before-Einstein epoch). In his construction there appeared a gradual transition from one inertial system to another in which there arose a special "most inertial" system of reference. Fock quickly caught him at an error in his reasonings and he lost a respect of the majority of theoretical physicists for good. This visible conciliation caused anger in many.

One can find more amazing the words of Bohr on philosophers in the West. The thing is, however, that the philosophy that reigned there was of positivist kind and Bohr did not agree with it. Let us also recall Einstein who wrote: 'I do not see any "metaphysical" danger in including into a system, as an independent notion, a *thing* (an object in the sense of physics) although even mentioning of a notion of "objective world" is suspicious in the eyes of philosophical police'. [Einstein (1965a)] "It is in the nature of things that we can not discuss these issues otherwise than with a help of notions created by us which are impossible to define". [Einstein (1965b)] And later: "... In our days there dominates a subjective and positivist point of view. Proponents of this point of view declare that consideration of nature as objective reality is an old-fashioned prejudice... In any epoch there is some dominating fashion and most of the people do not even notice the ruling tyrant".[Einstein (1965c)]

All this is aimed against philosopher-positivists with whom he failed to reach an understanding.

In this way one and the same science was accused by some philosophers in positivism and idealism as it was by us (which threatened scientists by real and heavy punishment); by others, for not accepting positivism as it was in the West (because this was not threatening with punishment scientists felt compassion towards philosophers). Of course, one should not say that philosophers never understood science and do not understand it now. This is seen from a history of philosophy and from what was happening here in this field in recent times.

W. K. Heisenberg

Chapter 8

HEISENBERG, Werner Karl (1901–1976)

8.1 Tragedy of Heisenberg*

When in 1961 Niels Bohr was in Moscow he visited FIAN three times and had long discussions with small (six–seven people) groups of its staff members headed by I.E. Tamm. I was among these people. Once there took place a significant episode on which I wrote a detailed note in my notebook in the evening of the same day, May 17. I have already described this[1] but would like to remind the reader what happened.

The conversation was on Werner Heisenberg, an outstanding scientist with world reputation, one of the creators of the physics of XX century, a Nobel Prize winner at the age of thirty, one who stayed in Hitler's Germany and was one of the leaders of the German 'uranium project' (which, as well known, did not bring any results). Bohr's statement about him, unexpected and even sensational, went beyond a simple interest in the personalities of Bohr and Heisenberg. In essence one touched upon a problem of the behavior of a scientist in conditions of a merciless dictatorship in a totalitarian state.

Everything began with a question to Bohr of whether Jungk did correctly describe a meeting of Bohr with Heisenberg who came to visit him in September 1941. What one had in mind was a place in the well-known book presenting a history of creation of atomic bomb for a broad audience [Jungk (1960)] which describes that an aim of Heisenberg's visit was

---

*A journal version was published in the magazine *Znamya* (Nmb. 3, 1989). Its German translation is in Duerr, H.-P., Feinberg, E., van der Warden, B.L., von Weizsaeker, C.F. (1992) *Werner Heisenberg*. Wien: Hansen Verlag.

[1]See Chapter 7.

to inform that Germany would not be able to produce an atomic bomb and that one should have persuaded English and American physicists to stop working on this as well. Heisenberg could not speak directly and his cautious speech did only frighten Bohr who stopped understanding anything after the first mention of the bomb. The conversation ended with nothing. This version is presented in other books [Irving (1960); Groves (1969); Snegov (1972)] and also in the book by Heisenberg himself [Heisenberg (1980)].

Here is what I wrote: 'Bohr even got agitated: "Heisenberg is a very honest man. It astonishing, however, how one is capable of forgetting one's views if he is gradually changing them... Heisenberg came in the autumn of 1941 when Hitler had occupied France and was rapidly progressing in Russia. Heisenberg persuaded me that a victory of the Nazis was imminent and that it was foolish to doubt it. Nazis did not respect science and therefore treated scientists badly. It was necessary to unite and help Hitler and then, when Hitler would win, the attitude towards scientists would also change. One had to collaborate with institutions created by Nazis".

Bohr lit the pipe and, keeping it in his mouth and puffing, looked at me with surprise. Because of this his face was strongly stretched, his brows hung on corners of his eyes do not hide immense somewhat yellowish whites and blue pupils. He was still very surprised.[2] "He thought that Hitler's victory was imminent! I could not say a direct "no" to him. I said that I could not decide on such an issue by oneself so I had to consult my colleagues." So, Bohr did not trust him. Jungk was right that they were secretive with each other. 'From what Heisenberg said we came to a conclusion that Hitler would possess atomic weapons. Otherwise why was the victory imminent?'... However, later neither Heisenberg nor Weizsaeker who came with him did return to this issue (either understood that I was outraged or the first defeats in Russia, near Moscow, influenced them). Gradually their views changed. I wrote about it to Jungk but this had no effect ...'

We began getting up and leaving. I approached Aage[3] who had arrived shortly before this and asked him: 'Do you think that one should not have relations with Heisenberg?' He denied it and referred to the good attitude of Heisenberg towards his father. He said that although Heisenberg was a nationalist he 'did not like nazis and anti-Semitism, etc.'

---

[2] Later Landau told me that in a private conversation with him Bohr, having had said the same, was not "surprised" but rather excited and angry.

[3] Aage Bohr (b. 1922), a son of Niels Bohr who accompanied his father. A theoretical physicist. Later a Nobel Prize winner.

So, in 1961 Niels Bohr made an astonishing statement: Heisenberg, "a very honest man," was in September 1941 persuading him that all scientists (and intellectuals in general?) should unite, help Hitler and thus ensure his good attitude towards science. Some historians even think that Heisenberg persuaded Bohr to participate in the German uranium project. This is, however, already nonsense. Aage Bohr did categorically state in 2002 that knowing the personalities, characters, political views of the participants of the meeting and also those of Weizsaeker this possibility should be excluded [Bohr (1957)].

8.2 Bohr and Heisenberg

Heisenberg was an outstanding physicist, a creator of quantum mechanics and, at least in the past, a close friend of Bohr. In Copenhagen he was at some time essentially living at Bohr's place and they were discussing until exhaustion, for days and nights, main difficulties of the new science. As a result of these discussions there appeared famous uncertainty relation of Heisenberg and duality principle of Bohr — two aspects of the same fundamental law of quantum mechanics. And now they met almost as enemies. As rightly noted by Weizsaeker[4] this was "a tragedy of an epoch." So, there arose two different versions of their meeting. Here there immediately arise several questions.

Did Heisenberg really suggest scientists to unite and help Hitler? How could in this case Bohr tell that Heisenberg was a very honest man? What was the true political face of Heisenberg, to which extent did he collaborate with the Nazi regime and why? How did he behave himself in Hitler times in general? And further: Why German atomic scientists and the Nazi state did finally fail to make an atomic bomb? And what is Heisenberg's "guilt" in it? Answers to these questions, with an exception of the first two ones that have not been discussed so far, differ widely.

There exists an immense literature on the German "uranium project," on the "Heisenberg problem." Even in the middle of the nineties in the

---

[4]Carl Friedrich von Weizsaeker, an outstanding theoretical physicist, b. 1912, a friend and disciple of Heisenberg. Answers to questions of an American historian of science A. Kramish (manuscript, 1986). I am grateful to Professor Weizsaeker for sending me a copy of this manuscript and for long, lasting for many hours, exceptionally open and interesting conversations that we had in March 1987 and especially in May 1991. His name will be often mentioned in the following text.

New York Times there appeared an article on two new books discussing the failure of the German atomic project and making diametrically opposite statements titled: 'Heisenberg — an Ignoramus or a Saboteur?" This characteristic contraposition runs through the whole literature.

In fact, as I will try to show, *neither of the two is completely true* and at the same time, in a certain sense, both statements are right. *The situation is more complex and subtle.* To understand it one should comprehend well what is a totalitarian state, what is the place of a person, in particular of an intelligent, in it, what is the psychology of such an intelligent in conditions of physical and ideological oppression. It is difficult for people from Western democracies to understand this.

A new boom arose in the first years of the XXI century. It was to a large extent related to an appearance of the play by Frayn *Copenhagen* [Frayn (1988)] (on this very meeting) that goes with huge success in different countries, to declassifying a part of Bohr's archive containing 12 (!) very similar draft variants of a letter to Heisenberg that was never sent to him and put on paper in 1957, i.e. 16 years after their meeting (and later, let me remind that Bohr died in 1962).

First, however, on the suggestion of Heisenberg to help Hitler.

Neither published materials nor evidence of those who knew Bohr and Heisenberg personally confirm that such a suggestion was made. Nevertheless Heisenberg was a nationalist, albeit a "proven anti-Nazi" as he was called by Beyerchen, a serious historian of science [Beyerchen (1977)], and did even right after the war many times strike with his statements his former friends that had fled from Hitler's Germany. Once in 1974 in a place of one such friend he said, as his official biographers, the well known physicists N. Mott and R. Peierls [Mott and Peierls (1977)] write (*they were not present there themselves*): "One should have left the Nazis to rule for some 50 years more, they would have become quite respectable." The authors remark in this respect that nobody collected and analyzed in details Heisenberg's similar statements but consider the one cited by them being indicative. All this does in their opinion witness that Heisenberg was completely incapable of understanding a position of an interlocutor (same is claimed by an outstanding Dutch physicist Casimir [Casimir (1983)], see below).

Of course the words of Heisenberg were quoted by his interlocutor and one could assume that they were somewhat distorted by him. For example, he might have said: "*If* Nazis had stayed in power for some

50 years more ..." Still such a statement provokes at least a benevolent bewilderment.[5]

Bohr's words that Heisenberg is "... a very honest man" could be understood as a recognition of directness and openness with which Heisenberg always expressed his opinion.

What do other people who knew Heisenberg closely say about his personality? Here is E. Teller, a Hungarian, German, then, after the Nazis came to power, an American scientist, an outstanding theoretical physicist who is customarily titled to be a "father of the American thermonuclear bomb" who hated both communism, that during Teller's childhood had for a short time established itself in its native Hungary, and Hitlerism. A person who was in no way inclined to soft statements, who accused an almost universally respected Oppenheimer, a leader of the American nuclear project (Teller accused him of an intentional delay in works on the hydrogen bomb). This caused a known Oppenheimer process and a decades long moral isolation of Teller among physicists (in 1987 Sakharov told in a conversation with me and Weizsaeker: "Teller is a tragic figure"). In his youth Teller was for three years working with Heisenberg as his, let us call it, Ph.D. student,

---

[5] However, if he had said "If the Nazis had ...", would this have been so terrible for an honest person who had been for many years living under totalitarianism and had changed his hopes depending on circumstances and in the case of Heisenberg was, for example, most of all afraid of an arrival of Soviet communism and its terrible vengeance? Was it not just naivety?

Let us recall one episode described in Chapter 5 on our academician M.A. Leontovich. This was an unquestionably honest, direct and brave man who was not afraid of shouting at the face of MVD general (who supervised FIAN where M.A. worked) angry words on his disgust to communist party which "plants anti-Semitism and keeps peasants in kolkhoses", etc.

Once in 1943 this fearless hater of the Soviet regime naively told me when, as already described, I was spending a night at his place and we "were talking for half a night about everything that tortured": "Do you really think that kolkhoses will still exist after the war?" My words that a system that wins never changes itself did not convince him. At that time many expected that after the people, at a price of losses unseen in the history, would save the country, the regime would be significantly softened as a reward to this heroic people. In reality what happened was of course the opposite. The destroyed half-hungry country was even stronger gripped in a bloody vice and terrible cruelty came to reign. And Hitler, contrary to the naive hope of Heisenberg, would certainly have not become more human "in 50 years" after his victory but rather would have accomplished "his task" — would have killed all Jews and other "inferiors" and made his gloomy power unshakable and slavery of conquered people unlimited.

How difficult is it for an intellectual sodden with democratic ideals to understand totalitarianism and its dopey, fooling power!

Let us however return to Copenhagen.

and consequently knew him closely. Here is what Teller said in his published memoirs (a quotation from the internet version of a broadcast of *Liberty* radio station on March 7 2002):

'Heisenberg was not only a brilliant physicist but also a person whose decency and feeling of responsibility I could observe many times. I cannot imagine that he supported Nazis by his own good will, even less that he did it with enthusiasm as Bohr's version declares. How could it happen that Bohr misunderstood him? Information that I have gathered leads me to a thought that Heisenberg went to Bohr for moral advice... The meeting consisted of two parts... Why did not Bohr tell about the second part of the conversation? (Please remember this important phrase, this will be elucidated below. — E.F.) The reason could be simple: as soon as Heisenberg said that he was working for his country Bohr stopped listening to him." (See more on this below.)

First of all, on discrepancies with respect to the discussed meeting of Bohr and Heisenberg. From one side are the words by Bohr in my notes, on the other, the statement of Heisenberg himself also repeated by others.

The 1941 trip to Bohr was planned by Heisenberg and Weizsaeker who were worrying for the future of Bohr and his institute in Denmark captured by Hitler (this was also understood by Bohr who wrote the same in one of the draft letters to Heisenberg). Finding an official motive for the trip was not easy. Finally it was found: A participation in the German–Danish conference organized by Nazis with an aim of bringing closer the cultural circles of both countries. Heisenberg went to Bohr and Weizsaeker waited for results of the conversation in the hotel. (Nowadays we know that Bohr's wife Margaret was decisively objecting against inviting Heisenberg but Bohr nevertheless insisted on his point of view and invited him.) Heisenberg returned in despair from the failure. In Weizsaeker's opinion Heisenberg approached the main issue too slowly and it was quite possible that Bohr misunderstood him. Weizsaeker writes: 'In reality Heisenberg wanted to say that physicists all over the world should unite so that no country would make an atomic bomb ... Now I think that Heisenberg made a mistake... He should have said right away: "Dear Niels Bohr, now I will tell you something that will cost me life if it becomes known to those who should not know this (*"die falschen Leute"*). We are working on atomic weapons (an extent to which this was true is discussed below — E.F.). It would be of vital importance for humanity if we and our Western colleagues would understand that all of us should work in such a way that a bomb would not appear. Do you consider this being possible?"

Why are then the descriptions of this meeting by Heisenberg from one side and Bohr (in my notes) from the other one so different?

There were three moments that made this conversation difficult.

First, from the very beginning Bohr saw in Hesienberg not a former close friend but a scientist collaborating with an inhuman regime, with a government that had not only killed millions of completely innocent people but had also occupied, crushed his native Denmark and many other countries. So albeit this cooperation was very restricted Bohr's attitude towards Heisenberg could not be the same as before. Even after 16 years he wrote in one of the drafts of an unsent letter to Heisenberg: "We had to be considered as representatives of two sides engaged in a mortal combat." Terrible words. Compare them with a position of Hautermans (see below). As we know now, Heisenberg's position was very different. It turned out that precisely in 1941–1942 he wrote a philosophical book *Order of Reality (Ordnung der Wirklichkeit)* in which there was the following remarkable phrase: '*Wir muessen uns immer wieder klar machen, dass es wichtiger ist, dem anderen gegenueber menschlich zu handeln, als irgendwekche Berufspflichten, oder nationale Pflichten oder politische Pflichten zu erfuellen.*' We should always keep in mind that it is more important to act in a human way with respect to others than to fulfill any kind of professional duties or national obligations or political obligations'. Due to the extreme importance of this phrase I give it both in original German and my Russian translation. The word *Pflichten* also means "duty". And this is told by a German to the very marrow of his bones, and broad masses portrayed Germans of that time as soulless formalists. Of course such a book could not be published in Nazi times. Heisenberg showed the manuscript only to his most trustworthy friends and it was printed only after his death. I am grateful to his son, Professor Johann Heisenberg (a physicist), who presented me this book and clarified some details related to it.

Second, as well known by friends and disciples of Bohr he spoke better than listened in general and was quite capable of "misunderstanding what was told to him by others." This was written by Peierls [Mott and Peierls (1977)], himself a disciple of Bohr who knew him very closely. The same was essentially told by Teller in the above-cited excerpt.

Third, a lack of caution of any of them could cost a life to both. They were talking when walking in the street in the evening because they were afraid of hidden microphones in Bohr's apartment but any of them could slip a word. Both were very tense (in one of the above-mentioned drafts

Bohr writes himself that a conversation took place on a background of "sorrow and tension").

Jungk describes this meeting as follows: "... From the very beginning Bohr behaved himself towards his former student and favorite in an extremely reserved and dry way." Heisenberg, in turn, "... began gradually and cautiously approaching an issue of the atomic bomb. Unfortunately Heisenberg failed to achieve a required degree of trust and candidly say that he and his group will do everything in their power to delay a construction of such a weapon provided the opposite side agrees to act in the same way." Because of caution in the conversation both were speaking through hints and each one heard what seemed most important for him. In particular, as told to me by L.D. Landau, a friend and disciple of Bohr, who after the war did for the first time meet Bohr in 1961 in Moscow and talked a lot with him at that time, the question of Heisenberg, "What do you think about construction of atomic weapon?", was unambiguously interpreted by Bohr as an attempt to learn whether physicists in the countries of anti-Hitler coalition were working on such weapons, what were their achievements, i.e. simply as an attempt of espionage.

He kept this conviction forever, as shown by his conversations with Landau 20 years after the Copenhagen meeting and also by the part of his archive published ahead of time in 2000 (not in 50 years after his death as decided at first by his family but 10 years before in relation to the arisen boom). The latter contains in particular similar variants of drafts of unsent letters to Heisenberg, without dates but probably written at the end of the fifties (a usual manner of Bohr who did even rewrite many proofs of a scientific article sent to print). Each of these drafts contained a key phrase written in a categoric tone, an aggressive phrase of the type: "I would like to know which German governmental body did give you a permission to visit me and conduct talks on the so heavily guarded classified problem?" In this phrase as such one hears a full conviction that Heisenberg was sent to him on an espionage mission. I do not know anything that supports this conviction.

In articles of other Western authors we find "terribly" sounding words on how after returning from Copenhagen Heisenberg presented to the Gestapo his report on the trip ("which unfortunately was not found") or that the Gestapo demanded a report from him. The authors (like Bohr) clearly consider this to be a hint that Heisenberg traveled with a task (perhaps, a spying one) from the Gestapo.

What naivety and lack of knowledge of customs of the totalitarian system are characteristic of these people!

I have to remind how this happened in another similar system, by us, when a scientist attempted to go abroad, for example at a scientific conference. Such trips, very seldomly, began after Stalin's death and there existed a definite official procedure.

At first administrative and scientific bosses decided on the number and list of delegates going to the conference. Of course this did already take into account whose questionnaire looked good. Then the one who wanted to go (or received an invitation from an organizing committee) had to fill in a giant questionnaire (including questions about his wife's parents), pass an inquisitor interview at a meeting of the complete partkom of his institute where the members of partkom asked tricky, sometimes scoffing questions. If everything went well this cross-interrogation resulted in a positive "characteristics" (the main document also containing such obligatory phrases as "morally stable, lives with a family" and the concluding one: "partkom bears responsibility for comrade (...)." For example, singles were in a difficult situation — they could not leave hostages in the country, they could have stayed abroad, become "non-returners." Usually they were refused).

In parallel the file was sent to "organy" (the analogue of the Gestapo). This was considered to be a secret but everyone knew about this. A conclusion of "organs" remained unknown, it was sent to the final body — the "travel commission" of the Central Committee of the party. One still had to pass an interview at the "travel commission" of the district party committee. This had to approve the "characteristics" of the institute's partkom. This commission usually consisted of strict old party members from Stalin times. They knew that they had to "pull and not let go" but understood nothing in science, its needs, and in particular in scientists.

If all the described stages were passed successfully one formed a delegation which was invited to the party CC for instructions. Here ignorant party officials conducted a strict briefing on how one should behave "abroad": avoid unnecessary contacts, avoid "provocations", never walk alone, avoid women (it is known how one instructor clarified: "those women are no better than ours, when you return home you go on vacation and there have fun"), etc. One also had to read a printed instruction on "behavior abroad" and put a signature confirming that you had read it, etc.

Then one wrote "technical instructions" where one listed the assignments: To present a certain talk, visit only the specified cities, in contact with foreign scientists to defend a policy of Soviet government, reveal an attitude of some foreign scientists to Soviet policy (these last assignments were of course accomplished by few "enthusiasts" of such things, professional agents, etc.)

And, after all this, during the last 1–2 days before the departure (so that there was no time to object to a refusal) everything was decided by the "travel commission" of the CC. Very often some scientists were crossed out at this stage without any explanation (recall that it was here where secret "conclusions" of "organs" did come). Quite usual was an expulsion from a delegation on the last day before departure. There were cases in which people were taken out of a plane. In such a way in the sixties in Moscow airport one returned *from an airplane's ladder* a well-known physicist, a disciple, colleague and co-author of Landau, Ya.A. Smorodinsky, who was going to Yugoslavia, had successfully passed all the above-described stages of "purgatory" and was *already going up the ladder*. In 1977, before my eyes, one did not let into a plane L.I. Dorman, a specialist in cosmic ray physics known in the whole world with whom I had been collaborating for decades. He had just passed through "instruction" a day before and was included into a delegation going to an international conference in Bulgaria. Organizers of international conferences knew this habit and took it into account. For example, in 1979 I was invited as a speaker to a conference in Bern devoted to Einstein's centennial. It was organized in such a way that each session was devoted to one of the Einstein's topics and had one speaker. On "my" session one however foresaw one more speaker (Weisskopf) because one could have at the last moment prevent me from coming and the whole session would have been endangered.

Each delegation did of course include a KGB agent (he was ironically called "a physicist in civilian," this took place in all the delegations: When, say, violinists went to an international competition "a violinist in civilian" was included into a group "on the same footing"). His report and evaluation of behavior of each delegate during the trip influenced a possibility of further ones.

After returning home each scientist *had to present a written report* (it is clear where it went) on an accomplishment of the "technical instructions." Without this report a financial report was not accepted: The money spent had to be in exact correspondence with established miserable norms. All excess money, independent of their source — had it been Soviet organization, organizing committee of a conference or some foreign institution that invited to present a talk — that had not been foreseen in advance was diligently taken away up to the last cent.

All these humiliations were slightly modified in the decades to follow but in its basis everything remained exactly like this.

In the very same years when Bohr was writing his angry draft letters to Heisenberg I got three refusals for trips to conferences I was invited to in a row (my form was "bad") and did already categorically refuse to fill in numerous forms for the fourth time when, miraculously, a head of the delegation ensured for me a permission "behind the stage." Probably it was done out of business considerations: Either one needed at least one theorist who could understand works of leading experimentalists or because I could give a talk in English which at that time was quite rare — improbable as it may seem nowadays. So in 1957 I went abroad for the first time to an international conference on cosmic rays in Italy. Such outstanding, irreplaceable scientists like V.L. Ginzburg, I.S. Shklovsky, L.I. Dorman were however not allowed to go there and I gave talks for all of them (altogether I gave 6 talks and got acquainted with Heisenberg).

Feelings of shame, shameless humiliation, bitterness accompanied each of us. It is difficult to understand how could we nevertheless behave ourselves in a relatively free and honest way. An impossibility of telling the whole truth to the foreign colleagues was awful but nevertheless many of them became our friends. I have to say that the most clever of them understood everything themselves (or were alerted by their own "organs"?). I remember how in 1964 at an international conference in Dubna we were sitting on a bench near a hotel talking with Viktor Weisskopf and he asked me: "Do you know who in your delegation is from KGB?" I did not know and he pointed that man out.

Is it possible after all this (and I have omitted many picturesque things — space is limited) to pay serious attention to "angry questions" of Bohr and the equally "meaningful" words on the fact that Heisenberg did allegedly present his report to Gestapo? By those who went through a long-term school of constant watching by secret punitive bodies, a school of fear, terror, humiliation directed and realized by vicious villains that caused not only fear but also disdain towards these half-humans, these suspicions and accusations by Bohr and some Western historians and journalists can only cause a smile. Happy were they who did not live through such horrible long years.

This book is in essence written about those who experienced all this but kept their personality, overcame occasional derangements and kept internal freedom.

Let us however return to the Copenhagen meeting and the consequences of Bohr's conclusions from it.

As I learned from Weisskopf when Heisenberg's coworker H. Jensen had learned about the failure of Heisenberg's meeting with Bohr he, by his own initiative, came to Bohr and told him about a low level of work on uranium in Germany. Bohr understood this as a crude provocation. After the war it turned out that Jensen described everything absolutely correctly.

We would hardly learn what Bohr and Heisenberg did exactly tell each other during their meeting. As many thought at the end of the XX century it is possible that both of them did describe facts accurately (an evaluation of the whole episode as of an espionage attempt is a subjective evaluation, not a fact) but each one paid attention to what he considered most important. As we will see below there was, however, also something sensational.

However, there exists one generic psychological law which by itself is sufficient to explain how two interlocutors can give two very different descriptions of the same conversation at an event in which they both participated. Let us clear this in more detail.

On July 12, 2002 in Boston, at Massachusetts Institute of Technology there took place a joint discussion of the Freyn play *Copenhagen* between physicists and the theater company that staged the play. They discussed a question of what did in fact happen there. From the audience there came a question on how could one understand how Bohr who was clearly outraged by what he heard and was in an aggressive mood and Heisenberg who without any doubt felt a shock because of the fact that Bohr misunderstood him could produce different versions of the conversation. Let us note that Bohr writes 16 years after in his notes, "I clearly remember every word," while Heisenberg in a letter to Jungk writes, "I may be mistaken," and says on multiple occasions elsewhere how difficult is it to remember exactly something that happened many years ago. So, Bohr had an ideal memory and Heisenberg a bad one? The answer was that Bohr was so sure of himself because he had doubtlessly begun telling all this to people like Aage Bohr who published *a part* of what he had heard long before the archive was made open and everything published agrees with the material in the drafts.

This description is however vulnerable. Something heard by an interlocutor (in this case Bohr) is remembered only after getting through the receptor system and is, in general, undergoing some transformation in the psychical sphere. Bohr himself did many times stress that an action of a "device" on an "observer" is most pronounced in psychical processes. In general one may think that it is hardly possible that someone has a right to say "I remember everything that was told exactly." One just has a right to say only "I remember for sure how did I perceive what I was told, how I understood this."

Bohr, irritated, feeling hostile towards Heisenberg (as a "representative of the opposite side of the two clenched in a mortal combat," recall these above-cited words of him from his notes) was not a good "device." Heisenberg, feeling sad because of the failure, was perhaps no better but in any case a *different* device and a psychological transformation of the "object" by two of them could strongly differ.

It is interesting that, as told to me in 1988 by a many-years coworker of Bohr, S.A. Rosenthal, when after the war he asked Heisenberg whether he had come to agree on resistance against constructing a bomb he answered: "This would have been a pure madness: Had the agreement taken place one would have immediately decapitated me after my return to Germany." Weizsaeker answered the same question as follows: "We were very naive." They probably wanted to realize their naive idea very much and were rushing around not knowing what to do, sometimes making dangerous silly steps or making it possible for the interlocutor to form an ambiguous interpretation of their words. Therefore when attempting to reconstruct history one should rely *not on words* that are often distorted when retold but on *deeds*.

There again arises a question: How to reconcile the opinion of Bohr (in Landau's words) that the visit of Heisenberg was an attempt of espionage with the statement of the same Bohr that "Heisenberg is a very honest man"?

It is quite possible that with time Bohr learned more about the anti-Nazism of Heisenberg, his uncompromising defence of science, understood that he was honest and open in expressing his opinion and did somewhat change his attitude towards him. Weizsaeker recalls that when in 1950 he met Bohr for the first time after the war and wanted to clarify him the essence of what Heisenberg had wanted to tell him back in 1941, Bohr interrupted him with the words: "Oh, let us not talk about this. I understand quite well that during a war a priority for everyone is a loyalty towards one's country. Heisenberg surely knows that I think so." An interesting statement. A strange one when one speaks about the country of Hitler.

At the very beginning of his manuscript Weizsaeker remarks how difficult is it to recall something that took place and was told 40 years ago. However the above-cited words of Bohr are plausible. Considering Heisenberg to be a nationalist but anti-Nazist he did in principle recognize his right on the "defense" position.

Of course this was by no means an uncompromising anti-Nazism. Jungk cites words of a German communist, a physicist F. Hautermans: "Each decent person when facing a regime of dictatorship should have courage to perform a state treason."

Houtermans himself did this. In April 1941 Laue informed him that a Jewish physicist Friedrich Reiche and his family would manage in two days, right before a final deadline for Jewish emigration, to leave the country and that he was ready to pass important information. Laue trusted Reiche fully and assured him that Houtermans, whom Reiche did not know well could also be fully trusted. They met and Reiche whom American physicists knew agreed to remember and pass an oral message from Hautermans. This was very dangerous for both of them. At the border Reiche could have been interrogated, his meeting with Hautermans who was under surveillance could have been observed and Gestapo methods of interrogation were also applied to the family. Nevertheless he agreed and carried out the task. The train was a special one. Its windows were blackened and it was forbidden to leave the train on the whole one and a half day journey to Lisbon from which Reiche went to the USA by sea. On a specially organized meeting he delivered the message of Houtermans to a group of leading participants of the Manhattan Project — Bethe, von Neuman, Wigner and others. The message included the following phrases: "A large group of German physicists is working, under Heisenberg's leadership, on uranium bomb... Heisenberg is trying to slow down the work as much as possible because he is afraid of catastrophic consequences of success. Nevertheless he is forced to comply with orders given to him and if the problem can be solved it will be solved soon. Therefore he [Houtermans] advises the USA to hurry up in order not to be late." They silently listened to him, thanked him and left. Only after 20 years did Reiche tell about this meeting and the contents of the message in an interview. A Russian translation of this interview is included into a book by V. Frenkel [Frenkel (1997)].

Of course one could doubt this amazing story that provides a proof that the work in Heisenberg's group was not going "at full steam." After all this was an oral story told 20 years after the event and could have been arbitrarily distorted. Fortunately there exists an independent written confirmation. A well known physicist, Ladenburg, who had left for the USA earlier wrote in a letter to the chairman of the uranium commission of the USA Briggs:

> You might be interested to learn that my colleague who came from Berlin through Lisbon several days ago brought the following message: A trustworthy colleague working in a technical research laboratory (clearly Hautermans from von Ardenne institute — E.F.) asked to inform us that a large group of German physicists is working, under Heisenberg's leadership, on a bomb project, that Heisenberg himself is trying to slow down the work as much as possible because he is afraid of catastrophic consequences of success. Nevertheless he is forced to comply with orders given to him and if the problem can be solved it will probably be solved in the near future. Therefore he advises the USA to hurry up in order not to be late.

In two days Briggs answered: "I am deeply concerned with the contents of your confidential message. If you learn more, please let me know." [Powers (1993)]

A coincidence of the main points of the text with what was told by Reiche, other details (Lisbon, etc.) do not leave a doubt in the reality of this story, in that Heisenberg and his group were formally obeying orders from above and were under extremely strong pressure which would soon be impossible to resist [Frenkel (1997)] but did in no way desire success. Precisely during these months in the early spring of 1941 Heisenberg and Weizsaeker were discussing with Hautermans his results that seemingly opened a way for constructing a bomb. In this early period a pressure on them was exceptionally strong (this was also understood by Bohr). However, a manuscript of the work of Hautermans finished on August 1 was still remaining in the safe of the institute's chief Ardenne, a minister of post, who considered himself to be a competitor of military ministries and was therefore remaining unknown.

For decades there also existed a widespread version that *at the end* of 1942 Hautermans managed, in addition, to arrange a transmission of a telegram from Switzerland to the USA: "Hurry up, we follow the trace." [Frenkel (1997)] Subsequent research did however establish that this was a myth generated by a mistake in the memoirs of Wigner (I am grateful to I.B. Khriplovich who clarified this misunderstanding to me). Indeed, this telegram does not fit into the sequence of events known to us. Fears of close success generated at the beginning of 1941 were weakening under an influence of arisen difficulties. Already in the spring of 1942 there had taken place a well-known meeting with Speer. On his question "Will you physicists be able to construct a bomb in less than nine months?" Heisenberg did truthfully answer "no." Hitler's government lost all interest in the whole problem and Heisenberg and others could quietly resume working on their scheme (formally with the same activity but in essence without a hurry) while many began working on diverse projects that had no relation to the bomb (see below).

A state treason was also in fact committed (albeit in vain) by Jensen. A visit of Heisenberg and Weizsaeker to Bohr was also close to this (especially if one takes into account a newly uncovered fact, see below). Paul Rosbaud,[6] who was closely connected to scientific publishing houses, personally knew all leading scientists and visited them in their laboratories was fearlessly providing to English intelligence information on scientific defense work in Germany. His spy pseudonym was "Griffin." There is also a popular story about him by S. Snegov [Snegov (1972)]. This was however a special case.

We have not however finished our conversation on the Copenhagen meeting. Here, in 2000, something extremely important and sensational was found.

On March 27, 2000 in New York University there took place a symposium 'Creating "Copenhagen" triggered by the play of Freyn that had produced much noise. At this symposium a short speech was delivered by one of the biggest theoreticians of the XX century, one of the main leaders of the American atomic bomb project, Hans Bethe — a man enjoying immense authority and respect. He told about previously unknown facts on the meeting of Bohr and Heisenberg that presented it in a radically new light and also about his own understanding of the history of the German uranium project [Bethe (2000)]. In fact he had already told about this before in a collection of essays devoted to Bohr's centennial (1985) but from there one could understand that a described paper went directly to the USA. We use the later publication.

It turns out that in the process of conversation with Bohr, Heisenberg drew a scheme of the device they were working on. However, writes Bethe, "he overestimated Bohr's knowledge in the domain of atomic energy" so that Bohr did not understood anything in the scheme and took it for the one of some variant of the atomic bomb.

---

[6]There is a fundamental work on him: Kramish, A. (1987) *Der Grreif*. Munich: Kinder Verlag.

However when Bohr came to Los Alamos in 1943 he reproduced this scheme as he remembered it (?!?) and showed it to Teller, Bethe and Weisskopf. They 'immediately recognized a reactor with many "control rods." Bethe asks: 'What did Heisenberg want to say by all this? Perhaps he was trying to say: "Look at what we are trying to build and you will understand clearly that this is a reactor and not a bomb" (as we know, this was really true). If so ... he probably tried to make Bohr a messenger of goodwill and wanted Bohr to persuade other atomic physicists to refrain from creating a bomb as well. According to what a journalist Thomas Powers writes [Powers (1993)] this message was somewhat later repeated by Wolfgang Gentner, another non-Nazi German physicist.[7] Bohr, however, did not understand anything again ... Still neither he (Bethe) nor anybody in the USA in general would have believed such a message in any case (this could have been a deliberate misinformation).

So, the story with this sketch helps us to understand that:

1. Heisenberg really worked only on the reactor and did *never work on the bomb* ("had no interest in building atomic bomb" [Bethe (2000)]). Bethe sees a most reliable confirmation of the reality of all this in the episode that happened when he, among ten German atomic scientists, was in 1945 interned into an English estate, Farm Hall. When Heisenberg heard about Hiroshima his explanation to the colleagues on how the atomic bomb worked was completely wrong. Only after two days, in a second lecture, he calculated and explained everything correctly;
2. By showing Bohr the sketch he gave out a most important state secret, i.e. committed a state treason by showing that he was not working on the bomb (I am deeply grateful to professor Bethe for confirmation of all this in a private letter);
3. If the all-knowing Bethe was right in thinking that a delivery of this scheme played a role of the "message of goodwill" then it is true that in the course of working on the atomic bomb moral problems were playing for Heisenberg an essential role. Please recall an above-cited remarkable

---

[7] As a digression let us mention that a friend and co-author of Bethe was after the capturing of Paris by Germans accompanying (as an interpreter) a representative of the German military ministry who came to inspect an excellent laboratory of Joliot-Curie that possessed, in particular, one of the two European cyclotrones. After that Gentner got an order to supervise this laboratory which for Joliot, one of the leaders of French Resistance, was a base also in this respect. The situation was completely non-standard because Gentner was a secret furious anti-Nazist. Soon he established a contact with one of the Resistance groups.

phrase from his philosophical work *Ordnung der Wirklichkeit* which he began to write exactly in the autumn of 1941. He did really naively think that there still existed a world brotherhood of physicists and that one could rely on human relations in preventing a development of the bomb. This was of course a striking naivety probably based only upon a passionate desire to prevent a nuclear war. The above-cited answers of Heisenberg and Weizsaeker to S.A. Rosenthal show that later this naivety was understood by them;

4. The paper with the sketch shown to Bohr is more important than a multitude of empty words and retransmitted rumors. This is "a material proof" of the anti-Nazism of Heisenberg and the absence of any desire of building an atomic bomb by him (in any case Bethe and Teller are of the same opinion).

Bethe adds: "In the middle of 1942 Albert Speer, Hitler's minister for weaponry, asked Heisenberg whether he could build such a weapon in nine months. With a clear consciousness he answered "no." He did not even know how much fissile material one would need for this. When friends in certain cases asked him about it his answer was not definite at all — from 10 kilograms to several tons.

'Why didn't he know it?' continues Bethe. 'Why did pure intellectual curiosity not force him to investigate the properties of uranium-235 with respect to fast neutrons? (These quantitative characteristics are of primary importance for constructing an atomic bomb based on uranium-235. — E.F.) He could indeed obtain a small quantity of it at Parisian or Danish cyclotrones. Nevertheless he never asked anybody to measure this." [Bethe (2000)] Perhaps, *he did not want to know.*

A possible answer to this question is given in his memoirs by a sober and unaffected by emotions Edward Teller. His opinion on Heisenberg as a man, a very high one, has been already cited by us. He resolves Bethe's bewilderment in accordance with this opinion. Teller thinks that as soon as Heisenberg understood that construction of the atomic bomb was in principle possible, that this inhuman horror was realizable, he constructed "a mental barrier" that did not allow him to think in this direction (let us remember this — E.F.). Teller says: "Information that I have gathered leads me to a thought that Heisenberg went to Bohr for *moral* advice. He told Bohr that he was participating in the Nazi project (in Teller's opinion the word "bomb" was here not mentioned). He added that fortunately it was impossible to make an atomic bomb in Germany and that he hoped that English and American scientists also would not succeed (I am citing from the above-mentioned Radio "Liberty" broadcast, see http://www.svoboda.org/programes/ep/2002/ep.030502.asp). Teller's indignation refers to this second part of the Copenhagen conversation: Why Bohr *did not tell anything about this second part of the conversation with Heisenberg*, i.e. on the story with the sketch of a reactor (Bohr also did not give a hint of this event in the conversation with us, FIAN theoreticians, in 1961)? Does one indeed have to relate it to the statements that were many times repeated according to which after the very first mentioning of a bomb Bohr "switched off" and did not comprehend anything? Even though he did not understand the sketch he nevertheless remembered it even after 2 years when he discussed it in the USA. As already told, most probably this could not be made known in Heisenberg's lifetime so that he would not have been killed by some Nazi fanatic for state treason.

Therefore, in spite of a full confidence with which Bohr speaks about the Copenhagen meeting in his draft letters to Heisenberg 20 years after it and states that he remembers every word one has to admit that much was not understood by him and much was misunderstood. Lack of understanding of the sketch of the device that Heisenberg was working on led Bohr to a false conviction that he was working on the bomb. For example in one letter he writes: "I also remember absolutely clearly our conversation in my office[8] in the Institute when you, in unclear expressions, were talking in a way that could only create by myself a firm impression that in Germany, under your leadership, everything was being done to develop an atomic weapon and you told that there was no necessity to discuss details because you knew them and had spent the last two years more or less exclusively in these preparations." (http://www.nbi.dk/NBA/papers/docs/d01tra.htm) At the same time we have just seen that competent people, Bethe and Teller, do convincingly prove that he *never* worked on the bomb. It would be ridiculous to believe that he knew all the details when he had not even attempted to learn them, did not know even the critical mass and before 1945 (before Farm Hall and Hiroshima) had not even attempted to learn anything about it. At his first attempt of explaining an explosion in Japan this complete lack of interest to building an atomic bomb did clearly show itself, he demonstrated a complete incompetence. Of course, he was a great physicist so that at the second attempt,

---

[8] By the way no conversation on Heisenberg's work in the office was ever mentioned. They talked in the street.

after thinking about the issue seriously, he explained everything in just two days.

Whatever our admiration of the personality of great Bohr, whatever feelings towards him had we experienced, one has to admit: All said questions much of Bohr who so confidently told about the Copenhagen meeting. Precisely universal admiration of Bohr, reverence for him and love to him was for many decades forcing authors to be silent on errors and apparent incompetence of the genial man and correspondingly to mistrust Heisenberg who was also sincerely devoted to Bohr, loved him, never publicly reproached him, never argued and justified his actions under a torrent of accusations, often completely absurd ones, falling upon him.

Let us therefore look in more detail at an important and complicated problem: A scientist, an intellectual in the conditions of a hard regime.

W. K. Heisnberg in 1934.

8.3 Heisenberg and Science During Nazi Times

Many physicists who emigrated from Germany (these were mainly Jews affected by racist laws but not only, let us recall Nobel laureates

E. Schrödinger, P. Debye, M. Delbruck) thought that those who stayed in fascist Germany did by this express their agreement on collaboration with Hitler, on supporting Nazism. Moreover they were convinced that everybody who stayed had to resign in protest against nazism. Heisenberg, however, explained his unwillingness to emigrate by the hope that despite of the fact that he would have to live in difficult conditions, constantly make compromises with the regime, he would still be able to protect German science, bring up young scientists and do what he could for the science not to degrade completely and recover after the war. He said that he got such inspiration from a conversation with M. Planck whose authority was very high.

Max von Laue, a famous scientist, a Nobel laureate (he died in 1960) did also stay in Germany and did not resign. He explained his reasons exactly like Heisenberg. In a conversation with Einstein in 1939 he added: "I hate them so much that I have to be closer to them." He also participated in the uranium project[9] (in particular in April 1945 (!) he was present at a desperate attempt of launching a self-sustaining chain reaction in uranium[10]). At the same time he enjoyed universal respect. When in the middle of the thirties a German physicist, P.P. Evald, visited Einstein before returning from the USA to Germany and asked whether Einstein wanted to pass a word to somebody Einstein answered: "Convey my greetings to Laue." Evald asked: "Maybe, to someone else?" and gave several names. Einstein thought for a while and repeated: "Convey my greetings to Laue." [Beyerchen (1977)]

It is known that Laue saved people. His position in science was firm and his behavior in conditions existing at that time holds as an example for scientists. He was not teaching in university and therefore did not have to, like Heisenberg, stretch out his hand and cry *"Heil Hitler!"* when

---

[9] His participation was in fact not active. He was first a deputy director of Berlin Physical Institute when Debye was its director and then, after Debye left, its director. This institute hosted a group working on the uranium project headed by Heisenberg.

[10] Strange as it looks German physicists thought that they were ahead of English and Americans and thought that if their attempt would have been successful then after Germany's defeat the allies would feel respect towards German science and create conditions for its development. In reality such an experiment was successfully carried out under leadership of Fermi in Chicago already in December 1942. (However, with all the unprecedented scale of this work in the USA one still needed 2 years and 8 months to build atomic bombs and drop them on Japan so even a success of the last German attempt would not have had military significance.) An amazing evidence of a complete failure of German intelligence in this most important domain.

beginning a lecture. Moreover, one says that when leaving home Laue usually had a briefcase in one hand and some package in the other so that he had the possibility of not answering greetings of his acquaintances in this fashion.

He did not make compromises and together with others withstood a Nazi baiting of relativity theory and quantum mechanics. For example he did not yield to persuasion of Hamburg Professor, Lenz, to organize a publication of a series of articles on relativity theory "to clean it from Jewish stain by declaring that its author was a Frenchman Henry Poincare thus making it suitable for the Third Reich." [Beyerchen (1977)] Let us also not forget that it was Laue who arranged a meeting between Reiche and Hautermans, i.e. he was an initiator of passing information on the state of uranium project in Germany to the USA.

The question "to go or to stay" was in essence not new for scientists. For the first time it probably arose in Russia in 1911 when 130 professors left Moscow University protesting against actions of an extremely reactionary minister of education, Kasso (allowing police to enter the university, mass expulsion of revolutionary feeling students, etc.).

Among these there were outstanding scientists, in particular a physicist P.N. Lebedev, an author of results of principal importance (he was the first to measure pressure of light and was nominated for a Nobel prize but soon died) in which he demonstrated the remarkable art of making experiments. Lebedev created the first school of physicists in Russia. He generously endowed with ideas talented young people who were from their student bench working with him and were growing not on "reproducing the known" but on self-made research. Those who knew Lebedev recalled that he did not sleep many nights in extremely hard contemplations on whether he should leave the university. Civil feelings and public opinion made him resign. For some time he was trying to continue his work with students in an apartment specially rented for this purpose on his own money but this was "no longer the same." His sick heart could not bear this and in less than a year, just reaching an age of 46, he passed away. After Lebedev left the university the physics there fell into decay. Students were now taught by professors who were far behind modern science. Things changed only in the middle of the twenties.

Contemplating about the consequences caused by the resignation of Lebedev from the university one involuntarily asks oneself a question: Did he do the right thing? Let us recall that academician I.P. Pavlov who was extremely hostile to Soviet power did not go abroad but until the end of his life (1945) was working in his laboratory. One recalls the lines by Akhmatova[11]:

> No, not under foreign skies
> Defended not by foreign wings
> There with my people have been I
> Where my people, alas, has been.

Therefore one should hardly condemn Heisenberg simply for staying in Germany as he was condemned by American, English and other Western scientists, especially those who had emigrated from Germany and Italy.

Like his teacher A. Sommerfeld, Planck and many others who stayed in Germany, Heisenberg opposed Nazi ideology which, as well known, recognized only applied (in narrow sense) physics, chemistry and mechanics while the role of fundamental knowledge was ascribed to half-mystical research in ancient German and Nordic mythology in general as well on anthropometric "foundations" of aryan race theory. Theoretical physics as such was considered to be a fruitless mental exercise and quantum mechanics and relativity theory to be creations of Jewish spirit.

In these conditions one should not forget and underestimate a courageous defense of science by Heisenberg (who, being a pure aryan, was named "a white Jew" by Nazis) and his colleagues. From the pages of SS edition *Der Schwarze Korps* there came direct political accusations of Heisenberg. An article told: "... One should deal with such people in the same way as with Jews." Heisenberg, one of the founders of the physics of the XX century, was not allowed to get a chair in Munich after Sommerfeld, who had strongly recommended his disciple, had retired. The chair was given to a mediocre specialist in aero- and hydrodynamics who reduced a course of theoretical physics to just (classical) mechanics.

This defense of science within the physics community took different forms. For example, one arranged a discussion with nazi physicists at which one was able to agree on a compromise resolution [Beyerchen (1977)]:

1. Theoretical physics with all its mathematical toolkit is an indispensable part of the whole physics.
2. Experimental facts summarized by special theory of relativity provide a firm basis. However, application of the theory of relativity to cosmic laws is not reliable to the extent that no new confirmation of its validity is required.

---

[11] *Requiem*, translated by A. Leonidov.

3. Four-dimensional representation of processes in nature is a useful mathematical tool but does not mean that new concepts of space and time are introduced.
4. Any relation between theory of relativity and a general concept of relativism (apparently the philosophical one — E.F.) is denied.
5. Quantum and wave mechanics are the only currently known methods of describing atomic phenomena. It is desirable to go beyond the formalism and its prescriptions to achieve a deeper understanding of atoms".

This document contains both banal truth included only with an aim of being able to withstand the stupidity of Nazi ideologists (points 1 and 4, the first phrase of point 5) and diminishing, to their pleasure, of new physics (end of point 5, the first phrase of point 2, theory of relativity is useful only as a systematization of facts but, according to point 3, do not change our understanding of space and time although in reality its grandeur is precisely in providing this new understanding of space and time).

It is not too pleasant to write about this but a physicist of my generation cannot fail to see how painfully similar are the statements in the compromise agreement to those vulgarizing and diminishing the role of quantum mechanics and relativity theory on which there also sometimes agreed our Stalinist philosophers who were attacking modern science from the thirties till Stalin's death. Of course, the basis for these attacks was not race ideas but a "necessity of protecting materialism from bourgeois ideology" but still the resemblance is striking (see Section 7.2).

Accusation in idealism was very dangerous and there were some physicists (fortunately very few) who out of fear or careerism accepted the vulgarization of science in the same way as some German physicists. Moreover, by us one often demanded from research direct and immediate practical benefits. One had to fight for the necessity of theoretical physics. Up until the war a research in nuclear physics was conducted in the institutes headed by S.I. Vavilov and A.F. Ioffe under a fire of critique for "a gap from the practical needs of people's economy" from the side of ignorant "ruling authorities."

Let us however return to the compromise document of German physicists. One cannot deny that it did play a useful role. Not only did it allow to keep in German universities the teaching of fundamental science (point 1), in particular of "vicious" new contemporary physics but, as it turned out later, even convinced some previously uncertain participants of the discussion so that they broke away with "aryan physics" of Lenard and Stark.

Besides that it was useful for students that in spite of having pro-Nazi sentiments did nevertheless probably understand the value of the new science.

Of course participating in such compromises was humiliating for scientists. Laue, Planck, Sommerfeld, Heisenberg could allow themselves to stay away but some had to do this in the interests of science and the young generation.

What was, after all, the political position of Heisenberg? First, it is impossible to understand it not taking into account the fact that German academic circles were, unlike the Russian intelligentsia, traditionally trying to isolate themselves from politics and considered it to be a low-level activity related to intrigues, etc. Second, one should not forget about a special psychological difficulty of an existence under totalitarian pressure which was characteristic of everybody, especially of those who tried to maintain a capability of thinking in an atmosphere of all-penetrating fear in Germany and in the USSR, who managed to keep one's personality (one should also take into account the milieu to which Heisenberg belonged). And, finally, and this is perhaps the most important point, the German people did in its vast majority follow Hitler.

8.4 Heisenberg and Western Physicists

Hitler came to power in a democratic way. He was elected by the German people who shrunk back from communism which by its policy of collectivization, hunger, terror and despotism had shown what it was able to do in Germany as well.[12] We (together with the whole world) consider Hitler to be an incarnation of devil (which he really was), a barbarian who crushed culture (and this was also true), killed millions of Germans, millions of people in countries conquered by him including six millions of Jews (yes, this is true as well). Cruelty, inhumanity, guile put hitlerism in line with the most abominable regimes in the whole history of humanity (all this is doubtlessly true). Why then the German people did in its mass already after the first period of his cruel reign rush with devotion and delight to his feet? Why did mothers reached out their children to him in affection and people were ready to die for him?

Evidently the explanation is in the fact that he led Germans out of the state of despair, hunger, hopelessness. Jobless got jobs, the shame and burden of Versailles ended when Hitler broke off with it, occupied the

---

[12] See Chapter 9 in this book.

Rhineland, began intensive armament. A young generation got an inspiring (albeit horrible) aim of enslaving other people. He did something that social democrats with their Weimar Republic could not do. So — why should one have pity for them when the Gestapo kills them (and "even more so" communists and Jews)?

Should one be surprised that masses of German people and even many intellectuals, in particular students, saw in this an accomplishment of a national idea of immense importance. Precisely because of this they wanted to shut (or at least half-shut) their eyes at the horrors of Nazism, to pay no attention to information on concentration camps, to believe that all this, together with the barbarian ideology of Nazi leaders, was inevitable "collateral damage," a side and temporary phenomenon, that one "cannot make an omelet without breaking eggs," that in the process of achieving the goals that are necessary for the nation the negative phenomena will weaken and finally disappear. How familiar is this to us Russians from our tragic experience!

This position is characteristic both for intellectuals and broad masses in any dictatorship *realizing big national goals* and admitting a usage of merciless, inhuman methods in the process.

Didn't the Soviet people tolerate the equal cruelty and crimes by Stalin that found their justification in a solution of the most important national problem of transforming a relatively backward agrarian country into a modern and powerful industrial state? In both cases skillful propagandists and demagogues just had to persuade people that there was no other way of achieving the great task. Because of this in our country one concealed and disparaged the immense progress in Russia after the reforms of Alexander II carried out within a very short historical period of about half a century.

One certainly needed a powerful, all-penetrating terrifying punitive apparatus. However, already Machiavelli had told in his tractate *The Prince* (*Il Principe*): "The prince must arouse fear in such a way that even if he will not deserve love he will avoid hatred because it is quite possible to scare and at the same time not be hated." Stalin, like Hitler, also could do this with a majority of people's masses.

It is of course also very important that Hitler got, in various forms, the support of Western capitalistic countries who considered his regime as a barrier against communism. Moreover, one French division was sufficient to force him to retreat when he moved troops to Rhineland and broke the Versailles Treaty. They closed their eyes on this instead. Here there was possibly an admixture of a fear that this would cause an unrest of German people in his defense. Nobody prevented him from arming himself. This armament and "fighting practice" acquired on its basis were successfully tested in a Spanish civil war in the thirties and a "non-interference" policy of Western democracies in Spain confirmed that Hitler was a good defense against communism. Let us however return to Heisenberg.

Heisenberg was a true German patriot (many were calling him a nationalist, we will talk about this later). His attitude towards Hitlerism could not be straightforward. From one side, of course, he felt disgust towards cruelties of Nazism, its barbarian ideology, was outraged by oppression of the intelligentsia and free thought, stupidity and cruelty of small and big Fuehrers. However, like millions of his countrymen, he could not fail to see that with Hitler coming to power there ended a period of despair of the German people that had lasted many years.

Of course for Hesienberg and a vast majority of other patriotically minded intellectuals everything was not that simple. It is no wonder that, being a patriot, he was all the time experiencing swings and in different periods expressed different points of view. The same thing happened to a majority of Russian intelligents, writers, "fellow travelers," etc.

Moreover, by the end of 1941 Heisenberg had been already living under Nazi regime for almost 9 years. The regime of crazy (in the literal sense of this word) terror and totalitarianism did habituate people, both in Germany and by us, to "double standards" in behavior and pronouncements, to the ability of hiding the truth even when seemingly saying it. Thereby although Heisenberg was working on the reactor, not on the bomb, objectively this was a preparation for it. Weizsaecker did in fact understand very early that in the reactor one also produced a "weapon," plutonium, and Hautermans made a full study related to plutonium (therefore they tried to hide these results). Therefore they could tell Speer that they were working on the bomb. The important thing though was *how* were they working on it!

It is still unclear how one should understand Heisenberg's words as repeated by Bohr (in 1961 in FIAN) that one had to collaborate with Hitler's institutions and then the attitude towards science would change. In this case fine details are of importance: How was it said? What was meant? This was however not for Bohr with his mood.

There is also nothing surprising in the fact that people that were at that time living in completely different conditions, in the USA, England, Denmark, etc., in the countries where there was no other large-scale national task but the only one — salvation from Hitler's aggression. People were correspondingly concentrated on feelings of enmity and hatred towards Nazism and could not find a common language with many intellectuals from totalitarian countries.

In conversation with physicists, especially the Western ones, one often hears a condemnation of Heisenberg's behavior at Nazi times motivated already by the fact that he was, after all, cooperating with authorities too closely, followed official instructions and always wrote "Heil Hitler" at the end of official letters and said the Hitler greeting.

It is difficult to judge what "too close a cooperation" means. We have already mentioned that such a person as Laue did also participate in the uranium project. Heisenberg (like Weizsaecker and also their friend and coworker Wirz) was a member of the Nazi party. This detail can be considered as being purely formal. After all Jensen whose travel to Bohr was described before was a member of the party but at the same time was far from being true to Nazism. Providing Bohr with absolutely secret information on the work on the reactor was in essence a state crime and the information of Hautermans passed (thanks to an initiative of Laue!) through Reiche does directly show that Heisenberg and his coworkers did not want a bomb although formally were obeying discipline and were doing something. In general, already from the above-mentioned names and conversations of ten leading German atomic scientists (interned to Farm Hall) taped by English intelligence, [Khriplovich (1993)] one sees that sentiments of a majority of real scientists (if not all of them) was sharply anti-Nazist but they were formally accurately fulfilling their duties. Fierce dangerous ideological attacks were falling upon Heisenberg and he, as has been already mentioned, withstood them.

As to the Hitler greeting it was obligatory and regulated by state law. Heisenberg was consoling him with the fact that he had to write official letters very seldomly. An oral greeting was of special importance only at the beginning. P.P. Ewald describes a colorful episode (I am citing from the book by Beyerchen [Beyerchen (1977)]): 'Plank, as a president of Kaiser Wilhelm Society (in some sense an analogue of our Academy of Sciences), ... arrived at an opening of an institute of metals in Stuttgart. He had to deliver a speech (apparently this was in 1934) and we were looking at Planck and waiting how he would master an opening ceremony because at that time it was already officially prescribed to begin such a speech with the words "'Heil Hitler!'...Planck stood on the podium. He lifted his hand a little but after that lowered it. He did this one more time. Then finally the hand went up and he said: "Heil Hitler." In retrospect we understand that if one did not want to endanger the existence of the whole Keaiser Wilhelm

Society this was the only thing that could be done." (this society, grounded in 1911, united a broad network of research institutes,; it was subsidized by government and private capital.)

With time this greeting transformed into a pure formality: A casual hand-wave, which is known to everyone from cinema, and the sputter of two cabbalistic words. In any case a participation in meetings with "applause transforming into ovation" after every mentioning of the magical, deified name which was also so typical for us in the Soviet epoch bore more significance because here there really arose a mass besotting, feelings of reverence, admiration, affection and devotion. We know it from our own experience.

Apparently in Nazi times Heisenberg experienced psychologically conflicting sentiments while politically he was in many respects unstable, possibly even not ripe enough. One known physicist who had fled from Hitler's Germany and was informed in this sort of question told me (later I got confirmation from another physicist who had stayed in Germany) that during the first years of the war Heisenberg wished Germany a defeat. However when he learned about the awful order that the Nazis established in the conquered countries, about concentration camps, etc., he was frightened of the revenge of these people in the case that Germany would lose the war and began wishing it victory. A hatred towards communism as another form of totalitarianism and a fear of its revenge were very important factors for him. However, the participation of Heisenberg in "Wednesdays" (see below) together with such people as General Beck shows that his notion of "victory" was not identical with that of Hitlerism and its ideas. At the end of the war soldiers were afraid of revenge as well. It is perhaps due to this that many of them who went through Soviet captivity and who were themselves witnesses (and sometimes participants) of Nazi cruelties, of attitude towards Soviet prisoners of war, have such a good attitude towards our country. They did not expect a human attitude and a (pragmatically calculated by Stalin) magnanimity. This brought political benefits straight after the war.

The milieu in which Heisenberg lived was in many ways different from that community that had existed before Hitler's coming to power which had been an international community of scientists devoted to science, creating new science, communicating with each other friendly and freely. Political and ideological split of the world also caused a split in the world of scientists.

Recalling the "chaos of the last years of the war" Heisenberg writes [Heisenberg (1980)] that there were only few impressions that made him glad. One of them became a part of that foundation on which he later based his attitude towards generic political issues. This joy was given to him by the famous weekly meetings on Wednesdays on which there met, played music, discussed various topics, the head of anti-Hitler plot of 1944, General Beck, a priest Popitz, a well-known surgeon Sauerbruch, an ambassador von Hassel, a German ambassador to Moscow Count Schulenburg who on June 22, 1941 had handed to the Soviet government a note on the beginning of the war (and, as some say, shed tears at this event), etc.

A well-known Soviet political journalist Ernest Genri who in the post-Stalin years published a lot on issues of Hitlerism and the fate of Germany told me at the beginning of the eighties that Schulenburg was "a conservator and nationalist but not a fascist. Two weeks before the invasion of Hitler's Germany he warned Soviet diplomats about it, in particular a Soviet ambassador in Germany Dekanozov, i.e. in essence committed an act of state treason."

In July 1944, on the way from Munich to Berlin, Heisenberg learned about a failed attempt on Hitler's life, the execution of Beck and the arrests of some of those with whom he had met on Wednesdays [Heisenberg (1980)].

When in 1943 Heisenberg visited in Holland his colleague, a well-known physicist G. Casimir, he tried to persuade him that Europe under German (evidently even Hitler's) leadership would be lesser evil than the communism of Soviet type and that only in this way could one protect Western culture. Not denying and not justifying cruelties and disgusting features of Nazism in general to which Casimir was referring when objecting to him he only said that after the war one should wait for changes to get better [Casimir (1983)]. As has been already mentioned the same hopes with respect to Stalinism in the case of its victory existed in our country as well.

At that time the democratic Western countries of the anti-Hitler coalition, especially those that had already been enslaved by Hitler, did in the first line consider the USSR as a savior. Although they knew much about the horrors of Stalinism this was by far not everything that became known later. They were incapable of putting Stalin on the same level as Hitler (although, of course, the wise and cynical Churchill understood that, using his expression, "they differ only by the form of their mustache"). A drafty Stalin demagogy that could so blatantly deceive even a writer, L. Feuchtwanger, who visited Moscow, supported by a heroic victory of the Soviet Army, spurred Western intellectuals to turn a deaf ear to information on the horrors of the Soviet system in the same way as Germans in Hitler's times did with respect to his cruelties. They did not know yet what would later happen to "Stalinized" Eastern Europe, to East Germany. Even when they seemingly learned about it, Heisenberg's swings continued to cause only their indignation, not understanding.

Casimir himself asks in his memoirs [Casimir (1983)] a question: Why did Heisenberg tell him all this? Calling to mind the possible reasons (apart from the one described by myself) he once again explains everything by the fact that Heisenberg was not capable of understanding an interlocutor, in this particular case a Hollander who hated Hitlerism. However, one could also derive an opposite conclusion from his words: a person from a Western democratic country is incapable of understanding the ideological swings of an intellectual who was for many years living in the horrible conditions of totalitarianism.

It is necessary to mention one more thing. The same Casimir writes that before the war he "was always admiring Heisenberg not only as a physicist. For me he was a representative of many things that German culture gave. He was a good musician and a good sportsman, knew ancient languages much better than I did." Later, however, a hostile attitude towards him began to dominate, there arose many accusations based on false rumors. For example, one told me on many occasions with peremptory confidence that during his visit Heisenberg was persuading Casimir to take part in the German uranium project. The book of Casimir's memoirs contains nothing like this. Moreover, when talking to me in Geneva in September of 1988 Casimir categorically refuted this rumor. I have come across many such rumors, false but widespread and in a majority of cases aimed against Heisenberg.

With time there began to appear information that Heisenberg tried to help victims of Nazism. A Polish physicist, E.K. Gora, who lived in the USA published in 1985 in an American scientific journal a letter titled "Saved by Heisenberg" [Gora (1985)]. In this letter he tells that when Wehrmacht regiments occupied Warsaw in 1939 he was warned about Hitler's order to exterminate the Polish intelligentsia. Gora appealed to Heisenberg and he saved him and was patronizing him for many years: he invited Gora to Leipzig, helped him get a job as a tramway conductor. This provided a status of "alien worker." He declared him to be a "foreign student" which allowed to continue education and carry out scientific work (the results of which were published in 1943 in a German journal). When arrested by the Gestapo Gora was soon released — as he thinks, thanks to Heisenberg. At the same time the above-mentioned coworker of Bohr, S.A. Rosenthal, told me in 1988 that once when visiting Warsaw Heisenberg stayed with his former school comrade and at that time Nazist hauptleiter of occupied

Poland, Frank. Perhaps this is true. It is unclear why he accepted an invitation from this terrible person. Surely not out of sympathy. Indeed, our prominent writers, even Babel, not speaking about Gorky, Aleksei Tolstoi, etc., were in close contact with first grade butchers Jagoda, Ezhov and other NKVD people.

Heisenberg never wrote and never told about his help to hunted colleagues. He probably thought that this would have been indecent because it would have looked as a self-justification.

Talking about attacks on Heisenberg everybody mentions only that the Nazis called him "a white Jew." However due to some reason there remains in the shade one place in his memoirs [Heisenberg (1980), p. 289]. In a conversation with Weizsaecker concerning the possibility of trusting some officials in the fall of 1939 he said: "... You know that just a year ago I was many times interrogated by the Gestapo and and it is painful for me to even recall the basement in Prinz-Albrecht-strasse where on a wall it was written in thick letters: "Breathe deeply and quietly." Therefore I cannot imagine myself such relations of trust."

Of course this is recalled by Heisenberg himself and his ill-wisher could doubt the truthfulness of these words but, I repeat, I heard myself in 1961 how Bohr called him "a very honest man" and Teller (see above) speaks about his decency and justice with full conviction. Let us add that his above-mentioned interlocutor Weizsaecker is still alive and can confirm or disprove the fact of Heisenberg's interrogation by the Gestapo.

In general it is gradually becoming clear how many "traditionally non-political" anti-fascist feeling scientists were connected with each other and strived to help those in trouble. For example when Hautermans[13] was being transported from the Soviet border to Berlin he asked one occasionally arrested German who was about to be released to find Laue and tell him just three words: "Hautermans is in Berlin" (according to another version this happened when he was already in a German prison and sent this German to Rompe and the latter went to Laue). Laue immediately began to act and in less than half a year ensured the release of Hautermans.

---

[13]F. Hautermans, a remarkable German physicist, a communist. In the thirties he emigrated to the USSR, worked in Kharkov Institute of Physics and Technics, was arrested by NKVD like a majority of other foreigners political emigrants and together with many other arrested German communists was handed out to Hitler after signing a pact with him. See above on F. Hautermans in relation to sending through Reiche to the USA an alarming message and also the book [Frenkel (1997)], the article [Khriplovich (1993)] and in the essay on Landau in this book.

I knew about this connection among the scientists from various memoirs and got a confirmation of it in a private conversation with the physicist Charles Peirou, one of the two captured French officers whom P. Rosbaud managed to free under a ridiculous pretext of the necessity of translating into French a scientific book. What is especially interesting is that Rosbaud ensured with a hero of French resistance, a known physicist F. Joliot-Curie, that after the war this work would not be considered as a collaboration with nazists. Until the end of the war Peirou worked in the laboratory of "bison" — N.F. Timofeev-Resovsky. He also confirmed to me that when a son of Timofeev-Resovsky, an underground anti-fascist, had been captured by the Gestapo Heisenberg had tried to help to save him.

One knows only about a few facts of such interaction of scientists but they are quite meaningful because they show that a pre-war community of scientists in Europe did not break apart completely. They became known only with time and now there remains only a few contemporaries and witnesses of those events.

One has to note that after the war Heisenberg was among 18 West German atomic scientists that published a manifesto in which they condemned the atomic weapon and declared that they would never participate in its construction.

Still, for a long time to come, Western physicists treated Heisenberg with dislike. Scientific contacts were of course resumed — he participated in many conferences. His relations with Bohr did also resume although the unpublished drafts of his letters to Heisenberg show that Bohr's wound did not heal until his death. Heisenberg and his wife came to Copenhagen and the two couples took long strolls together. Heisenberg even lived in the Bohr's dacha in Tixwell. They were together in Greece in 1956. There resumed close relations with an old friend W. Pauli, an outstanding theoretical physicist. In their time they were together creating quantum field theory and now they were discussing new scientific problems (Pauli, a Jew by nationality, permanently living in Switzerland, in 1940 had left for the USA, and after the war was again living in Switzerland for 5–6 years in a row).

Doubtlessly it was not easy for Heisenberg to feel estrangement or just tense attitude towards him. Viktor Weisskopf told me that when in 1961 he came to the European Center for Nuclear Research (international organization CERN) for five years as a director general he saw this isolation of the German representative Heisenberg. He decided to do away with this and organized a reception in his honor. Weisskopf was a wise man.

In August 1959, for the first time after the war, in Kiev there took place a big international conference on high energy physics. Among a hundred of foreign scientists there also came Heisenberg. Almost all of them lived in the hotel 'Ukraine.' A girl at the registration desk, not knowing with whom she dealt, put Heisenberg into one of the rooms on the upper floor together with several Soviet journalists. The next morning Heisenberg approached I.E. Tamm and timidly asked for help: The water did not come to the last floor and he could not wash. Tamm was outraged, disturbed and everything was of course momentarily settled.

Of course Heisenberg did not know how he would be met in Kiev. He was not just a man from Hitler's Germany. He surely knew that our philosophers and some physicists adjusting to them had stigmatized him as an ideological foe, "bourgeois idealist," physicist of the "Copenhagen school". For them he was not one of the creators of quantum mechanics about whom Landau told with delight: "A man turned out to be capable to understand something that is impossible to imagine," but a carrier of ideological evil.

Heisenberg's fears turned out to be exaggerated, in the scientific community one could not feel anything like this. As I have already mentioned I first met him in 1957 at a conference on cosmic rays in Italy and for some reason, not at all political, could expect his animosity (with D.S. Chernavsky we had discovered that in one of his works he used the classical approach where according to his uncertainty relation this was impossible) but there happened nothing like this. He was lively, with enthusiasm, for a long time telling me about his fundamental new work that was his passion at that time (it aroused a great interest in Moscow. Landau told me with admiration: "At the age of 57 he puts forward such a brilliant idea!" But then some drawbacks were revealed and the excitement was gone. It was this theory that Bohr said afterwards: "This is of course a crazy theory but it is not crazy enough to be right.").

During the Kiev international conference in 1959 a young talented physicist from my group in FIAN, G.A. Milekhin (who passed away before his time in three years), told Heisenberg, with my help, about his work. In this work he found an explanation of the fact that two theories of the same important process of particle interaction at high energies (a so-called multiparticle production) independently developed by Heisenberg and Landau, both very attractive but seemingly different led to different results. Milekhin explained that two theories can be transformed one into another, that they are in principle equivalent and the difference in results is explained by that in a choice of a subsidiary element of the theory that should be independently based on other considerations.

The theory of Landau was more developed, more popular and Heisenberg did not hide his joy that everything was clarified. His round face shone. There were other interesting discussions. Then he presided at the plenary session at which I made a "rapporteur" talk (i.e. the one summarizing all the talks at the conference on the same topic, multiparticle production). It seemed that everything was returning to its previous state. Nevertheless Rudi Peierls answered in my presence a question of his friend Landau "Rudi, what is your attitude towards Heisenberg?" by "I am not inclined to forget the past as fast as some." "I thought so," said Landau in an approving tone.

Soon after that I learned from Weisskopf about the above-described episode with Jensen — on his trip to Copenhagen to Bohr with an information on the level of German works on uranium, on the fact that after he returned to Germany Heisenberg, who learned about the voyage, did not give him away but just reproached him in a private conversation. When I told Landau about this he said: "This establishes the limit of Heisenberg's indecency." In reality this is first of all a manifestation of their, Heisenberg and Jensen, anti-Nazist views. At the same time Landau himself was before 1935–1936 of radically pro-Soviet views and did only afterwards understand that "Stalin betrayed the course of Lenin" and was not essentially different from Hitler. This was written in an anti-Stalin leaflet in the writing of which he participated and paid for by a year in prison and torments. He was miraculously saved thanks to the cleverness and courage of P.L. Kapitza.[14] However, also after that he did things that gave no less ground to accuse him of "collaborating with Stalin" than those for which Heisenberg was accused of collaboration with Hitler. I mean for example Landau's participation in calculations on the effect of hydrogen (and, as became clear at the end of the nineties also atomic) bombs that took place many years after the end of the war. He did everything honestly and not "in a slipshod way" (see above). For that he got the highest state awards: a title of the "hero of socialist labor," Stalin prizes, a permission to become an academician (for which he had doubtlessly had a right for a long time but CC had not allowed it). This was disgusting for him but he could not refuse. Only after the deaths of Stalin and Beria he said "That's it, I am no longer afraid" and stopped this activity but there was also something else.

---

[14] See Section 10.2 in this book.

The conversation with Peierls and the reaction of Landau on my story about Jensen took place 14 years after the war. R. Peierls who married a Soviet citizen at the beginning of the thirties knew what the Stalin regime was better than many other Western physicists. At the beginning of the eighties I had several discussions with him on Heisenberg. In 1988 at a conference in Copenhagen I insistently pressed him for a summary short answer at the question: "What are you condemning Heisenberg for?" He could only say that Heisenberg was completely incapable of understanding a position of somebody staying "across the hill" (i.e. exactly the same as what Casimir had said).

Perhaps the fate of Stalinized Eastern Europe, in particular the crushing by Soviet troops of the Hungarian uprising in 1956, Czechoslovakia in 1968 and many other after-war events, exerted their influence and the attitude of Peierls towards Heisenberg did somewhat soften.

8.5 Tragedies of the Epoch

Summarizing what has been said we come to the conclusion that the political behavior of Heisenberg was to a large extent determined by a natural care for the fate of German people and the country itself. Such a deep patriotism can of course in some cases transform into nationalism and even chauvinism, in a conviction on the superiority of your nation over all other ones. Despite of the fact that German culture was for centuries open for fruitful mutual exchange with the cultures of other people such things did happen in the history of Germany. For example, right after the beginning of the First World War a wave of chauvinism overwhelmed the country without facing any obstacles. All recent loud anti-war resolutions of congresses of the Second International (in which a leading role was played precisely by a strong German social democracy), passionate promises to begin, in the case of the war, a general anti-war strike in all the fighting countries turned out to be momentarily forgotten.

The same thing did even more strongly manifest itself in the Hitler years. Did such transformation of patriot into nationalist however happen to Heisenberg? We have no grounds for such an assumption.

Science is in general about the only spiritual sphere of true internationalism. This role can be played neither by religions, each of them working out their own norms of moral, nor even by art. Peoples do often have to overcome some barriers to understand the art of other people. I dare to give a hackneyed but especially bright example of an opposite role of color in painting: In China the funeral color is white, not black. Therefore a painting of Malevich, *Black square* (on white background), is perceived by a Chinese brought up in national traditions in a completely different way than by a European. Simply saying, grief encompassed by joy or joy encompassed by grief.

On the contrary, a scientist fully understands everything done by his colleague on the other side of the globe. It is not surprising that before Hitler, before the time when Stalin's totalitarianism acquired its full strength, in the world science there dominated a sprit of mutual understanding, benevolence and openness that was so well described by Charles P. Snow [Snow (1977), p. 134] a writer and a physicist by education and initial professional activity: "Scientists that entered science before 1933 remember an atmosphere of that time. ... I would risk to provoke your irritation ... and say that the one who was not doing science before 1933 would not experience joys of a life of a scientist ...

The world of science in the twenties was as close to an ideal international community as it is possible in principle. Do not think that scientists composing this community were belonging to a superhuman race or were free of usual human weaknesses.... The thing was that the scientific atmosphere of the twenties was filled with benevolence and magnanimity and people who had dived into it were involuntarily becoming better."

World shocks brought by totalitarianism and the following wars did also split the world of science. Heisenberg, perhaps, was naively keeping a feeling of the unity of people of science from previous years and therefore "did not understand the point of view of people across the hill" who, above all, hated nazism and had insufficient understanding of life under totalitarianism and everything related to it. He was still freely telling what he thought and was at the same time rushing about and what he felt and hoped for was very different in different periods which caused the indignation of former friends.

On the occasion of Dirac's death (in 1984) one published extracts from his letter exchange with Heisenberg from the time when Dirac, with some delay, joined the process of creation of quantum mechanics. Although they were exact coevals. Heisenberg who had at that time been an author of ground-breaking papers could be considered as *maître* with respect to Dirac. Reading of this letter exchange is pure delight. In their relations there is so much of openness, of benevolent attitude of "senior" Heisenberg to the previously unknown colleague, laconic and reserved Dirac, so much mutual trust. They freely exchange ideas, communicate to each other not only the results already obtained but opinions, intentions, hopes (all this can certainly also be seen in the communication of other, more senior people, e.g. of Bohr, Einstein and Born).

However, with national socialism coming to power they found themselves in completely different political conditions. It is true that Dirac participated neither in politics nor in the development of nuclear weapons. His sympathies did however reveal themselves in one fact in which, like it was also characteristic of many other Western scientists outside Germany, one sees a misunderstanding of inhuman lying dictatorship and totalitarianism in the Soviet Union. In November 1937, a year of the maximum of incomprehensible Stalin terror, a scientific journal of the Academy of Sciences of USSR issued a special volume dedicated to the twentieth anniversary of the October revolution. For this volume Dirac wrote a special article [Dirac (1937)] and did therefore praise this very revolution.[15]

At the same time Heisenberg was torn by contradictions. From one side there was a denial of national socialism. From the other one there was a satisfaction from the fact that for the German people there had ended a many-year period of despair, unemployment, hunger, humiliation brought by the treaty of Versailles. From the third one there was a fear of a terrible vengeance that, he feared, would come to Germany, in the case of its defeat, from the people that it enslaved. From the fourth one there was a bitterness from a destruction of the wonderful international community of scientists, the atmosphere in which he had grown up. And, finally, a natural desire of protecting science, the foundation of his life, from Hitlerism and of staying with his people. He had to make compromises doing this and who can tell exactly what are the borders for these? All this was the typical fate of an intelligent in the conditions of barbarian despotism when he does not dare to make an unequivocal choice of a patriot, but not a nationalist. The same situation was in the Soviet Union under Stalin dictatorship as well as during later decades.

The widow of Heisenberg wrote a book of memoirs and gave it a title *Internal emigration*. In the preface to this book Viktor Weisskopf, who has already been mentioned many times, a remarkable theoretical physicist, a student of Bohr and Pauli who in 1937 emigrated from Europe to the USA, a person enjoying an immense scientific and moral authority in the

---

[15] In one Western book I found that Dirac was simply called a communist (of course he was not a party member). In Section 3.3 I have already written that apparently he influenced his friend Dirac with his propaganda.

physics community, writes that Heisenberg strived to create "an island of decency." This is perhaps one of the most exact definitions of his political position.

Nevertheless we have seen that after the war Heisenberg, together with many other German physicists, faced misunderstanding, reproaches and even hostility of physicists from democratic countries.

This was of course not only Heisenberg's tragedy. This was a tragedy of the epoch. However, those who experienced the same fate under another equally disgusting dictatorship (here, of course, I have in mind Soviet scientists) would seemingly have been able to understand all this. Nevertheless this did not happen. Heisenberg met understanding of just a few.

It was also surprising that these same Western physicists did not show any malevolence towards Soviet physicists who often went for even more humiliating compromises with "their" dictatorship while many of their statements sounded worse than those by Heisenberg.

When in 1956, after the 19 years of "iron curtain," many Western physicists came to Moscow for a conference and there immediately resumed friendly relations with those who had survived through this time and there appeared many new ones. There was no hint of any kind of reproach towards Soviet physicists. Why? Was it indeed just because during the Second World War our countries were allies and Stalin played a role of their savior from Hitler?

Fate and the position of Bohr were different. He hated Nazism and this hatred did not cause ambivalent feelings in him — everything was clear. He was in constant close contact with the Danish resistance and through it, with allied military bodies. When the time came Bohr was brought to Sweden and then, through England, to the USA where he took part in the development of atomic weapons. However, when this work and the war were approaching their ends he got painful thoughts about the post-war structure of the world with the bomb. He appealed to Roosevelt and Churchill insisting of communicating the "secret" of a bomb to the USSR in order to keep an alliance forged during the war (nobody in the West knew that in the USSR there had been already launched an active and successful work on developing atomic weapons and that in fact there was no secret at all).

However, experienced politicians played with Bohr like with a ball, throwing him from one to another. Everything had been done, scientists had given them the weapon which was now at their full disposal so that these crackpots could be just kept away.

This was always the same, when American physicists appealed for not dropping the bomb on the already barely breathing Japan and here when Sakharov protested against the technically useless but carrying mortal danger superbomb tests.

In this way Bohr and other scientists come to their eternal tragedy — a full negligence from the authorities for who they had forged military might. This is a very different tragedy which Heisenberg and his friend escaped, but this is also a tragedy of the epoch.

8.6 Why Hitler Did Not Get an Atomic Bomb

This question is still of interest to many since Jungk [Jungk (1960)] put forward a statement that Heisenberg and other physicists connected with him did meaningly sabotage its creation. As was already told there continue to appear books in which one reads that from one hand they did not guess how to do it and from the other did really hinder the work. Once again, I will try to show that neither the former, nor the latter provide a full explanation of the issue, the situation here is more complex.

In the literature one can find mention of many other possible reasons. Here is the first of them: Hitler forbade development of weapons that could not be soon used in the ongoing war. In 1941 German physicists, primarily from the Heisenberg group, came to a conclusion that for creating a bomb one needed material and human resources which were impossible to obtain during the war and in 1942 informed the leadership that they could not produce a bomb in nine months as the leadership wanted. In these circumstances Hitler's order could be interpreted as a permission to stop working on the atomic weapon. However, had the physicists passionately wanted to create it they would not have calmed down so quickly. Indeed, the duration of work could not be known exactly and they began working earlier than others, in 1939, when material and human resources were still available (during the next two years they were only becoming bigger due to conquering many countries).

Another opinion states that German physicists did not have enough understanding, did not know a lot (this is also true but *why* did it happen? See Bethe's bewilderment concerning an absence of interest to this issue by Heisenberg, see above) and their collective was weakened by mass deportation and emigration of big scientists from Germany. Indeed, Germany lost 15 Nobel Prize winners only in the field of physics and chemistry and big physicists in general who played a leading role in the creation of atomic weapon in the USA. Still in the country there remained many strong scientists who understood the most important elements of the problem very soon. Having a wonderful chemical industry capable of solving complicated problems of chemical purification of materials and of constructing complex machines and mechanisms, with the amount of stocked uranium (at the end of 1940 Germans even had somewhat more uranium than Fermi in the USA 2 years later), etc. Therefore the situation could in no way be considered as hopeless. One also points out at the general situation with science in Nazi times. Science and scientists were humiliated, Hitler despised them. Only on September 1943 did Borman issue an order forbidding to draft scientists to military service and use them for other military needs that were not related to their main speciality. (In the USSR such a decision was taken after just four months after the beginning of the war.) As a result of this order some scientists were returned from the front but this was already in vain. There was no single governmental scientific management. Research on the uranium problem was carried out by disjoined groups that were competing with each other. For example, even the last attempt of launching a self-sustained chain reaction in April 1945 failed only because the group of K. Diebner did not give to that of Heisenberg its stock of heavy water and uranium (cf. [Irving (1960), p. 318]).

An important role was played by the striking self-confidence of leading German physicists including Heisenberg himself. In essence they thought that if they faced unsurmountable difficulties then their Western colleagues would not be able to do anything at all because of their generic backwardness in science.[16] At first they did not believe information on the first American bomb dropped (see below). Thinking of creating in the future an energy reactor they, as has already been told, up to the last moment, April 1945, concentrated all their efforts on getting a self-sustainable chain reaction in uranium which had already been successfully launched in Chicago by Fermi in December 1942, so had the last attempt of German physicists been successful this would not have had any military significance anyway (we have already discussed this).

Of interest is the following fact. A bomb could be created in two ways, either from plutonium or from the isotope ^{235}U which had to be somehow

---

[16]This is in fact not so surprising. Before the war Germany was a center of world science, German was a common language for all the scientists in the world, main scientific journals were published in German. Almost all the main creators of American atomic bombs were formed as scientists when working in Germany for many years.

extracted from natural uranium containing it in a very small proportion. This last task was very difficult. German physicists did unsuccessfully try six methods of isotope separation but neglected precisely the one that was used in the USA and the USSR. Perhaps this happened because the best expert in this method, a Nobel laureate Gustav Herz, was not involved in the project because of his nationality. As a participant of the First World War decorated by an order, the Jewish Herz could stay in Germany, the race laws were not applied to him. Nevertheless he was not allowed to participate in such a secret project. As a result the whole problem of extracting the isotope of uranium-235 was considered in Germany to be of secondary importance. At the same time in the USA and the USSR this method of isotope separation was developed and used with success. Let us note that in the Los Alamos center where an American bomb was being constructed immigrants, at that time still citizens of European countries that were enemies of the USA (they had not yet lived in the USA for five years and so could not become an American citizen), were playing the leading role. These were the brilliant scientists E. Fermi, L. Scillard, E. Teller, V. Weisskopf, H. Bethe, J. von Neumann, E. Wigner and many others. In the USSR where an anti-Semitism planted by the state and the party was in full swing was not implemented in the uranium sphere.

All this did of course play its role. However I would like to make a special emphasis on still another main, decisive circumstance that is not so evident but does, in my opinion, really explain the failure of German physicists in creating the bomb.

While categorically denying a statement that all or almost all German physicists did *consciously* sabotage the creation of the atomic bomb one should tell about a different thing. A success of a scientific work does not exclusively depend on conscious decision. Each scientist — a mathematician, a physicist, a chemist, a biologist, a physician — knows very well *that an achievement of something really essential and difficult is possible only at a price of full concentration of intellect and internal forces,* only by full immersion, *by passionate desire of achieving a goal* and this is already determined by subconsciousness. Were German physicists overwhelmed by such a desire?

They could not but experience disgust towards Hitlerism. Many were anti-Nazis. (It is characteristic that zealous Nazis that were present among the physicists, in particular apologists of "Aryan physics," did not take any part in the work on the atomic problem. Something similar was used by us as well: Those few qualified physicists that joined the Soviet party philosophers in a baiting of "idealistic" quantum mechanics and theory of relativity did not do anything for the creation of atomic weapons.) Please recall what Bethe and Teller told about this in relation to Heisenberg (they surely knew what they were talking about). Heisenberg and many other German scientists could not fully distance themselves from the moral aspect of the problem. They were convinced that Americans and Englishmen were far behind and therefore it was them, Germans, who had to decide whether the awful weapon was to be made.

It is known that Otto Hahn (who discovered uranium fission) did understand what this could lead to. He morally suffered and was close to suicide, dreamt about drowning all the uranium in the ocean but was nevertheless participating in the project. By the time of the atomic bombardment of Japan main German atomic physicists were interned in England in the estate Farm Hall. Having learned about a dropped bomb, about the death of hundred thousands of people, Hahn was so horrified that his friends worried about his life and did not leave him alone before they were convinced that he had gone to bed and was sleeping.[17]

Could Max Laue, who hated Nazism with all his soul, wish Hitler to get a bomb? He wrote to his son in 1946 [Beyerchen (1977)]: "In the process of all the research on uranium I always played a role of an observer who was often, although not always briefed, kept updated by his colleagues" (see the text and footnote on page 325). Nevertheless he was all the same considered to be one of the main participants and it was not accidental that he was interned, together with others, in Farm Hall and participated in the desperate attempt of launching the fission process in the April of 1945.

Heisenberg and Weizsaeker write that they felt themselves relieved when in 1941 they became convinced that it would be impossible to create a bomb in the wartime Germany and there is no reason of not believing

---

[17]Unfortunately the discovery of fission is related to unpleasant personal conflicts. The very fact of fission in the process of neutron absorption by uranium nucleus was established in a joint work of Otto Hahn and Strassmann. A talented Lise Meitner also took some part in this work. Understanding of the process of fission was achieved in a subsequent work of Lise Meitner and (her nephew) Otto Frisch. It seems to me that the Nobel Prize should have been given not only to Hahn, as it happened, but all the three scientists. Meitner took this injustice very heavily. Unfortunately her feeling of offence led to a sharply negative attitude towards Hahn (as well as to other physicists that remained in Germany) that found its reflection in attacks on him in the literature. As far as I can judge these attacks, in any case the majority of them, are unjust. I am writing this so that readers who will run across these attacks would have this story in mind and make a corresponding evaluation of the sharply negative attitude of Meitner to Hahn (and to Heisenberg and his colleagues).

them. Indeed, from the very beginning they restricted themselves to work on reaching a self-feeding reaction. In the published materials there is no hint that during this period they were thinking over a construction of the bomb or perform any calculations on it (that a bomb should be "of the size of a pineapple" was everybody's knowledge for an already long time). Both Heisenberg and Weizsaeker state that in principle they knew everything about major theoretical issues needed for constructing a bomb, but in the autumn of 1941 became convinced that to build a reactor and then obtain a quantity of another fissile transuranium element (plutonium) required for making a bomb or extract from natural uranium a corresponding quantity of ^{235}U one needed efforts that were impossible in the wartime Germany. As is clear from what has been told this was to a certain extent an exaggeration. They knew principal issues but did not know and *were not even eager* to get exact information on the most important qualitative characteristics of substances and particular processes — and it is impossible to do anything without it [Bethe (2000)].

A statement on the fact that the main principles were known to them is quite grounded (if one does not take into account an error of an experimentalist W. Bothe. Because of this error graphite as a moderator of neutrons was rejected and all the consequent work went along an incomparably more difficult path, we will return to this once again below). It is confirmed, in particular, by a very detailed unpublished work of the already mentioned F.G. Hautermans finished in the August of 1941. It was done in Berlin in a private institute of Professor Manfred von Ardenne who sent me a typewritten copy of this work.[18]

Repeating myself I dare to remind nevertheless that this was a man with an active communist past and communist views, from 1934 he worked in Kharkov Institute of Physics and Technics. In the years of Stalin terror he was arrested and after signing a pact with Germany in the April of 1940 handed over to Hitler. In July 1940 Houtermans was released with a ban on working in governmental organizations. Laue, and also Weizsaeker,

---

[18]Hautermans, F.G. *Zur Frage der Auflosung von Kern-Kettenreaction.* Mitteilung aus dem Laboratorium Manfred von Ardenne, Berlin-Lichtnfeld (Manuscript, 1941. I am very grateful for Professor Manfred von Ardenne for sending me this important document). Done within eight months (January–August 1941) the work of Houtermans demonstrates a full understanding of the physics of the problem. Pessimistic conclusion on practical impossibility of making a bomb in Germany was based only on considering heavy water and not graphite, as in the USA and the USSR, as a moderator of neutrons in the reactor (a consequence of incomprehensible error of W. Bothe).

helped him to find a job at the institute of von Ardenne (an interesting fact giving additional evidence on mutual understanding and trust of true scientists having anti-Nazi views even in the conditions of that time: two noblemen, von Laue and [baron] von Weizsaeker provided maximal help to the convinced communist Hautermans).

The work of Hautermans is a special phenomenon. In the introduction to Farm Hall documents Charles Frank (I am using this opportunity to express my gratitude to Dr. Horst Kant who kindly sent me a full text of this edition [Frank (1993)]) calls it excellent [Frank (1993)]. According to [Frenkel (1997)] in the first month of 1941 Hautermans informed Weizsaeker about it in a confidential conversation but added that he "keeps silence on everything having to do with constructing an atomic weapon." Weizsaeker and Hautermans who also understood a possible role of plutonium as a second kind of a weapon in addition to that based on ^{235}U (an extremely important fact! As is known one of the two possibilities was used in the bomb dropped at Hiroshima and the other in the bomb dropped on Nagasaki) decided that it was undesirable to inform the authorities on these possibilities. Then they had a meeting with Heisenberg and, according to Powers [Powers (1993)], decided to hide these results from their bosses (if this is true such a decision can indeed be considered as an act of sabotage). According to [Frenkel (1997)] copies of the work of Hautermans were sent to a number of participants of the project: W. Bothe, K. Diebner, O. Hahn, F. Strassman, H. Geiger and some others. This is obviously in poor agreement with an intention to keep everything in secret.[19] One tells in addition that Hautermans sent the paper to the censorship committee where, in accordance with his expectations, it lay until the end of the war. Something here needs further clarification, in particular because the name of Hautermans was never mentioned in the voluminous notes on Farm Hall's long discussions of the physical foundations of the issue. Several authors point this out. This is strange. Here one can suggest three equally unlikely reasons: A spirit of competition of different groups, a desire of hiding the work as deeply as possible and a wish of not showing the authorities close connections of a convinced communist with a top secret project. When reporting at a meeting of scientists, military and state officials in 1942 Heisenberg mentioned

---

[19]The choice of the addressees is somewhat funny. Bothe did not work on the weapon but measured physical constants. Diebner's credit as a scientist was low, he was a half-official. Hahn was totally horrified by the possibility of constructing a weapon. Strassman was an expert on inorganic and nuclear chemistry. Could it be that this was one of the ways of hiding the work?

Weizsaeker, not Hautermans. This can however be understood as an unwillingness to mention a communist.

Having come to a conclusion that it was not realistic to construct a bomb during the war and, as we have seen, not having a passionate desire of constructing it, Heisenberg and his colleagues directed all their efforts towards construction of the energy reactor. Later Heisenberg told [Heisenberg (1980)] that they had overestimated the difficulties (let us remember this important detail) but at that time had been glad that they had not had to intensify the work related to the construction of a bomb and to give corresponding recommendations to the government. They felt moral alleviation but the thought that atomic weapon could be constructed in another country worried them. Therefore Heisenberg and Weizsaeker decided, in their words, to visit Bohr (which was not at all easy but they managed to find an above-described official reason for the trip), describe to him a status of the work and ask to talk to physicists in other countries so that they would decline to participate in constructing an atomic bomb.

There also exists, however, another explanation of the position of German atomic physicists (who, if talking about leading scientists, were in their majority, as we have seen, anti-Nazis). According to this version they were not at all striving to persuade the Hitler leadership to urgently intensify the work on atomic weapons but kept its interest only at a level allowing them to save young scientists from the front, etc. Scientists understood that if they promised to construct a bomb and would not succeed then all of them would lose their heads. It is possible that such a reason did exist but it would be a mistake to consider this as the only one. Indeed, to a certain extent they seem to be right: Even Americans who began building giant plants without waiting for a confirmation of correctness of calculations related to the construction of a reactor with natural uranium (this was provided by the first self-sustained chain reaction on the experimental installation of Fermi in the December of 1942), who spent 2 billion dollars, exactly 1000 times more than German scientists on their uranium project (an estimate by Weizsaeker which is most probably correct) got the bomb only after the end of war in Europe. Thus *formally* all of them did their job quite conscientiously. Formally — yes.

Here are, however, three (or even four) facts. I think that they render irrelevant all infinite discussions on what and how did this or that physicist, historian or journalist *say*, how Bohr understood or misunderstood Heisenberg. All this is words, words, words and we shall discuss deeds. The first of this fact is the following one (I will have to repeat myself a little).

The failure is to a significant degree, if not entirely, related to the above-mentioned fatal mistake of Bothe. A remarkable experimentalist, a Nobel laureate, in the January of 1941 he was measuring, seemingly accurately, the most important physical characteristic of carbon nucleus — a diffusion length of thermal neutrons in graphite. Half a year before an experiment of the same Bothe had given a value of 61 cm. One had expected that in a specially purified graphite one would get at least 70 cm but Bothe got a value of only 35 cm.

From this there followed that a parasite absorption of neutrons by graphite was inadmissibly large so that it could not be used in a reactor as a neutron moderator. One had to orient oneself to heavy water which was much more difficult to get. It was produced only at one plant in Norway that consumed an immense amount of electric energy. Norwegian patriots managed to destroy the plant and the stock and then a transport with the earlier produced water (many of them died in this action) which had a heavy impact on the German uranium project.

Bothe, however, made a heavy error. Something (perhaps a contamination of nitrogen from air) was not taken into account. In the USA and the USSR a first choice for reactors was to use precisely graphite, not heavy water. If Bothe would not have made a mistake a remarkable German chemical industry would have easily fulfilled an order on ultrapure graphite. However, in an incomprehensible way neither Bothe nor anybody else had doubts in the correctness of his measurements. Neither he nor anybody else in Germany did repeat them. Hans Bethe writes on this: In Germany one has a very big trust to authority. Bothe was a recognized authority and one believed him unconditionally.

At the same it is clear how a scientist having a deep interest to the problem would have acted. He would have purified graphite again and again, would have changed experimental setup, would have squabbled the problem. On the contrary, here Bothe and all other physicists did with a striking light-mindedness believe in his absurd result: A path length of neutrons in deeply purified graphite (35 cm) was smaller than in badly purified (61 cm) i.e. a parasite absorption of neutron in more contaminated graphite was less than in the purified one. The error of Bothe, more exactly — his behavior with respect to his result would be unpardonable even for an experimentalist at the beginning of a career. Were it not for special circumstances, a relation to the entire bomb problem, bomb for Hitler, if this mistake had been made in peace times by, say, a Ph.D. student working on a task given by Bothe then one should not have any doubts — the same Bothe would laugh at him and force him to find a reason of an absurd interrelation between the recent and past measurements.

My late friend, a remarkable physicist Gabriel Semenovich Gorelik, did tell me almost half a century ago: "What is a real experimentalist? He comes to a laboratory in the morning, takes his working place, switches on experimental devices and suddenly notices that a speckle of a mirror galvanometer (at those times one of the main devices — E.F.) is deviating two centimeters less than two weeks ago in conditions that seem the same. He would not be quiet before finding out why." On the contrary, Bothe was quiet from the beginning.

Moreover, in 1944 this work was fully published as done in 1941 by Bothe and Jensen [Bothe and Jensen (1944)]. Here one sees a thoroughness of the work and the same erroneous value of 6.4 ± 1 millibarn for the cross-section of absorption of neutrons by carbon nuclei was given. A correct number known to everybody who worked in this field is 3.5 millibarn.

The first value makes usage of carbon impossible while the second one is quite suitable and carbon was used both in the USA and the USSR.[20]

It is interesting that at the beginning Fermi also thought that carbon was unsuitable but a remarkable physicist, Leo Szillard, who was working with him was by his first education a chemist-technologist. He knew about possible contamination of carbon by, say, boron and carried out an immense work to persuade industrial farms to purify graphite better. In this way

---

[20]There is in fact one more mystery related to Bothe's error. It was also noticed here by us. In 1942 Diebner (a head of a group competing with that of Heisenberg) and collaborators published in a German journal a voluminous compilation of measured (by different authors) cross sections of interaction of neutrons with nuclei for almost all the elements of the Mendeleev periodical table. For carbon it gives a value *close to the correct one* with a reference in parentheses "private communication of W. Bothe" (cf. *Physikalische Zeitschrift*, v. 43 (N2), S 440–465, 1942). What does this mean? Did Diebner take one of the early values of Bothe obtained before the last erroneous experiment of 1941 that disproved the previous ones (indeed, in the article in the footnote on page 440 the authors tell that they used data on all experiments published before 1940) but the Bothe data were taken from "private communication" (they were probably considered to be secret and were unpublished). However, due to some reason this value that rehabilitated carbon was not used in later attempts of German atomic physicists. This can be seen, in particular, from the above-described sudden suspicion of Wirtz that arose in April of 1945. It was however sufficient to attentively read one nondescriptive phrase from the paper of 1944 [Frank (1993)] to put everything on its place: In the footnote 2 on page 754 it says: "As communicated by Mr. K. Diebner an incorrect reference to our result in *Phys. Zs.* v. 43, p. 442, 1940 did probably arise due to a misprint." A funny episode. Indeed, the difference of these two numbers in the article published in 1944 did cause the serious interest of many physicists.

everything was finally in place. Germans, on the contrary, did not have Szillard.

Everybody else also had their share in what happened. They immediately trusted Bothe on such a super-important question and were relying on this absurd result in all the work on the uranium project. A shame for physicists. Happiness for humanity.

Could it be that the story with Bothe has a simpler explanation? In a well-known fundamental book *The Making of the Atomic Bomb* by Richard Ross one reads on p. 345: "Nothing in the existing documents points out that the error was deliberate. It is still worth noting that Walther Bothe (a Nobel laureate! — E.F.), a protege of Max Laue, was removed from a position of the director of the physical institute of the University of Heidelberg in 1933 (i.e. in the first year after Nazi's coming to power — E.F.) because he was anti-Nazi." This affected him so strongly that he was sick for a long time and underwent a treatment in a sanatorium. When he finally recovered, Planck gave him a position in a physical institute of the Kaiser Wilhelm Society, also in Heidelberg. Nazis were however continuing to haunt him and even accused him of fraud and deception.

Only in April of 1945, several weeks before the capitulation of Germany, when because of a lack of heavy water the installation was surrounded by a graphite "coat" and neutron multiplication, contrary to all expectations, turned out to be much more significant than one thought, did K. Wirtz get a suspicion that Bothe had made an error. It was, however, too late. Indeed, had Bothe not made a mistake in the January of 1941, the critical test would have been accomplished at least one and a half years before Fermi. Who knows whether in this case the Nazi leadership had decided to give the necessary 120,000 workers for building powerful reactors and then a plutonium bomb (of course following the advice of physicists)? Indeed, at that time almost the whole of Europe was in Hitler's hands and Germany possessed an immense economic power.

As has been already told, in the USA the bomb was constructed two and a half years after the Fermi experiment (although building the reactor started somewhat earlier). This means that in principle Germans could have built the bomb by, say, the beginning of 1944. Of course, a relative scarceness of scientists (as compared to the USA) would have hindered the task. And, possibly — this is the most important thing — that furious pressure that experts in America (as well as by us) did demonstrate would not have been present.

Now to the second fact. From numerous published memoirs of participants of the 'Manhattan Project' we know how scientists and engineers working in the USA were extremely afraid that Germans (whose science and technics were traditionally considered by them as the strongest in the world) could be ahead of them in building the bomb. All these people were entirely occupied by the atomic problem. At that time they were not haunted by moral aspects — one had to save humanity from becoming enslaved by Hitler. They could not even think about doing something else but working on the construction of the bomb.

Let me remind that Soviet nuclear physicist G.N. Flerov did at the end of 1941 get an idea that the USA had carried out secret work on the uranium problem because publications of all (almost all) experts on nuclear physics *completely disappeared* from American scientific periodical journals. Flerov immediately addressed Stalin with this conclusion and his activity played a significant role in resuming our nuclear research in the most difficult period of the war.

And what were the leading participants of the German uranium project doing?

In June 1943 Heisenberg signed as an editor a collection of scientific papers called *Cosmic rays* that was published several months later. Published in honor of the 75th anniversary of Sommerfeld this collection, albeit being quite valuable from a purely scientific point of view, was devoted to questions that had *no* relation to the uranium problem. From 15 articles in the volume 12 (!) are written by the leading participants of the uranium project: 5 (!) by Heisenberg himself, 2 by Weizsaeker, 2 by S. Fluegge, and 1 by K. Wirtz, E. Bagge and F. Bopp. It is even more remarkable that the articles were written on the basis of talks of their authors given at a series of seminars in Kaiser Wilhelm Central Physical Institute where Heisenberg was appointed by military officials to head the uranium project during the summer semester of 1941 (!) and the winter of 1941–1942.

In the same 1943 Heisenberg publishesd two papers that marked a beginning of an entirely new direction in fundamental quantum theory of fields and particles (on a theory of S-matrix). They have no relation to practice, even more so to reactors or the bomb. In the most acute period, from the end of 1941 till the end of 1942, he writes a philosophical book of 300 pages, *Order and Reality*. In a conversation with me in 1991 Weizsaeker added that in the summer of 1941 (!) and also later they regularly gathered at a seminar on biophysics and he himself worked on cosmology and astrophysics. In the summer of the same 1941, in the same institute, there took place a colloquium on physics and chemistry of proteins (Heisenberg and Weizsaeker were among the speakers). On May 5, 1941 Heisenberg went to Budapest with a talk "Theory of Colors of Goethe and Newton in the Light of Modern Physics", etc.

Nothing like that could happen in the USA where scientists were working on the bomb feverishly and without interruption.

Soviet scientists were occupied by the same task. During the war they were also afraid that Germans could overtake and then, during a period of cold war, considered a necessity of ensuring the parity of forces crucial for keeping peace. Also here the work went along without deviations to other topics and with incredible concentration of forces. At that time in a scientific collective headed by I.V. Kurchatov people even did not digress to defending theses. When the first hydrogen bomb was successfully tested in 1953 Sakharov was elected to the Academy of Sciences (together with some other leading participants in constructing a bomb who were not bearing a honorable title of an academician). In the process it turned out that he was only a candidate of sciences — one had not thought on scientific degrees, one had thought on vitally important things. To maintain due order one had to hastily supply him with a doctorate. It is interesting to know if anybody could find a single paper by Kurchatov, Khariton and their collaborators written at that period?

And, finally, the third fact.

As has been already told, in the August of 1945 the leading German atomic physicists were interned to the estate Farm Hall stuffed with covered microphones. They were under English military guard, their conversations were taped and in 1993 transcripts of these tapes were published [Khriplovich (1993)]. However the edition [Khriplovich (1993)] does not only contain these transcripts. The majority of tapes were analyzed by a special team and a resulting synopsis was sent to higher officials every one or two weeks. Reading this edition is extremely interesting. General Groves, an administrative head of all American work on atomic bomb, did earlier already quoted excerpts from these tapes in his book *Now It Can Be Told About It* [Groves (1969)] (now we know much more from [Khriplovich (1993)]).

The very first transcript contained a question by Diebner. "How do you think, did they install microphones here?" and a self-confident answer by Heisenberg: "Microphones? (Laughing). No. They are not so diligent. I am sure they have no idea about real Gestapo methods." Therefore the tapes fixed statements about which they were sure that nobody could hear them.

When German physicists learned about the first bomb dropped on Japan there began furious debates and mutual accusations. The atmosphere became hot and here there came a voice from Weizsaeker: "I think that the main reason for our failure lies in the fact that a majority of physicists did not want it out of principle. *Had we all wished Germany a victory we would have certainly succeeded.*" (italics are mine — E.F.) Bagge understood these words straightforwardly and told: "I think that the statement of Weizsaeker is absurd. Of course one cannot exclude that this was true for him but one cannot tell this about everybody".

Weizsaeker, however, certainly told not about a conscious unwillingness but stressed an internal, subconscious protest. Hahn replied to him: "I do not believe this but I am nevertheless glad that we failed," and Wirtz, meaning the bomb, told "I am glad that we did not get it." [Khriplovich (1993)] Rephrasing the above-cited words of Fazil Iskander on the boy Chik ("he knew it but he did not know he knew") we say that "German physicists did not want to build the bomb but did not know (at least a majority of them) that they did not know it." This is precisely what Weizsaeker did say.

It was impossible to solve the grandiose problem of the creation of atomic weapon with such sentiments.

Therefore everything brings us to a conclusion that patriotic German scientists who in their majority were anti-Nazi or in any case not pro-Nazi were formally carrying out their work on the uranium project conscientiously. Nevertheless the biggest scientists made uncomprehensible miscalculations, simply errors, erroneous estimates of the entire situation and of the choice of direction of their efforts and, most importantly, displayed astounding passiveness. They were working "in a slipshod manner." One gets convinced in this from many above-described examples of what they did and even more so, perhaps, from what they, with their extremely high scientific qualification, did not do. Somewhat repeating ourselves, let us summarize.

The first cardinal error was an erroneous measurement of parasite absorption of carbon, graphite, made by Bothe in January 1944. He was defending this erroneous result even in 1944 when he published an earlier secret paper. We have discussed this at length. Because of this error one did not consider one of the two main ways of creating an atomic weapon — an accumulation of plutonium in an uranium-graphite reactor and creation of plutonium bombs. This could in principle be compensated by production of heavy water but this could not be accomplished due to actions of Norwegian partisans and English pilots. Let us remind that Weizsaeker got an idea about plutonium very early and Houtermans did also understand it himself and worked it out in detail. The easiness with which Bothe himself and other atomic physicists accepted the erroneous result is difficult to understand (following Bethe [Bethe (2000)]) as a national habit of believing authorities. A rude Russian expression "devil-may-care attitude" does suit here much better. Let us note in passing that Bothe was in very good relations with a prominent figure of the French resistance, F. Joliot-Curie.

The second stunning phenomenon is a full refusal of Heisenberg to work on the problem of a bomb as such. Bethe convincingly proves: If even in 1945 after learning about Hiroshima his first calculation of the bomb was completely wrong and only after several days he made another, correct, calculation then he had *never* worked on the bomb [Bethe (2000); Khriplovich (1993)]. Having come together with his collaborators to a conclusion on the impossibility of constructing a bomb he not just restricted himself to an attempt of launching a self-sustained chain reaction (from which, even with gigantic efforts, there lies a many-years extremely effort-consuming way) but, like many of his collaborators, worked on various serious scientific problems that had *nothing* to do with the uranium problem (see above). Where was his genial intuition? Bethe wonders why did not he, even from intellectual curiosity, study an impact of fast neutrons on uranium-235 (an atomic bomb works precisely in this way). For example Fermi, when justifying uranium studies, said that "first of all this is good physics." For Heisenberg this was apparently an "uninteresting" physics. How could this be? For him, a patriot? Carelessness? No, apparently he did not want to work on it. Something inside was hindering him. The moral aspect caused a feeling of disgust (recall his remarkable phrase from the book *Order and Reality*, p. 311). At the same time he made official reports on uranium problem in governmental bodies and did it formally, completely honestly.

We have already discussed another immensely important fact (fourth or fifth?) that suddenly surfaced only very recently. It turns out that when Heisenberg came to Copenhagen to Bohr in the fall of 1941 he covertly sketched on a piece of paper a scheme of an installation at which his group concentrated all its efforts during the second part of their conversation. Because of his technical incompetence Bohr did not understand anything. Having been sure in the espionage goal of Heisenberg's arrival he did not tell anybody about this second part of the conversation (at the same time, as already discussed, he could not do it in order not to reveal the state crime of Heisenberg, not expose him before the Nazis). However, when Bohr arrived in America in 1943 and drew this scheme from his memory to Teller and Bethe these great specialists did immediately recognize a scheme of a reactor, not a bomb. Heisenberg wanted them in the USA to know that he was not working on the bomb! So, what is this — a negligence? Alas, this is rather giving out a state secret to an enemy country, a state treason. Teller was indignant: Why did Bohr not tell anything about this second part of the conversation? And Teller made a conclusion: After he got scared by the first word of Heisenberg about the bomb he "switched himself off." Hardly. Had Bohr switched off he would have forgotten and he did not, even after two years.

And, finally, the fact that had already been mentioned. When in Farm Hall the German physicists learned about American atomic bombardment only few of them felt shame that they had not built the bomb. We have cited a decisive phrase by Weizsaeker: "Had we all wished Germany a victory we would have certainly succeeded."

Summarizing, we come to a conclusion that already the four above-described facts, together with many other ones (these are *facts*, not *words* but, as once already called, "material evidence" if one uses a juridical terminology) do explain a failure of the German atomic project. Physicists, first of all Heisenberg and his friends, were German patriots but completely rejected Nazism and in their majority did subconsciously (and some, like Houtermans and, to a large extent, Heisenberg, consciously) did not want the bomb to be made.

This point of view also meets the understanding of some other authors. For example when in 1988 I told Weisskopf about my above-described understanding of the issue he showed me in response a foreword to the memoirs of Heisenberg's widow that he had written. In this foreword Weisskopf draws attention to the words of Heisenberg (that I have cited before) that the participants of the German uranium project *overestimated* the arisen diffculties. At the same time, as justly written by Weisskopf, when a researcher does passionately want to achieve some result he always *underestimates* the difficulties. Therefore they (most probably, only subconsciously) did not have much interest in the success.

Even more direct was a statement by Teller who, as has been mentioned, knew Heisenberg closely and was of very high opinion of his personal moral qualities. In Teller's opinion as soon as Heisenberg realized that an atomic bomb could in principle be constructed he created a "mental barrier" that did fully switch him off from practical thoughts on this horrible weapon. This point of view of Teller is very close to that described in this essay.

As the reader remembers, Teller stresses that all information collected by him witnesses that Heisenberg worried about moral aspects of the usage of a horrible weapon. He went to Copenhagen to learn an opinion of the authority that was highest for him but he went there in vain. He was apparently worrying not about the fact that when making a bomb he was making it for Hitler. He was horrified by the perspective of atomic war as such.

Chapter 9

What Brought Hitler to Power? And Who?

Russian bolshevism came to power as a result of revolution and bloody civil war. On the contrary, Hitler came to power in a democratic way. He won the elections and President Hindenburg charged him with a formation of a government. Cruel oppression of opposition forces began later, at the beginning these were the actions of the "legal" government. The provocative arson of Reichstag gave into the hands of the "legal ruler" Hitler immense power and he began using it in a barbarous way.

What did however impel German people, one of the most advanced peoples in the world, with its deeply-rooted social democracy and very strong communist party, to throw itself into the arms of a merciless obscurantist and deify this tyrant in the first years of his reign? One has however to recall in which terribly horrifying state had Germany been living in the preceding years and what an awful choice its people faced in 1929–1932 because of events that were by no means happening only in this country.

What did in fact happen between 1928 and 1932? In 1928 the Nazis got at the elections to Reichstag only 2.6% of the votes while the socialist democrats together with the communists got 40.4%. In 1932 Hitler supporters gave 37.4% (at the end of the same year when Reichstag was dissolved and new elections were declared this was 33.1%) while worker's parties, social democrats and communists had almost the same result of 36.4% (in December this was 37.3%).

Of course the reason is primarily seen in the world economic crisis of these years. The crisis particularly strongly hit Germany which in addition suffered from the consequences of the First World War. For Germans these were years of despair. First of all this was a time of horrible permanent unemployment (in 1932 more than 6 million unemployed) which in proletariat milieu reached 45% and even those who were employed did often work for only a part of the week. A dole, in fact a small one, was paid for a limited time and afterwards one could count only on public charity equivalent to 1.75 dollars a week. The country was depressed by moral humiliation and shame of the Treaty of Versailles, almost all revenues from export went to indemnity paid to countries that had won the First World War.

All this did at those times coexist with overcrowded luxurious restaurants in the center of Berlin. At the same time right there, as witnessed by American journalist Knickerbocker [Knickerbocker (1932)], people fell on the asphalt from hunger. He saw all this at Alexanderplatz. Here is how he describes a small restaurant in the working, primarily communist, district of Berlin Wedding which one called German Moscow:

> Behind iron bars protecting a buffet stand there were wooden trays with roasted horse-flesh and pairs of sausages from horse-flesh. The customers were hungry. They sat at their tables staring at horse-flesh behind the bars. It was dinner time but they did not order anything ... From forty customers only two did at least have something on their table: In between an old man and a freezing woman there stood a mug of malt beer. He made a gulp, put the mug on the table and began staring at horse-flesh ... *And this was not happening in Russia* (italics are mine — E.F.). At the door of the restaurant *At Old Fritz* there was a picture of Friedrich the Great. The restaurant bearing a name of the greatest of Hohenzollern's should be German even in the Red Wedding.

From 1929 till 1932 Germany was rushing about looking for a way out. Both the Nazis and the strong communist party promised to find it. In between these poles there were also social democrats, more numerous than communists, President Hindenburg and nationalists of old Prussian type. All of them had mass paramilitary organizations — for the Nazis this was Sturm Abteilung; for the communists, Rot Front; for the social democrats, Steel Helmet. Their noisy meetings, aggressive street marches, bloody clashes belonged to everyday elements of life.

Other Western countries followed the development of events in Germany with anxiety. The reason was not only in the fact that, as was written by Knickerbocker, because of American loans to Germany 'every American citizen — a man, a woman, a child, has a direct interest in the fate of Germany to the size of 33 dollars.' Much more important was that (as written by the same author) a communist Germany meant an appearance of the Red Army in the Rhineland.

An appearance of the communist Germany was quite real. In 1930 only in Berlin where there was half a million of unemployed, for communists there voted 739,235 people (for social democrats, 738,094) while for national socialists the number was almost twice smaller — 395,988. Roughly speaking the difference was in that "communists are those who never had anything while national socialists are those who had something but lost everything," [Knickerbocker (1932)] but life was unbearable for both.

In 1931 Knickerbocker traveled all over Germany, talked to representatives of all parties and the fears described in his book were well grounded. Already in the twenties one could see (I saw it myself) at festive demonstrations in Moscow a slogan "Soviet sickle and German hammer will unite the whole world." Communists in Germany used this slogan as well, it was also known in the USA. In the twenties and even in the thirties many of our komsomol member and people of Komsomol age in general wore *jungsturmovkies* — khaki-colored shirts with open collars girded with a belt and trousers or a skirt of the same color supplied with a leather shoulder-belt. This was a form of a German Union of red war veterans and its youth organization headed by the leader of the Communist party Ernst Thaelmann. A slogan "Red front from Vladivistok to Rhein" was declared at meetings in Germany itself, in the USSR and the USA. People faced a terrible choice.

An economic crisis by itself could not however explain the growth and victory of Nazism. Indeed, the crisis did primarily hit industrial regions and the working class. At the same time the two working class parties taken together remained remarkably stable. During all the years of the crisis their share of votes oscillated between 36 to 40%. However one justly points out at another important circumstance that helped the Nazis — a split in the worker's movement, an isolation of communists that did radically weaken the worker's front. We will return to this later.

Even with this taken into account it remains, however, unclear what was the source of strengthening of the Nazis that increased the number of votes for them 14 times if two working parties taken together did keep their stability so that the above-described extremely difficult conditions of life did not influence the worker's choice. This issue was studied in detail in the literature. Let us point out, in particular, at the fundamental analysis of an outstanding English historian Alan Bullock [Bullok (1994)] which was recently published in Russian. Bullock, also presenting results of other researchers, shows that a growth of a number of votes cast for Nazis had its origin in the middle classes, small businessmen, craftspeople and, let us stress this, peasantry, where Hitler's followers were extremely active — in 1932 the

Nazi "party organization did reach every village" [Bullok (1994), p. 292]. "The main source of new votes for the Nazis were voters that had earlier supported "bourgeois" parties — nationalists, liberals and progressive liberals" [Bullok (1994), p. 263].

Bullock quotes the conclusions of an American historian, T. Childers: Already in 1930 Nazis "captured important positions in each of the main middle classes" [Bullok (1994), p. 267] (these voters constituted 40% of the whole electorate). "The leading party, the nationalists ... were progressively losing votes in favor of Nazis *in rural regions.*" (italics are mine — E.F.) (ibid). A Canadian historian, R. Hamilton, tells: "More than a half of recognized votes at German elections were cast in villages and small towns with populations not exceeding 25,000 people." [Bullok (1994), p. 267] A detailed analysis of voting in different German provinces shows that in those years Nazis always got many more votes in rural regions than in industrial ones [Bullok (1994), p. 264]. Why?

Amazingly neither Bullock nor, apparently, other historians did pay attention to the fact that in this atmosphere of "fear, discontent and despair" [Bullok (1994), p. 269] a development of events was suddenly perturbed by a powerful external factor that moved the scale into a definite direction: in the Soviet Union there broadened a total collectivization with all horrors accompanying it — dispossession of kulaks that also hit a significant percentage of middle class, with expropriation of property and sending the "dispossessed" families to uninhabited woodland northern regions and to Siberian taiga. The very procedure of sending to exile was so awful that many died from hunger and diseases on the way there. How many was it?

Winston Churchill writes in his memoirs that in the process of his many-hour friendly conversations with Stalin in his home in Moscow in the August of 1942 there also took place the following conversation:

> 'Please tell,' I asked, 'are the hardships of this war influencing you personally as heavy as an enforcement of a policy of collectivization?' This topic did immediately energize Stalin.
>
> "No," he said, "the policy of collectivization was a terrible struggle."
>
> "I thought so that you consider it hard," I said. "Indeed, you were dealing not with several thousand noblemen or big landlords but with millions of small people."
>
> "With 10 millions," he said raising his hands. "This was something awful, this lasted for four years (evidently from 1930 till 1933 — E.F.)...

"These were the people you called kulaks?"

"Yes," he answered, not repeating this word. After a pause he noted: "All this was very ugly and difficult but necessary."

I asked: "What did happen?"

He replied: "Well, many of them agreed to go with us (i.e. were herded into kolkhozes — E.F.). Some of them got land for individual exploitation in Tomsk or Irkutsk region or still further to the north but the majority of them were unpopular and were killed by their own farm hands." There came a sufficiently long pause [Churchill (1955), p. 493].

The figure mentioned by Stalin is by no means an exaggeration.[1] If one believes this story by Churchill (and why should not one believe him? Churchill himself was a "tiger" and in his speech in the English parliament after Stalin's death praised him) then a four-year war against his own people was for Stalin more fearful than that with Hitler's Germany even in its most difficult period when Germans began their fight for Stalingrad.

A hasty forced collectivization was accompanied with the killing of peasants who were herded into kolkhozes with their livestock. For example only for cattle from 1928 till 1932 the number went down by more than 1.5 times. The level of 1913 could not be reached even at the beginning of the war with Hitler. Productivity of agriculture as a whole went down as well. Even afterwards when it was equipped with technology and by 1941 the number of tractors, lorries and combine harvesters reached almost a million, wheat production from one hectare reached 7.3–8.6 centners while in old Russia with its plow and wooden harrow it was only somewhat smaller: 7.0 [Big Soviet Encyclopedic (1969), People's Economy (1987)].

The state began a merciless confiscation of "excess corn" and other agricultural products from peasants. One took from peasants the last and left them to die in order to support export, get hard currency for buying machines needed for a simultaneously unrolling unimaginably accelerated industrialization. In villages there was a true hunger, it even came to cases of cannibalism. The hunger moved to the cities, but there it was less terrible: One introduced (up until 1934) a card system of food rationing with small rations.

---

[1] In general he did make the event look better than it had really been. He did not mention, e.g. that in Ukraine, Povolzhje and in other regions many millions of peasants who had not been sent into exile did die from hunger (only for Ukraine one quoted the figure of 5–7 million and more).

All the resources were oriented towards industrialization. However, also here a "volitional" (or, rather, "voluntaristic") decision of Stalin on a sharp, 1.5 times and more, increase of targets for the end of the first five-year plan (1932) did have destructive consequences. As a result of this a previously carefully planned and grounded five-year plan that had been going very well, even faster than planned, carried out before this decision (oil industry had reached the targets planned for the end of the five-year period already in 2.5 years, a construction of Dneproges and other grandiose construction projects had been proceeding very well) went to pieces.

All interrelations and balanced proportions were broken. For example, the five-year plan target for smelting of cast iron for its last year, 1932, was initially 10 million tons but then it was increased to 17 million tons. However, because of a chaos that arose in the economy, even the original target was by far not reached. In spite of great efforts, in spite of the fact that this target was placed into a focus of attention (on the first page of *Pravda* was published, on a daily basis, data on a quantity of metal smelted during the previous day) in 1932 one smelted not 10 but only 6.2 million tons and only in 1940 the smelting reached 15 million tons [Big Soviet Encyclopedia (1969), People's Economy (1987)].

Often even when plants (for example, the key Stalingrad tractor one) were launched, "for the report," in time they were afterward stopped for finishing the construction, completion of installation of equipment, putting everything in order and putting everything to work which took a lot of time.

Of course this allowed to begin, sooner or later, a construction of modern machines, transport, aviation, diverse defense technics, but not at all at the pace prescribed by the "volitional" decision. Besides that one had to pay for all this by the disappearance of goods for the population. Normal shops were shut down and "special distribution centers" were opened (one also had to pay for goods there) for certain categories of the population. A sharp need in hard currency needed for buying abroad equipment for new plants forced to open a chain of "Torgsin" shops — formally for the needs of the numerous foreigners in the country but in practice for everybody who possessed and could bring there valuables including engagement rings, old dental crowns, silver spoons and icon frameworks. In these shops one could buy both food and goods of highest export quality. All this made the situation even more difficult in the moral respect.

Once I heard a *chastooshka* (with a feeling of horror with respect to a simple working woman who sang it):

> In Torgsin-shop,
> Cheese and butter, sausage.
> And in Soviet shop,
> Stalin goggles eyes.

Portraits of Stalin were hung everywhere — even above empty shelves of groceries.

One needed to find somebody to blame for the chaos in the industry, construction, transport. There began a search for "saboteurs," in particular among technical intelligentsia. Before the revolution Russian engineers were considered to be among the best in the world — on par with the Belgian ones. There also existed more detailed estimates: The best designers were Belgians, Russians, French; the best technologists, Germans; the best transport engineers, without any doubt, the graduates of the Petersburg institute of communications. It was not by chance that it was them who built the Great Siberian Railway and in general the whole vast network of Russian railways. It was not by chance that after a loss of the Russian military fleet at Tsushima, in just ten years before the First World War, they constructed a new quite modern fleet. Outstanding engineers did a lot for the recovery of industry after the civil war. It is sufficient to name an author of Dneproges design, I.G. Aleksandrov; a constructor of VolkhovGES, G.O. Graftio; V.G. Shuhov; E.O. Paton; a ship builder, A.N. Krylov. Personally these people were not touched but how many remarkable ones were placed into a category of "saboteurs" and were killed.

There began fantastic court processes — of Promparty, of Menshevist center, etc. (in 1930–1932). Newspapers were full with reports from them, with incomprehensible confessions of the accused in carrying out improbable crimes — from organizing an underground government headed by a brilliant engineer, Professor Ramzin to "conscious sabotage" that even for example showed itself in the fact that, as one professor confessed at the Promparty process, he had deliberately approved a project of Ivanovo textile factory with a height of ceilings in workshops of five meters while according to his public repentance one had been able to restrict oneself with four meters.

At the process of the Menshevist center, which to a large extent was a process of the top managers of Gosplan who had prepared a remarkable first five-year plan, a head of the metallurgy department repented his crime: 'I,' he said, 'did deliberately diminish the target for smelting of cast iron in 1932 by restricting it to 10 million tons while now we know that we will smelt 17.'

These were those above-mentioned mythical numbers. Ramzin, an organizer and director of All-Union Institute of Heat Technics and "the head of Promparty," his "minister of foreign affairs" academician Tarle and some others went out with long terms of imprisonment but a majority of the "enemies of the people" were executed under a loud approval of numerous mass meetings and demonstrations. "Unlawful methods of carrying out investigation" (a soft official definition for the horrible evil deeds of that time coined in a period of much later Khrushchev times) did their job.

Why am I telling all this in detail? Because (and this is extremely important) all this could not be hidden from the outside world. At that time the country was sufficiently open to foreigners. In connection with industrialization, construction of plants, buying machines and equipment abroad at construction sites and plants there worked many foreign specialists and highly qualified workers. In particular, there were very many Germans. (During my diploma practice in 1932 at Kolomna locomotive plant I saw myself German workers and technicians equipping a giant workshop for the production of diesel engines for submarines. I also saw how bitterly were the Soviet workers of this plant and their families starving.)

This phenomenon was a mass one. Foreign journalists traveled around the whole country. Although the foreign specialists themselves did have a special supply a sharply falling level of life of Soviet workers, peasants, rank-and-file intelligentsia could not remain a secret. In a German newspaper there appeared, e.g., the following cartoon: A Russian peasant stands in trousers made of cabbage leaves, with the text: *Kohlhosen* (a play on words: *Kohl*, cabbage; *Hosen*, trousers). It is especially important to underline that the German population was especially well informed about what was going on.

First, already from the times of the Rapallo Treaty (1922) that had broken a blockade of Soviet Russia and had brought much profit for both sides there took place a wide development of economic and many other kinds of cooperation of the two countries that were related to prolonged stays of German citizens in our country. A German ambassador to the USSR, von Dirksen, wrote in his memoirs *Moscow–Tokyo–London* that when he came to Moscow in 1926 he found out that in the whole country there worked 5000 German specialists. They knew very well what has happening in the country — up to what was going on in the regional party committees.

Under the Versailles treaty the German military forces were sharply restricted. Military aviation was totally forbidden. However German pilots were getting mass training at Soviet airfields, tankmen and officers of other combat branches, at Soviet firing ranges, in special Soviet military schools. The secrecy of all this was indeed a secret of *polichinelle* (i.e. something that everyone knows).[2]

Second, and this is perhaps more important, from the times of Katherine II who had invited German peasants and generously given them land, in Russia there existed many prospering German villages, "German colonies" (they were called like this even in Soviet times) with schools in German, Kirchen, etc. They were especially numerous in the Ukraine, in Crimea and the Volga region. From 1924 till 1941 on the left bank of Volga opposite to Saratov there existed an autonomous republic of Germans of the Volga region (with the capital in Engels) with a population of 600,000 people from which 64% were Germans. I myself had a chance of living for four months in one such "island of Germany" in Crimea. Later I learned about the horror that collectivization brought there.

Of course, dispossession of kulaks, devastation, exile to the north did not leave any of these islands intact. In Germany there enrolled a loud company running under the slogan "Brothers in need." Moscow newspapers did of course "rebuff" this "mendacious anti-Soviet provocation." However, the anti-Soviet and anticommunist campaign in Germany grew and included, in particular, collection of money and food for the Germans in the USSR. It went so far that, as written by the West German historian Niclaus, in 1930 the German government handed a special note of protest to the Soviet ambassador in Germany N.N. Krestinski [Niclaus (1966)] (this knowledge about the hunger in the USSR possibly explains the words "And this was

---

[2] Pilots were training in Lipetsk. Based on declassified NKVD documents, one finally openly told everything about this on TV on January 28, 1998. German pilots were wearing Red Army uniform, their groups changed once in several months. They were mastering the Dutch Fokker aircrafts and other models with which they later, during the war, bombed our country. They were served by Soviet personnel (naturally, numerous): medical staff, cooks, waitresses, cleaners, etc. (later they were all executed besides one man who occasionally survived who was present at the TV broadcast). It is noteworthy that during the war Germans did never bomb Lipetsk. The TV journalist ascribed it to sentimental recollections that captured them. However, the rumors spreading at that time explained everything simpler: In Lipetsk during the years of education of German pilots there were born many purely Aryan-looking children. It was them who were protected by German pilots. However, in the course of education they were allowed to fly over Voronezh, Tambov and other nearby towns. They learned these from the air very well and during the war did mercilessly bomb them.

not happening in Russia" in the above-quoted citation from Knickerbocker in which he describes a hunger scene in a Berlin restaurant in Wedding [Knickerbocker (1932)]).

In his book [Niclaus (1966)] Niclaus analyzes the crisis in the relations of the two states in 1929–1930 caused by the inner policy of Stalin that made it impossible for the German government to continue the course of Rappallo. Representatives of the German government during numerous meetings with the Soviet ambassador, N.N. Krestinski, as well as the German ambassador, H. Dirksen, during meetings with narkom of foreign affairs, M.M. Litvinov, did declare that collectivization and the campaign against religion, shutting down and demolition (demolition of the Cathedral of Christ our Savior!) of churches, persecution of priesthood, prohibition of peal in the towns caused common indignation in Germany which especially related everything that was happening with the fate of German "colonists." A flood of protests addressed to the German government and the president of the republic coming from all places in Germany and the anti-Soviet campaign in the press made the position of German government in the Reichstag very difficult.

In these conditions, was it difficult to persuade German peasantry and other intermediate and undetermined layers of the German population that had previously voted for nationalists and other right-wing parties that had carried out the course of Rappallo as well as starving workers that it was not from Russia, not from communists wherefrom they could expect salvation?

In such a way everything that was happening in the USSR during these critical years pushed significant masses of German people to another pole. This was however not all. It was no less important that in this decisive historical moment the left wing forces were split. Communists were in complete isolation because Stalin used no other words for social democrats but social fascists and any collaboration with them was considered to be a treason that could only weaken communists. Only when in France there arose a powerful people's movement "Popular Front Against Fascism and War"[3] that in 1936 came to power and really blocked a road to power for fascism in France did Stalin and consequently Comintern recognize a

---

[3] On July 27, 1934, one and a half years from Hitler's coming to power, this people's movement was shaped by the French Communist Party and United Socialist Party of France signing an agreement on the unity of actions. Then FCP suggested to form a People's Front. In its organization committee, besides communists and socialists, there also took part other left wing groupings. Besides that, one restored a unity of trade union movement.

necessity of collaboration with all left forces.⁴ For Germany, however, this was too late. The Popular Front later existed in the form of the Resistance during the war.

As a result of all this the German people did make a choice for Hitler. How could this happen to a nation that possessed ancient culture, that gave to the world great scientists, writers, artists, people of religion and philosophers, with the nation that gave founders of scientific socialism?

However, this people was suffering unemployment and hunger at the level that threatened to lead to extinction. It experienced not only physical sufferings but also a feeling of humiliation. Talented heads and crafty hands were doomed to inactivity in the presence of remarkable industry. They were helpless when they needed to save their starving children. All this could not but result in a general animosity. There remained only one question: would this animosity turn into a "sacred rage" which Block saw in *The Twelve* and which, in his words, was led by Christ "with the bloody banner" (also bloody!) or into a troglodyte rage of "superpeople" capable of killing babies and burning books, into a rage inspired by a fuehrer with his Auschwitzes and mobile gas chambers. External reasons that were described did away with the uncertainty and the choice was made.

Stalin's guilt with respect to mankind is immense. *With his crazy internal policy* (cruel total collectivization, incompetent acceleration of industrialization, hunger and horrible terror) he did not just ruin agriculture in our country, physically exterminated many millions of peasants, created chaos in the industry and hunger in the country. It is by all this *and also by incomprehensible isolation of German Communist Party from other left wing forces he objectively opened to Hitler a way to power, pushed in his arms first of all peasantry as well as other middle classes and as a result provoked the Second World War.*

---

⁴At the 13 meeting of the Executive Committee of Comintern (November–December 1933) one recognized a necessity of unity of a worker's front but it was only the Seventh Congress of Comintern (July–August 1935) that proclaimed a full-scale policy of unity of all the left forces, i.e. the policy of the "People's Front."

L. D. Landau

Chapter 10

LANDAU, Lev Davidovich (1908–1968)

10.1 Landau and Others: '... Verklaerungen und Neubegruendungen ...'*

I was introduced to Landau by Yuli Borisovich Rumer right after I had graduated from MSU in 1935.

Rumer, who had returned from Germany at the beginning of the thirties after several years of working with Max Born, read us a part of a course on theoretical physics. He was elegant, behaved himself relaxedly, gave lectures in a clear manner, somehow effortlessly. He openly told that he was learning himself: He had graduated from university as a mathematician. Once I met him at the department with *Optics* by Planck in his hand (a finger holding a particular page). "I am learning physics," he rapidly told me with a smile, passing by with a springing gait.

Without feeling shy he could answer a student's question: "I do not know, I do not understand this, I will try to give an answer next time." He was charming, brilliant, benevolent.

Due to occasional circumstances I got personally acquainted with him when I was still a student. Once, in 1933 (or 1934?) I visited him at his dacha. When seeing me off to the station he suddenly said: "I would very much like to go to Kharkov to work in Landau's group." (As is well known since 1932, when Landau was 24, he headed a theoretical department in the Kharkov institute of physics and technics, UPTI). At that time I still knew nothing about Landau besides what my one all-knowing friend had

---

*'Elucidations and New Substantiations' (translated from German). At that time the main language of physics was German and the main journal was *Zeitschrift fuer Physik*. In our country, in Kharkov, there began a publication of *Physikalische Zeitschrift der Sowietunion*. One often encountered these words in titles or subtitles of the papers.

told me in 1930–1931 about a talented trinity in Leningrad — G. Gamov, D. Ivanenko and L. Landau, who liked to make jokes teasing people around, especially elder and respected ones. He described details with a thrill while by myself these childish outbreaks did cause nothing but irritation.

I was surprised and asked Rumer: "Well, is Landau very clever?" Rumer just tossed his beautiful head and stretched "U-u-u..." This could not but cause interest. By this time Rumer was one of the founders of quantum chemistry (together with W. Heitler, F. London, E. Teller, E. Wigner) and knew many people.

During a defense of my diploma paper which drove me to despair by the lack of contents — there is a witness who could confirm these words (written in 1988 — E.F. [Feinberg (1988)]) — there suddenly came excessive praise (this did not change my own evaluation). Soon after the defense Rumer called me: "Landau has come, he stays at my place. Come, I want to introduce you".

When I came to Rumer's small room in Tverskaya-Yamskaya (Gorky) Street densely packed with occasional pieces of furniture, he asked me to wait: "Dau is taking a shower." (As everybody knows, in Landau's environment one used short nicknames: Landau — Dau; Rumer — Rum; Pomeranchuk — Chuk). In a few minutes Landau slowly came in drying his wet hair with a towel. Rumer said: "Dau, here is Evgeni L'vovich, he wrote a good paper, talk to him."

'OK,' said Landau somewhat lazily, 'go ahead, just so that there will be none of these *Verklaerungen und Neubegruendungen*." We sat, facing each other, at a small (due to some reason marble) table and I could only pronounce, before being interrupted, the first phrase: "The work is on quantum-mechanical theory of stability of crystalline lattice." However, hardly did I plot on a piece of paper a curve (of the type of the potential in two-atom molecules) and explain: "It is known that such a curve displays a dependence of an energy of crystals on lattice constant," Landau immediately exploded: "Where did you take it from? Nothing like that is known. In the best case we know some points around the minimum if one takes into account data on compressibility. Everything else is invented."

I was struck dumb. I did not even realize that I did not need the whole curve at all, knowing the vicinity of its minimum was sufficient. Attempts of justifying myself with words like "Everybody writes this, for example there," caused only new indignation: "What of it! For example, one draws Sargent curves," (here he mounted the hobby-horse of that period. Everybody who dealt with Landau knew that he always had some favorite subjects for scoffing. At that time one of them was Sargent who said that if one plots experimental data on beta-decay of various elements (vertically — lifetime; horizontally — decay energy) the points group around certain curves corresponding to a degree to which the transition is allowed). "There is no Sargent Kurven (Sargent curves), there is Sargent Flaeche (Sargent plane), the points are uniformly distributed in the whole plane," roared Landau. It continued in the same fashion[1]: "What else do you have?"

In what followed I only managed to murmur several unconvincing phrases — particularly because, as already mentioned, I myself did not see anything really essential in what I had done.

Soon everything was over. Then there followed a brief quite friendly conversation on side topics (we were both from Baku and this gave food to the conversation on the town of our youth, on a common friend, etc.) and I left in a state of a shock.[2]

After that we were in contact many times and participated in common discussions, in particular when after an arrest of Landau, a year of imprisonment and his subsequent release in 1939–1941 a "group of Landau" (several people) and a group of Tamm (also several people) did sometimes get together on Fridays in FIAN and IFP (Institute of Physical Problems where Landau worked) for informal discussions of physics.

All the same, I usually did not feel at ease and was mostly silent. Only once (four years after our first encounter) did I dare to put in an

---

[1] Later the quality of experimental data was significantly improved, the grouping of points around certain lines became more visible and the lines themselves got a more detailed explanation. In short, Landau's roar was misdirected.

[2] All of the written above was written by me and published, in 1988, in the volume *Recollections on Landau* [Feinberg (1988)] and only then did a hidden motive behind this episode become clear. My younger colleague of many years in FIAN, I.M. Dremin, who had graduated from MEPHI (Moscow Engineering Physics Institute) *at the end of the fifties* told me that he and other students-theorists knew about my meeting with Landau very well (25 years after it!) — knew even more than I did.

It turned out that before our meeting somebody (possibly Rumer) had told Landau that I had written a diploma paper which had carelessly (due to the lack of experience) been broadly titled "Intermetallic Bonds." Landau had reportedly replied: "Such a theory could have been created by a theorist of Tamm's class (this is true — E.F.). Feinberg does not have such a class therefore the work is wrong." Thus Landau had decided (Dremin used a student's lexicon) "to rub Feinberg's face across a table" which he had done. Wherefrom, however, could Landau know my "class"? I guess, from the two small papers cut out from my diploma work sent to the Kharkov *Physikalische Zeitschrift der Sowietunion* (and published there) with which he was naturally tightly related. Of course, my diploma paper should have been titled in a much more modest way: "To the Question of the Theory of ... " or "Remarks on the Theory of ... ". Then Landau would possibly have not objected.

objection. At that time one thought that a muon discovered in cosmic rays was that very nuclear meson (pion) that according to Yukawa's hypothesis was responsible for strong nuclear interaction. Speaking about it Landau mentioned with disappointment that because of inapplicability of perturbation theory to strong interactions calculations with this meson are impossible, it is somehow of no use to a theoretician. I timidly put in: "No, it has an electric charge, one can calculate various electromagnetic processes." Landau, who was walking, stopped, thought for a while and slowly said: "Well, this is probably true."

We continued for a little while and soon there appeared a few corresponding papers — by Landau himself, Landau and Tamm, Ya.A. Smorodinski (a disciple of Landau). I was awarded by the reprints of two papers by Landau. Something in his attitude towards me had possibly shifted. Still we were very far apart from each other for a long time to come. I did already understand very well who was Landau as a physicist but years had to pass before I was capable to discuss with him physics tête-á-tête, describe my work without panic (although always with some anxiety) and defend my point of view.

Let us now discuss this little story. Two moments are of interest here: (1) What did "none of *Verklaerungen und Neubegrundungen*" mean? (2) Was Landau really such a beast who was capable of paralyzing with a few words a theoretician who had approached him with a question? (By the way, he was formally only 4 years older than me, at that time, in 1935, he was 27, I was 23 and Rumer was 33, but, as seen from the above, a formal comparison of age did not mean anything. Landau himself told that a reference point is not a date of birth but a year of publication of a first scientific paper. He published his first paper at the age of 19, in 1927. Therefore, by this count he was 8 years older than me). Let us start with the first question.

At that time in our country (also abroad) there appeared quite a few papers on theoretical physics which did not contain any new results but did just ruminate again and again various more or less fundamental elements of quantum field theory or theory of relativity. Dau could not stand this because he was a man of action. The result could be small but had to be new and reliable. Here there also played a role that, in my opinion, Dau held himself personally responsible for a state of theoretical physics in our country. The mere fact that he was outraged by every embellishment of the situation in our physics could serve as an indicator of this.

For example, in 1936 in Moscow, in an overcrowded giant audience of the then existing communist academia in Volkhonka there took place a general assembly of the Academy of Sciences devoted to an activity report of Leningrad Institute of Physics and Technics. For many years the institute belonged to the People's Comissariat (Ministry) of heavy industry (*Narkomtyazhprom*) and was constantly under attack for dealing with "problems far apart from reality" (like nuclear physics) and in such an atmosphere its founder and director Abram Fedorovich Ioffe delivered his report.[3] The outstanding role of the institute and Ioffe himself in the development of Soviet physics is well known. He also did a lot personally for Landau when he worked at his institute.

Nevertheless Landau and also Aleksandr Il'ich Leipunski, both young and well acquainted with the world level of science due to their work abroad, came out with a ruthless critique of the work of Ioffe and the institute. They pounced upon a too optimistic evaluation of the state of our physics that was given by Ioffe.

The speech by Landau [Ioffe (1936), pp. 83–86] was remarkable. He began it with the words: "Whatever are the drawbacks of Soviet physics there is no doubt that it exists and is developing and it seems to me that the very existence of Soviet physics is largely due to A.F. Ioffe." Nevertheless after this, having had once again stressed the achievements of A.F. Ioffe, he vehemently attacked his report. He derided Ioffe for his statement that we had 2500 physicists and told that in their mass these people "play a role of laboratory assistants and do not possess any significant knowledge," that "... if counting together with physical chemistry one could name about a hundred of real physicists and this is far too little," etc. He criticized many works of Ioffe for being erroneous and his position for praising rank-and-file works, for ascribing to our physicists discoveries that in fact are replications of the work done abroad, for "propagation of style that can only be characterized by the notion of boasting". All this "is harmful, corrupting Soviet physicists and not contributing to their mobilization to a giant work that lies ahead of us," etc.

Reading it nowadays, one can not but acknowledge that with all the unheard sharpness of this speech the 28-year-old Landau spoke from state,

---

[3] Perhaps to somehow defend himself from such attacks Ioffe had a separate section of the report called 'Problems of Socialist Technics' that included about 30 applied problems on which the institute was already working or was going to work. Among these there were some valuable ones but also such problems as "windowless construction, rational form of windows and heating systems" (what was meant was a replacement of metallic heating batteries by ceramic ones), etc. I recall very well one "problem" that seemed especially striking to me — "painting floors indoors white to save electric energy on lighting." Apparently he was in a really difficult situation (cf. [Ioffe (1936), p. 77]).

civic point of view, with a feeling of pain towards our physics. Of course he did somewhat exaggerate. For example, by that time two works that many years later got Nobel Prizes had been done: A discovery and theory of chain reactions (N.N. Semenov) and a discovery of Vavilov–Cherenkov radiation (an example of an extremely exact, difficult and reliable experimental investigation the scarcity of which Landau also lamented, which was in a year after this session theoretically explained by I.E. Tamm and I.M. Frank). He also knew, of course, about a discovery of combination scattering (Raman effect) by G.S. Landsberg and L.I. Mandelstam that did not get a Nobel Prize only because of misunderstandings and about many other remarkable works. Nevertheless he was essentially right: all this was insufficient for such country as ours and, most importantly, there were too few qualified physicists.[4]

When Landau left the rostrum and, when heading through listeners sitting on the stairs of amphitheatre, passed by me, I told him: "As long as the present academicians are alive, you will not become an academician." He smiled awry. He was not in a mood to joke. Probably it was not easy for him to decide to speak against A.F. Ioffe. Still he had to do it: He felt a problem of development of our physics as his personal one.

This feeling of responsibility also showed itself when he, together with Evgeni Mikhailovich Lifshitz, created their many-volume course on theoretical physics, brought up his school based on his own system encompassing all stages of development of a theoretical physicist. Those who visited him at home after the automobile accident of 1962 remember how he repeated: "When I get better I will work on the school program in physics — it has to be changed totally." Of course there exist skeptics and cynics who would prefer all this to a desire to dominate, head all the theoretical physics. However, if an element of ambition that is necessary for any researcher and active person was present here, it was not the dominant one. It is sufficient to recall that when ranging theorists in "classes" by their achievements he gave himself a modest class. In general, in all his scientific activity there was so much honesty, soberness, conviction that even when he seemed to underestimate some of our outstanding physicists this was a consequence of his own scale of values based on deep conviction and in no way of a spirit of competition.

Let us however return to "*Verklaerungen und Neubegrundungen*".

---

[4]The speech if A.I. Leipunski [Ioffe (1936), pp. 72–83] was equally clear in its critical part but much more balanced in its tone and more objective in its estimate of the role of different schools in our physics.

Of course, it is inadmissable to reject the works of this sort completely. In any case, at that time what was most important was the creation of new physics, not these works. Besides that, working on such problems should not be fruitless as well. For example, a theory of electromagnetic showers in cosmic rays was created not by Landau and Rumer but by H. Bhabba and W. Heitler from one side and R. Oppenheimer and H. Sneider from another. However, although the latter did extract some basic consequences from their theories, using them was extremely awkward. Landau and Rumer put the theory into such a clear and convenient form (and in some sense this was also "*Verklaerung und Neubegrundung*"!) that this form became canonical. After these works one recalls its predecessors only to pay them a tribute.

Let us turn to another inference of my first encounter with Dau.

Bursts of accusations towards Landau in the cruelty of its attitude towards theorists that wanted to hear his evaluation of their work have not stoped until now (let me recall — this was written in 1988 [Feinberg (1988)]). It is true that Landau did not soften the sharpness of his evaluations and this often hid one's self-esteem hard. He was sometimes unforgivably hard in public conversations even when an object of his statements was a worthy person and Landau's attitude towards him was good. At times this was simply insulting. However — how generous was he in giving advice to his disciples! I remember especially well one case when in my presence he suggested a simple and transparent way of calculational implementation of a physical idea that was not easy to grasp and this idea helped his disciple to write a whole series of papers (together with his student who, on the basis of these works and their continuation, also became a well-known scientist). And advice was given easily, "in passing," the name of Dau in the work of his disciple was later not mentioned and Dau himself never mentioned this advice although later we had many conversations with him related to this problem.

It is worth recalling how in his speech at the joyful celebration of his fiftieth birthday Dau said: "Some think that a teacher steals from his students, some, that students steal from their teachers. I think that both are right and participation in this mutual stealing is remarkable."

Based on my experience, in particular on the above-described first encounter with Dau I understood the main thing: One should not approach him with something unfinished, not completely (to the extent one can) understood, with something that you can not defend as convincingly as he attacks. Later I got many confirmations of the honesty of his critique. If, as a result of a discussion, it was possible to prove him wrong, he was ready to acknowledge that he had erred. However, I never heard him (although I was told that such things did happen) saying in plain words: "Yes, I was wrong." This was however a manifestation of some childish feature of his personality that could only cause a smile.

Here is one characteristic case. For a long time Dau refused to accept a notion of isotopic invariance. He cursed it not choosing expressions. Nevertheless several years after the appearance of this concept when it was already broadly acknowledged, at Landau's seminar a colleague gave a talk on a published paper in which a speaker simply could not avoid discussing isotopic invariance. Approaching this point with fear he strained himself in expectation of another scoffing. Suddenly one heard a quiet voice of Landau asking with interest: "Well, this is something interesting, tell us about this in more details." The seminar exploded with laughter and Dau continued interrogating the speaker as if nothing unusual had happened. If one recalls that Dau himself selected the papers for discussing at seminars when looking through journals (sometimes reading the whole articles) one could assume that he had understood his mistake earlier.

This is however enough of this childishness. In essence his honest attitude towards science displayed itself here as well. Let us return to something more essential. Why, after all, did he not soften his critique? I think, because he always discussed things "as equals." He was apparently assuming that his interlocutor was "a grown up person," had to have his opinion and be responsible for his words.

The authority of Dau was extremely high and, perhaps, he should have remembered this more often and have been more careful with this dangerous weapon, understand that as a rule the conversation was in fact not proceeding in an "as equals" fashion.

It seems to me there was another psychological reason for this unrestrained behavior which I will discuss in the second part of these notes.

I am turning to one quite painful question that arose because there were one or two cases in which a negative attitude of Dau towards a work an author described resulted in the corresponding work (an important one!) not being published and later repeated abroad. The priority and sometimes the accompanying honors were lost. To which extent, however, Landau is to be blamed for this?

First of all I should refute inappropriate and completely false statements (that I sometimes hear!) that a negative attitude of Landau meant a ban for publication. This is completely wrong. Let us even put aside a purely formal detail: Dau was not a member of the editorial board of any journal — and there were several of these. I heard many times (I do not recall — from either Dau himself or E.M. Lifshitz) that Dau thought that it was admissible to publish anything (provided, of course, there was no direct mistake) if there was no contradiction with quantum mechanics and theory of relativity.[5]

There was however something else: An authority of personal attitude which some did not have enough internal strength to withstand. So, one not yet quite experienced theorist came to a certain bold idea and even realized it in numerous calculations of physical phenomena in which predictions of this theory could be verified experimentally. However, his colleagues (by the way, disciples of Landau) completely rejected the basic idea as absurdity and nonsense.

Then he dared to go to Landau. Landau understood everything right away, told that there was nothing absurd about the idea, the world could look like that but he personally did not like the idea in question and that such a world did not look attractive to him. Is Landau guilty that after this discussion this theorist did not dare to send the paper for publication, that the discovery was shortly afterwards made abroad and brought fame to someone else? In fact everything was honest.

As was already mentioned Landau published his first paper when he was 19. From his point of view everybody who talked to him were people that was a "grown up enough" person. That theorist did, in Dau's opinion, have to decide whether he published (risking, in the case of being wrong, to be laughed at) or not. He, that theorist, made a wrong decision (and honestly acknowledged this [Shapiro (1988)]). Landau talked to him in an "as equals" way here as well. I think that in his place a softer person, I.E. Tamm, would have said: "I do not like this idea but this does not mean anything, you should publish the paper anyway." Dau did not say this although he was inclined towards something similar in other questions. He did after all write about cinema: "Despite of a certain flatness of my statements I am very far from enforcing my artistic taste upon anybody. I can assure that if I were a director on movie distribution I would gladly show the movies that would be very bad from my point of view provided there existed an audience that would enjoy them." [Livanova (1983)]

---

[5] Of course this should not be understood too elementarily as a ban of constructing a more general and perfect theory than quantum mechanics or theory of relativity. After all in 1959 Landau himself proclaimed a necessity of abandoning hamiltonian quantum mechanics, spoke on radical changes that abandoning local theory would bring, etc.

However, to have this movie on a screen a film director who created it should make a decision that does not take into account a critic's dissatisfaction, show enough conviction, after all — boldness to counterweight his opinion to the authority of even Landau himself. Many physicists did not have this courage. Dau did not take this into account and one could just regret it.

Igor Evgen'evich Tamm whose esteem of Landau was exceptionally high and who was in perfect personal relations with him in spite of a striking difference in characters[6] and a 13-year age difference (in essence they belonged to different generations) gave, as witnessed by V.Ya. Fainberg [Fainberg (1995)], the following instruction to his disciples who were going to discuss with Landau some scientific question: 'Do not pay attention to "general" remarks by Landau (of the type "this is nonsense!"; "this can not be!" etc.). However when Landau tells something concrete on the work — become very attentive and do not miss anything!'

The excessive influence of Landau's authority in such cases was not so much a fault of Landau but rather of those who did not dare to oppose their own opinion to this authority. In theater one knows an elementary truth — if an actor plays a role of an emperor then the emperor is played not so much by him but rather by his entourage. People around Landau did not always recognize that Landau spoke to them considering them to be independently thinking persons. An inequality of levels, a gap between a talent of a rank-and-file theoretical physicist and an exceptional, extraordinary talent of Landau could be compensated only by one thing — a serious attitude towards the problem under discussion, a well-prepared line of reasoning, a responsible attitude towards one's point of view. Under these conditions one could withstand a real storm.

Once after Dau's seminar he, Chuk (I.Ya. Pomeranchuk) and I stayed in the conference hall to chat about different things as we often did. Dau

---

[6] Here I could not help remembering, with affliction, words of Vyach. Vs. Ivanov, our brilliant philologist, to whom I feel infinite respect and sympathy, cited in the above-mentioned book by A. Livanova: 'Neither among his disciples nor among his ill-wishing colleagues were people that were even distantly comparable with him [Livanova (1983), p. 219]. First of all and most importantly all of our other theoretical physicists did not necessarily belong to the two above-mentioned categories. Landau enjoyed great respect and sympathy of L.I. Mandelstam, I.E. Tamm, V.A. Fock, their schools and many other physicists as well. As to "comparability" what is meant are perhaps personal features (indeed, a philologist would not judge upon professional quality of theoretical physicists). It is unclear to me how could one compare such differing personalities as Landau, Mandelstam, Tamm, Fock and others. All outstanding people are special in their own way. Their characteristics "span different coordinate axes."

asked, 'Well, what's new in FIAN?' I answered, 'Well, Ter-Mikaelyan has finished his dissertation. A curious result: It turns out that if an energy of an electron is very *big* then its bremsstrahlung in a crystal displays an interference structure so that it feels a crystalline lattice.'

'You wanted to say if the energy is *small* and the wavelength is large?' asked Dau.

'No, it should indeed be large — exceeding hundreds of millions of electronvolts.'

'What a nonsense!' exclaimed Landau.

Chuk began stroking soothingly a lapel of my suit and said: "Zhenya, Zhenya, you should understand it yourself, this is impossible." However, Ter-Mikaelyan was at that time a Ph.D. student and I was his advisor so I continued to insist.

'There could be a small effect, under the logarithm,' admitted Dau.

'No, the effect is strong. Its essence is in the fact that it is not a wavelength that is important here but a formation length, an inverse transferred momentum which grows with energy'.[7]

After short wrangling of this sort Chuk left and me and Dau were walking in the garden around a flower bed on a former tennis court and Dau continued a furious but quite constructive critique in his style: 'You did not check a role played by inelastic scattering, it is essential.'

'We have checked it, it is not essential.'

'But thermal oscillations of lattice nodes will wipe out everything!'

'No, it turns out that they will not.'

And so on, and so forth. It is clear that it was not easy to keep up with Dau and explain why this or that factor was not essential. I listened to all of his remarks with anxiety. After all his intuition was very strong. In a state of this anxiety I left for home.

At 7 p.m. there came the first telephone call: "No, there is no such effect. The issue is in general a difficult one but mathematically everything is elementary. One could maximally get 10%." We had talked, my anxiety got stronger but I did not see a mistake. Around midnight the telephone rang again. And suddenly: "Yes, of course, the effect exists. It certainly does. Everything should however be done in a completely different way."

I felt relaxed and sat down to write a letter to Ter-Mikaelyan to Erevan. Miraculously the letter survived: All this happened on July 17, 1952. I listed

---

[7]Nowadays this is trivial, at that time it was unexpected. 'Formation length' is sometimes also called 'coherence length.'

nine final statements of Dau related to the details of the problem and added my own comments: These points are worrisome, in these Dau is clearly not right, this we know ourselves and concluded by "You have to think over the main points at which Landau aimed (or jumped) at again and again to either find a mistake (I would not believe it) or strengthen your position".

In three months Ter-Mikaelyan reported his work at the seminar in an atmosphere of complete agreement and in nine more months Dau and Chuk sent for publication their well-known remarkable work on *bremsstrahlung* in amorphous media in which they themselves used the above-mentioned effect of the growth of formation length with energy.[8]

I recall this episode as an example of scientific honesty and feeling of responsibility of Dau. The discussion was on phenomenon belonging, so to say, to his "household", he was responsible for keeping it in order and could not be quiet until he understood everything for himself. Needless to say that by Dau (and certainly not by me and Ter-Mikaelyan) there was

---

[8]There was an unpleasant misunderstanding related to this work which I will now describe because here as well there revealed themselves characteristic features of the personality and behavior of Landau, in particular his honesty in science.

Dau never wrote his papers himself, this was delegated to coauthors and in the case he was a single author the paper was written by his closest disciple, coworker and friend E.M. Lifshitz. This was done along the lines pointed out by Dau and in constant discussions with him. In the discussed case there wrote Pomeranchuk. The paper begins with a key phrase "When considering radiation process in media one can establish that at large enough energies the B.-H. (Bethe–Heitler — E.F.) theory is no longer applicable." This phrase does in essence summarize the main message of the work by Ter-Mikaelyan (let us recall how this statement on inapplicability of this theory precisely at these energies was met by Dau and Chuk with disbelief). These works gave rise to a big direction of research.

However, neither in this phrase nor in the list of references the name of Ter-Mikaelyan was not even mentioned. This was clearly a consequence of oversight. Indeed, before sending the paper for publication Chuk had given a manuscript to Ter-Mikaelyan and had asked for comments. Ter-Mikaelyan had fully approved the paper but had felt shy to tell that one had to refer to his work. As a result most of physicists did think afterwards that the work by Landau and Pomeranchuk *preceded* that by Ter-Mikaelyan and that it was them who first understood the role of formation length growing with energy.

Afterwards it got even worse. At that time, as a result of struggle with "low-bowing before Western science", our scientific papers were published only in Russian and only in our journals (in Armenia in Armenian, etc.). However, a well-known American theorist F. Dyson who knew a little Russian noticed the name of Landau and, after reading their paper, understood that there should be an effect related to crystalline lattice discovered by Ter-Mikaelyan. Not knowing about Ter-Mikaelyan's work he, together with his colleague Uberall, published a paper repeating the result of Ter-Mikaelyan. Having read it Dau became very anxious and wrote a letter to Dyson and sent Ter-Mikaelyan's papers. When there arrived an answer of Dyson confirming the priority of Ter-Mikaelyan Dau read it aloud at his seminar. Several years afterwards Dyson visited Moscow and Dau introduced him to Ter-Mikaelyan and myself.

no aftertaste left and we even became closer to each other. However, it was not easy to withstand such an attack and I seldom risked having such discussions.

Soon (at the end of 1952 or beginning of 1953) Landau and Pomeranchuk made a work in the field of multiparticle production at high energy. I was also working on this together with D.S. Chernavski who had just graduated from MEPHI and had come to FIAN. In their work which was described to me by Chuk in details (earlier he had already given a talk on it at some seminar and I suspect that the paper had already been written and possibly sent for publication) they made a principal mistake (quasiclassical approach was used outside of its applicability domain). Chernavski and I had already found the same mistake in similar works by W. Heisenberg and H. Bhabba. Indication of this mistake made an impression on Dau, their paper was not published and I got a flattering offer to work together with him and Chuk. Due to some reasons I refused and was probably right.

Already in a month Landau himself made a very difficult large-scale work — a remarkable hydrodynamic theory of multiparticle hadron production in collisions of nuclei at very high energy. Dau called me to tell about this work. Sitting next to him at the famous sofa I admiringly followed how he, covering page after page with formulae, was telling about this work that has charmed me for decades to come. It was clear that with his remarkable technique I had not been able to keep up with him. Later he said on many occasions that this work was more difficult for him than all other ones.

However, the world did not pay attention to this work for 15–20 years thinking that non-quantum hydrodynamics 'inside nucleus' is an inadmissable oversimplification. Nevertheless me and other members of our group in FIAN (S.Z. Belen'ki, D.S. Chernavski, G.A. Milekhin who died very early, I.L. Rosental, N.M. Gerasimova and others) were of very high opinion about this work, extracted new results from it, showed its exceptional usefulness in explaining experimental data. Only when after a long-dominating opinion (also of Landau himself) that one should abandon quantum field theory its authority was re-established and after improvement of experimental technique that happened during the same period and allowed to study phenomena at much higher particle energies and prove the correctness of predictions of hydrodynamical theory it became a widely accepted foundation in treating the processes it aimed to describe. Hundreds of papers were published.

Let us however return to the above-cited instruction by Tamm. He himself did many time leave without attention 'general' remarks of Dau (and, as turned out later, was usually right. This happened, for example, when Landau scoffed introduction of 'isobars,' or resonances, on the same footing as other particles. These were introduced by Tamm when considering pion-nucleon interactions in 1952, see Section 2.1. Nevertheless all of us knew how exceptionally valuable was a furious critique by Landau, his concrete advice and did not miss anything when he spoke.

However 'general' remarks did not necessarily contain offending words. I have already described a conversation with him in which (as far as I remember it, I am not totally sure) he used the word 'nonsense.' Even if it was so, this was the only time when I heard such a word from him in my address. He always spoke to me in a completely correct manner although he was at times offensively sharp not only towards his disciples and friends, even in the presence of other people. Why did it happen so that I was lucky?

Dau could not specially select me among his entourage as a physicist. This is obvious. Possibly the reason was that after the memorable acquaintance in 1935 I never began a discussion of physics with him if I was not sure I could withstand a storm. He did twice reject my ideas — in both cases softly, although in disproof he gave only 'general considerations' (albeit quite correctly).

Once when I thought that there should be new levels in atoms he said, 'I do not think so, it seems to me that the levels we know are sufficient.' I did not publish this work, not because of this remark of his but because after experiencing tortuous difficulties for two months I could not eliminate painful drawbacks in the derivation. 20 years later a right way, at much higher level, was found by D.A. Kirzhitz.

On another occasion his remark, also made in a soft form, sounded really unconvincing ('I think that at high energy the dimensionful constant would disappear and then everything is wrong'). I was however supported by Chuk who participated in our conversation ('Dau,' he said, 'you are not consequent: if you believe in your own hydrodynamic theory you should acknowledge that Zhenya is right!'). I was sure that I was right, published the paper and am very glad I did it.[9]

---

[9]In 1959, on the basis of the above-mentioned hydrodynamic theory, I came to a conclusion that in such a picture there should exist a supplementary 'direct' production of photons and dileptons (electron-positron pairs). However at that time Dau had doubts concerning foundations of quantum field theory (this explained his objection) although he still considered his hydrodynamic theory to be right and cherished it to the end. In the last two decades direct photons and dileptons were probably observed and became a focus of intensive investigations.

L.D. Landau in 1961 (looking at Bohr).

However my caution in choosing a topic for discussion on physics does not provide a sufficient explanation. I think the reason is different. Due to some reason Dau turned to me with a different side of his personality. I will elucidate this in the next section.

10.2 Two Landaus

In his remarkable article on Landau, Evegeni Mikhailovich Lifschitz writes that in his young years Landau was shy and this brought him many sufferings but with time, due to self-discipline and duty with respect to himself, was able to 'bring himself up and turn into a person with a rare ability of being happy.' [Lifshitz (1969), p. 430]

How did he achieve this? I would dare to make a statement that might seem excessive: He created for himself a character, a mask and got used to it to an extent that it became natural for him. Unfortunately this mask was not passive and controlled his actions and statements. In my opinion it was precisely this mask that dictated the harshness of behavior that caused bewilderment (this is that additional cause of incontinence in Landau's statements that I have mentioned before).

It happened however that Landau took off this mask and there appeared another person, another Dau — a soft, sensitive one. I saw this Landau many times. However this never happened 'in public' — only in the presence of one or two people related to him by something beyond physics or simply close to him. What has been said does not mean that I do not consider the words of E.M. Lifshitz, 'Beyond his harshness there hid a scientific impartiality, great human heart and human kindness"' [Lifshitz (1969), p. 434] to be just a nice phrase (at the same time I would replace the word 'impartiality' by the word 'honesty' — he did have partiality but I do not consider this to be a negative quality). This means furthermore that I understand and accept the words by Petr Leonidovich Kapitza: 'Those who closely knew

Landau were aware that behind this briskness in statements there hid, in essence, a very kind and responsive person always ready to come to help to somebody who was unjustly offended.' [Kapitza (1981), p. 389].

The mask did however play its role here by causing damage to these very qualities. As a matter of fact, challenge and bullishness covering shyness are not so unusual (especially with teenagers and young people in general). By Dau these were (meaningly?) developed to consequent entirety. Harshness, mockery, even rudeness, boy-like behavior constituted necessary elements of a character he adopted which was in the first line visible for those with whom he contacted, especially in his young years. Although to a weaker degree, this remained with him until the end.

He was different when relaxing at a moment of tiredness or speaking about something serious for him personally or, e.g. contemplating about verses (but not when reading them aloud in some foreign language to make an impression, to boyishly show off, to flaunt with memory and command of a language!), when listening to poetry.

At the end of the forties and the beginning of the fifties I would often read him aloud verses of Mandelstam, Pasternak and others which at those times were distributed in the manuscript form and he diligently wrote down those he liked, for example Pasternak's *Hamlet*. I remember with which quiet seriousness and concentration did he rewrite from my school copybook (which is lying before me now) a penetrating long verse by Olga Bergoltz of 1940 (who was, simultaneously with Dau and for almost same time, in imprisonment)[10]:

> No, not from our empty books,
> The kind of beggar's bags,
> You'll understand how hard,
> How impossible did we live.
> How loved we, bitterly and crudely,
> How in this love were we deceived,
> How when interrogated, with teeth clenched,
> Did we renounce ourselves.
> . . . .
> Oh, days of shame and sorrow!
> It can't be so that, even by us

---

[10] *Autumn*, translated by A. Leonidov.

> Was human anguish not exhausted
> In starless marsh (or "mines"? — E.F.) of Kolyma!

With an ending:

> But if, crumpled from pain
> You come across this verse
> As if in deserted field, adust
> From fire a circle dead
> But if of our suffering
> Cold smoke will you feel
> Well, recall us by standing
> As we are silent meeting you.

In such cases this was a different Landau.

There was of course also no need in the mask when Landau lectured or gave a talk. Command over material and consequently over audience were complete. In his speech or gesticulation there was no trace of briskness or tension — only seriousness. The naturalness was genuine, not put on.

I will conclude with one especially memorable episode which again, like at the beginning of these notes, is related to Rumer.

As is known in 1938 Landau and Rumer, as was told in those times, "went from the physical sheet of Riemann surface to the unphysical one" i.e. simply speaking, were arrested by NKVD. Thanks to the civic courage, cleverness and perseverance of Petr Leonidovich Kapitza, Landau returned home after a year (see below).

Rumer, on the contrary, "surfaced" only ten years later in the distant Enisejsk (at that time this was the end of the world, albeit with a pedagogical institute in which he began working). He lived there in exile for three years — with a wife and a child born there. The then president of the Academy of Sciences, Sergei Ivanovich Vavilov, managed to ensure his transfer to Novosibirsk. However, just at the moment this happened, at the end of January of 1951, Vavilov died before he could ensure Rumer's work there and Rumer found himself in a "suspended" state — without a passport (with an obligatory visit to the local NKVD office every fortnight), without work, living almost entirely on his friend's help (he made some money only with translations which his Novosibirsk friends gave him after signing contracts with publishing houses in their names).

It happened so that in the summer of that year I flew, on a scientific trip, to Yakutsk. At that time this flight made an overnight stop in Novosibirsk. When it was announced, I was agitated. I went to the city. I called Moscow, asked for his address (from the last Rumer's letter), rushed to find him but he was not at home. With difficulty, after some adventures, I found him by phone at some friends of his. We met at the boulevard near the central square, kissed each other and began making plans — what could be done, how could we help him? At that time Rumer was passionately carried away by his work on "five-optics" (a variant of unified field theory) which he had begun already in imprisonment and considered being so important that viewed working on it as a sufficient ground (in the eyes of officials) for his transfer to Moscow.

After returning to Moscow I hastened to go to Dau and put the following note on his desk: "I have seen Rumer." He said: "Let us go for a walk".[11] We went into the garden and walked and walked discussing Rumer's fate. Dau was serious, sorrowful, somewhat perplexed and was repeating all the time: "What shall we do? What can be done?"

Finally, petitions to CC, if I am not mistaken, of Rumer himself and one of the officially recognized big scientists did their job. After some time Rumer got an invitation to come to Moscow for a discussion of his work. Soon, on some early morning, Dau called me: "Zhenya, come over, Rum has come, he is at my place." When I came to Dau to his famous room with sofa on the first floor Rumer was sitting at the table in the corner, at the window, and having breakfast (I even remember he ate fried eggs). Dau, thoughtful, quiet was going back and forth across the room. Approaching Rumer he touched his shoulder and said softly, even tenderly something like: "Rum, please have more."

In such a way after one and a half decades — and what one and a half decades they were!, three of us met again with permutation of characters. This was simultaneously a joyful and a sorrowful meeting.

The scientific discussion of Rumer's work took place in the Institute of Geophysics in Bol'shaya Gruzinskaya (perhaps because an entrance to this

---

[11] Around 1950 I occasionally learned that all of us underestimated a level of technique and a scale of eavesdropping even in domiciles (for example, by reflection of infrared rays from window glasses vibrating from the sounds of conversations in a room) and warned Landau about this. Soon he and Lifshitz thanked me: The meaning of an always inaccessible room at the end of the living block began clear to them. How Landau did afterwards coordinate this new knowledge with his stormy "private life" — this is not clear to me.

institute was free). This was an important moment in Rumer's fate. Theoreticians expressed an opinion that in the course of difficult explorations in theoretical physics this direction that was developed at a very high level could not be neglected and that it had to be supported even despite there being no guarantee that this direction would lead to overcoming difficulties in the theory of particles. (Landau did not attend the discussion. He did not believe in this direction and was organically incapable of telling a lie or even a half-truth in a scientific dispute.)

All this changed the life of Rumer. He did not move to Moscow but started working (first at the pedagogical institute, then at Novosibirsk Institute of radiophysics and electronics). However, soon died Stalin, everything changed and he even became a director of this institute. When later near Novosibirsk there appeared an Akademgorodok he moved there.

So now when someone tells me about harshness and ruthless behavior of Dau I recall him — soft and repeating with a pain in his voice: "Rum, have some more."

When these memoirs were originally written I showed them to E.M. Lifshitz, the closest friend of Landau. In response he read me a letter which Landau wrote in the summer of 1946 to his wife in a minute of tragic discord. In this letter there is so much tenderness, thoughtfulness, deep feeling, so much care and desire to keep at least a good memory on their happy days forgone that I am sure that if one crossed out names and showed the letter to someone who knew only the "usual" — mocking, joyful, "merciless" and extremely rational — Dau nobody would believe that its author is precisely the "usual" Landau known to everyone.

I have to say that Evgeni Mikhailovich expressed disagreement with the word "mask" that I used in my essay because of having a shade of something dishonest and insincere while Dau was always honest. Simply, beginning at some age, he started allowing himself to display features of his character that had previously been suppressed by shyness and lack of self-confidence. I do not know. Could be, could be ...

Nevertheless it seems to me that everything written by me is correct and, in any case, is not in essential disagreement with such a point of view as well. Let psychologists and, in general, more perspicacious people decide.

I place here a photo of Dau. It was taken at a moment when he was looking at Bohr, whom he loved so much, after a separation of a quarter of a century during the Bohr's visit to Moscow in 1961.

Yes, two Landau.

10.2.1 *Landau, Kapitza and Stalin*

A curious combination of names in the above subtitle is neither occasional nor of little significance. New times have revealed striking earlier completely hidden and unknown sides of the fate and behavior of Landau and Kapitza related to the personality of Stalin. The materials shedding light on these are the investigation file of the arrested Landau and a number of letters of Petr Leonidovich Kapitza to Stalin and other "less important" leaders. The Landau's file was first published in Gorbachev's time in the journal *Izvestia CC KPSS* (v. 3, p. 134, 1991) but a real professional analysis was performed by a physicist and historian of Soviet physics, Gennadi Gorelik. As a staff member of the Institute of History of Technics and Natural Sciences AS USSR he was allowed to study the files of arrested physicists in the KGB Archive (below I will use his results) [Gorelik (1991), (1995)].

The letters of Kapitza, astonishing by their courage, wisdom and feeling of internal independence, were kept by their author in deep secrecy. He understood how cruelly would have the addressees avenged themselves for a "public disruption of their prestige." These letters were published by a referent and secretary of Petr Leonidovich of many years and now a director of his archive, Pavel Evgen'evich Rubinin [Kapitza (1990)].

One should however begin with the political position of Landau and its transformation over time.

Due to some reason only few know (or recall) that in the twenties and in the first half of the thirties Landau, a son of a well-to-do oil engineer, was sincerely and demonstratively pro-Soviet oriented. When he was abroad he wore a provoking red shirt (one told that he did not wear a red suit only because he was told that this was a uniform of waiters in posh restaurants) and did not hesitate to make lots of statements in the corresponding direction. Many years later he told on many occasions, also to myself, that he was a materialist and a supporter of historical materialism but at the same time openly ridiculed Soviet philosopher-diamatchiks[12] who prescribed concrete "scientific" points of view to scientists working in natural philosophy. Like lots of other intelligents Landau was convinced in advantages of the Soviet system over the capitalist one.

In this relation it is worth rereading memoirs of H. Casimir who in 1931 was working at Bohr's institute in Copenhagen simultaneously with Landau and was in general in frequent contact with him. Excerpts from these

---

[12]Specialists in dialectic materialism.

related to Landau were published [Casimir (1990), pp. 150–160]. Casimir tells that Landau "... was a revolutionary ... He believed in extermination of all prejudices and privileges if these are not recognition of real achievements." Of special interest is however a complete interview given at that time by Landau to the weekly *Studenten* (N 22, March 12, 1931) reproduced in these memoirs. Then Landau gave a talk for a student society on March 16. Two leading Copenhagen newspapers wrote about it.

Answering in the course of this interview to a question on education in Russia, Dau said:

> Organizing new educational institutions is very expensive and at this moment main efforts should certainly be directed towards developing a healthy system of socialist production. This taken into account it is all the more amazing that nevertheless one found resources for expanding the system of universities and scientific institutes that took place in recent years as well as substantial sums that are distributed every year as scholarships for students.

When he told that "... due to a higher level of education, intelligents get higher loans" there followed a question characteristic of left wing Western studentship: "Does it not automatically lead to a class structure of society?" Dau immediately replied:

> Yes, one can say so, but this structure is fundamentally different from the division onto proprietors of means of production and workers in the capitalist world (a marxist! — E.F.). Directors, managers and people at high technical level at a Soviet socialist factory get their loans from the state. Therefore administration cannot be interested (as it often happens at capitalism) in exploitation of the working force aimed at getting maximal revenues for the owners of means of production. ... In Soviet Russia there is no exploitation of majority by minority, each person works for the well-being of the country as a whole and there is no irreconcilable contradictions between workers and administration, they are at one with each other.

To one question he even responded as follows:

> It is clear that the state should withstand attempts of sabotaging the work on social (possibly, an inaccuracy of an interviewer, the word "socialistic" does rather come to mind — E.F.) development which are, from time to time, undertaken by some emigrants that return to the country with the only goal of sabotaging the five-year plan.

He did even still believe in a myth about "sabotage".

There is no doubt that Dau did completely honestly tell what he thought at that time. This is also confirmed by an open character of presenting his point of view (it was known to anyone who was at least to a certain extent close to him) with respect to a question on situation with liberal arts and in general with disciplines "not related to social development" (again "social", E.F.). He said: "...In my personal opinion too much money is currently spent on pseudo-sciences like, e.g. history of literature, history of arts, philosophy, etc. ...Metaphysical trinkets are not of any value to anybody but idiots specializing in them." Answering one question during the talk he said: "It is necessary to make a distinction between sensible and senseless fields of knowledge. The sensible ones are mathematics, physics, astronomy, chemistry, biology, while the senseless are theology, philosophy, particular the history of philosophy, sociology etc."

It was very dangerous to say something like this about philosophy and sociology in our conditions at those times. This shows that Landau spoke absolutely freely, without a shadow of conformism.

Such was Landau in the twenties and beginning of the thirties. Only the growth of Stalinist terror did perhaps lead him to begin experiencing doubts. Still, even in 1935 when after Kirov's assassination on the last pages of newspapers there appeared, almost every day, long lists (of 30–70 names) of "enemies of the people" unmasked and executed in one or another regional center he published in the newspaper *Izvestia* a long article on Soviet science. In this article he stressed that the Soviet system was more favorable for development of science than the capitalist one. The argumentation was, in particular, in the fact that talented people appear uniformly in different strata of society but in a capitalist country a scientific talent of somebody from a low stratum can seldom be developed because of a necessity of paying substantial money for getting higher education while in the Soviet country science is accessible to everyone.

As found out by Gorelik he wrote this article following an insisting recommendation of his friend of that time, a little known physicist M.A. Koretz. It is possible that a desire of "insuring oneself," to push away suspicions in "political untrustworthiness" did play its role here. Still, knowing Landau, one can be sure that he wrote what he did really believe in.

A discernment was however coming very fast. Horrible court trials of the middle of the thirties and the grandiose scale of terror could not but exert influence on Dau with his extraordinary soberness of thought. As a result there happened something unbelievable.

During those years we knew (of course unofficially) only about several cases of direct protest against the Stalin regime. For example, on Ryumin (secretary of Krasnopresnensky district party committee) and his group; on the group of young people, "young avengers," for imprisoned and executed parents. The very fact that the Landau's file is published only now shows that there were of course many more such facts albeit still their number was very small.

So, now we learn that in April 1938, after the terrible March Session of CC that called to still bigger "strengthening of alertness," Koretz came to Dau and said that one should publish an anti-Stalin leaflet and distribute it during the demonstration on May 1. Does Dau agree to look through the prepared text and give comments? And — Dau agrees under a single condition: He should not know the names of the participants of this action. This condition is easy to understand. Dau was always afraid of pain and was probably afraid that when tortured he would give out these names.[13] He did look through the leaflet. He made comments and it was passed for copying. But ...on April 28 Landau, Koretz, as well as Landau's friend Rumer,[14] were arrested.

Somebody had betrayed them. Suspicions of the members of the group fell on one of its members, K., whom I will not name first because this suspicion was not proven (below, at another example, I will show how terrible could be the outcome of such unproven suspicions). Second, when the war began this man volunteered to the front (those suspecting him thought — to redeem his fault by death) and perished. The only ground for suspicion was that it was to him Koretz, who had known him from childhood, handed the text so that K. and his friends (who in the words of K. had been striving for action) would produce copies on a self-made amateur hectograph (this was indeed done although the quality, in Koretz's words, was bad, this was for the most, art waste). Who the one guilty of leaking the information was remained unknown.

This leaflet shatters by courage, full understanding of horror reigning in the country and direct comparison of Stalin and Hitler. It is in Landau's

---

[13] When Landau and Rumer were arrested they occasionally met in the jail due to a laxness of escort. As told to me by Rumer many years later Dau anxiously asked him: "I suppose they would not beat, would they?" He was still naive. He himself was indeed not beaten.

[14] In fact Rumer had nothing to do with the leaflet. As found out by Gorelik an accusation of his involvement was waived in the course of investigation.

file. Gorelik did hold it in his hands. It is worth reproducing it here, even if not completely:

> Proletariat of all countries, unite!
>
> Comrades!
>
> The great cause of October revolution is meanly betrayed ... Millions of innocent people are thrown into jail and nobody can know when does his turn come ...
>
> Comrades, don't you see that the Stalin clique has made a fascist revolt?! Socialism remains only in the pages of newspapers that only lie. In his furious hatred to real socialism Stalin has reached the level of Hitler and Mussolini. Ruining the country to keep his power, Stalin turns it into a light prey for beastly German fascism ...
>
> Comrades, organize yourselves! Do not fear of NKVD butchers. They are only capable of beating prisoners, catching innocent people who have not done anything, steal people's property and inventing absurd court processes on nonexisting conspiracies ...
>
> Stalinist fascism holds only due to our lack of organization.
>
> Proletariat of our country which threw away power of tsar and capitalists will be able to overthrow the fascist dictator and its clique.
>
> Long live the May 1 — the day of struggle for socialism!
>
> Moscow Committee of Anti-fascist Workers Party.

(No such party did in fact exist.)

It is noteworthy that the leaflet speaks about the betrayal of the cause of Lenin and the October revolution the truthfulness of which the authors did evidently continue to recognize. Only 7 years had passed after his bolshevist statements (see p. 393). How much more did Dau have to endure, suffer through, think over in order to say in 20 more years the following words (intercepted and fixed by KGB in a top secret report to CC): "It is clear that Lenin was the first fascist."

One had to live through that terrible time when many millions of people were exterminated without any guilt on their side simply for establishing an all-embracing atmosphere of fear (this is precisely the meaning of the French word *terreur* — fear, horror), unconditional submission to tyranny and its ideology to understand the courage of the "naive" authors of the leaflet, independence of their thinking. At that time a vast majority of intelligents were trying to adjust themselves to the "system," find at least some kind of justification for what was going on.

Everything did mix up in people's heads: Crazy fear and ideological heritage of former liberal socialistically minded part of the intelligentsia; a sight of demolition of peasant foundation of the predominantly agrarian country — and real achievements in liquidation of mass illiteracy, in development of education and science; poverty, mass hunger of population — and real progress of industrialization.

There arose and spread a general besotting, even salutary self-besotting. When clever Il'f ridiculed a besotted intelligent who had finally "recognized metro as an achievement" one could have asked him in response: "Who are you laughing at? You are laughing at yourself." This mutilated psychics also gave birth to such terrible phenomenon as squealing — not only for self-salvation or career but sometimes really on "ideological grounds."

Of course this leaflet combined a soberness in evaluating the situation, a clarity of perception so characteristic of Landau with political naivety. This was in essence an appeal to a revolutionary mutiny. However, a mutiny without an organizing strong center, a party is doomed to failure. It was not without reason that Lenin, "a genial theoretical and practical expert on seizing power" (as written by an eser Chernov in an obituary devoted to him) was in the first line building a party with rigid organization and discipline.

And still, at that time — such a leaflet! One can say that the arrest of Landau and Koretz was one of the extraordinary, unique cases in Stalinist times when people "were imprisoned for something real," for a deed they had accomplished (it is also not excluded that their betrayal, a squeal, was also based on "ideological grounds").

What happened in the following, horrible as it was, also contained elements of farce. Some puzzling questions do not find answers even now, in the first line — the main question: Why Landau and Koretz involved in the case (and also Rumer) were not exterminated, crushed neither immediately nor later? Why, as was habitual in those days ("as was proper") one did not do the same with their closest relatives, friends and simply acquaintances? Indeed, none of them was sent to exile, fired from work or at least deprived of election rights (the life of these "deprived" was also horrible then).

Strange as it seems, some friends of Landau (I know two such ones) categorically deny, despite of clear evidence, an authenticity of the leaflet and the whole story related to it. Their argumentation goes as follows: (1) "I was such a close friend that without any doubt Dau would have told me about it;" (2) "This is in complete disagreement with the political sentiments of Dau."

The first argument does not hold because after the arrest and release Dau was extremely careful in his statements on what he had lived through. He was, as usual, sober and understood that each of his interlocutors could tell it to someone else and this would have led to serious consequences both for him and Kapitza with respect to whom he felt an extreme responsibility.

The second argument also does not withstand critique. If one takes two historical periods, one in the twenties and the first half of the thirties when, as described above, Landau was a passionate marxist and a supporter of Soviet power and the second one, after his arrest and release when he called Lenin the first fascist, then at a borderline between these periods, before the arrest, his intermediate position expressed in the leaflet is very natural: Lenin was good, the October revolution was a positive phenomenon but these were betrayed by the fascist Stalin. Taken together all this pictures a consequent evolution of his views from his younger years till those before prison. He was regaining sight gradually.

Let me add one more fact that I have recently learned. Koretz stayed in prison for many years (his term was expanded) and after his release (this was already after the death of both Dau and E.M. Lifshitz) did himself tell about details of the story with the leaflet to the widow of Evgeni Mikhailovich, Zinaida Ivanovna Gorobetz-Lifshitz.

It is possible that the distrust of the two above-mentioned friends originated simply from the fact that they were offended by him not telling them anything. Dau, however, was astonishingly good at keeping secrets. His closest friend V.L. Ginzburg learned from declassified and published documents that Dau had been involved in the atomic project already in 1946 (and possibly had become an academician in 1946 thanks to this) only in 2002.

Landau was released in a year but this is a fantastic story in itself related to Kapitza and we shall talk more on it later. Koretz did spend his term (and added years) in a camp and lived to Gorbachev times. Gorelik met him and his daughters (now living in Israel) and interviewed them. Rumer worked in an aviation *sharashka* — an in-jail, albeit privileged, construction bureau headed by another prisoner, the famous A.N. Tupolev and where there also worked other *zeki*, S.P. Korolev, V.P. Glushko and other outstanding people of science and technics; then he was sent to exile (we have already discussed it in these notes).

Rumer told me about a phantasmagoric episode when Beria organized for all the remarkable people in the *sharashka* a "friendly dinner" and did himself serve the "guests" with a plate of *pirozhki*. Softened by all this an aircraft designer Bartini, an Italian aristocrat, "red baron" (or marquise) who had come to the USSR to create Soviet aviation potential raised and said: "Lavrenti Pavlovich, we, together, are having a friendly dinner, talking to each other — so I would like to tell you absolutely truthfully and sincerely: I am not guilty of anything at all." Beria, in the same friendly tone (admittedly even clapping him on the shoulder) answered: "Of course you are not guilty. Were you guilty we would have executed you. And you, do it — put a plane into the air and you are free." All these people were however imprisoned for specially fabricated false accusations while here there was a real fact of the leaflet.

At the same time the although disciples of Landau of the first, "Kharkov" generation, future academicians I.Ya Pomeranchuk, E.M. and I.M. Lifshitz, A.I. Akhiezer as well as V.G. Levich, who adored Landau, did bring a suspicious seal of closeness to an "enemy of the people" were more or less normally continuing their scientific work and career (their fidelity to their teacher is indubitable and was not in any way sullied).

Relatives of Landau as well as his Moscow friends were also not touched. For that epoch all this was highly unusual and even strange.

The very course of investigation that was thoroughly studied by Gorelik does also cause bewilderment. In the course of the one-year imprisonment Landau was interrogated many times with a usage of a so-called conveyor interrogation widely applied at those times: An interrogated person is standing all the time, he is allowed neither to sit down nor even lean on a the back of a chair, or sleep; the bright light of special lamps shines into the face and the interrogator is asking question after question and fixing the answers. This lasts for many hours. After several hours an interrogator is replaced by another one and the interrogated person is still standing and answering questions. How did a physically weak Landau endure this? This is simply a miracle. One also widely used various psychological methods.

He broke down and began delivering evidence only in August. In the protocol of interrogation of August 3, 1938 one sees that at the beginning of interrogation he was still completely denying any kind of anti-Soviet activity but when shown a leaflet written, as he saw (and told it) by the hand of Koretz and Koretz's confession in the existence of an anti-Soviet organization and Landau's participation in it, he gave up completely. He wrote horrible fables on how "after beginning with anti-marxist positions in the field of science" (!) he had become close with a group of "anti-Soviet minded physicists" already in Leningrad, had developed more hatred after the arrest of his father condemned for ten years for sabotage in 1930 and then, in Kharkov, had become close to a group of leading physicists which, inspired by growing hatred, had at the end become an anti-Soviet organization carrying out sabotage in the institute (pushing out from the institute allegedly incapable but in reality useful people, etc., see below) and in this had come to the authorship of the leaflet.

It was important, however, that all accomplices named by him were either already executed or arrested. Still, he also confirmed nonsense, for example that his disciples Lifshitz and Pomeranchuk knew about his views but did not know about the existence of the organization while he was preparing them for entering it. He was also allegedly planning to rely in future upon P.L. Kapitza and academician N.N. Semenov as anti-Soviet "activists."

All this gibberish was repeated in the evidence written by his own hand on August 8.

As has been already mentioned, all this was interwoven with complete farce. Due to some reason investigators were trying hard to get from him a confession on "sabotage" on two issues that, with the leaflet and written confession in anti-Soviet activity, seemed superfluous and completely unimportant.

No, they wanted (and got it): Landau confessed, in addition, that (1) he performed a dedicated company on discrediting dialectic materialism and, therefore, fought against the party's ideology; (2) working at the Ukranian Institute of Physics and Technics (UPTI) he was an active proponent of its division into two separated institutes — that of applied physics (at UPTI one carried out a large amount of work on these issues including the defense-related ones) and that of fundamental research (besides the work of his theoretical department this included large-scale experiments on physics of atomic nucleus that at that time was far from having applied significance; in particular in 1932, only with a few month's delay, one repeated a famous experiment of Cocroft and Watson on separation of atomic nucleus; one conducted unique experiments on low-temperature physics, etc.). By doing this the accused was in a 'sabotaging' way trying to separate fundamental research from the needs of socialist development.

This was this "sabotage activity" (using precisely these words) in which Landau did confess after months of torment. All this could seem ridiculous were it not so cruel. How could one explain this tragic farce?

We find a key to its explanation in such unique material as letters by Kapitza and this equally unique activity in saving Landau (and before that — V.A. Fock).

Kapitza was working in England in the laboratory of Rutherford for 13 years. Using special equipment designed by himself and using new methods he invented he obtained first-class results. In 1934, at the age of 44, he was already a scientist belonging to world's elite, was a member of the Royal Society (an English equivalent of the Academy of Sciences). Many times he came to spend summer in the USSR but once, in the fall of 1934, one forbade him to leave the country — he was needed at home.

At first Kapitza was enraged. As Rumer told me right at that time he was going to quit physics and began working in physiology with academician I.P. Pavlov known for his anti-Soviet position. Later, when one promised to create absolutely extraordinary conditions for his work, he did however yield.

Following his detailed instructions, in one year an institute and a block of two-storied English style apartments for scientific staff and a villa for himself were built, all this in an old park in Vorob'yovy hills. One bought and brought from England his unique laboratory installation and allowed him to use director's rights in his special style.

Nevertheless he was still feeling himself insulted by this forceful act of deprivation of freedom. His origin and upbringing (his father was a military engineer, a general, built forts in Kronstadt), a long life in a free democratic country, his position in world science developed in him strong feelings of self-dignity (simply put — human dignity) and internal independence that were in complete disaccord with life in the USSR.

When the building and organization of the institute were almost complete he wrote a letter to Stalin. In this letter he expressed his displeasure both with respect to a progress of construction and an attitude of state officials towards him. Distrust and suspicion emanating from all of them, including those of highest rank, insulted him. There arose such an atmosphere in which Soviet scientists were afraid of contacting him. The main point is however not in this but in *how* was this letter to the almighty tyrant written:

> It was very heavy for me when more than a year ago one unexpectedly forbade me to leave and sharply cut my scientific work in a very interesting place. Afterwards I was treated very badly and these months in the Union were the most difficult in my life. I understand that there are reasons behind moving my work here but up to now I do not understand what was the purpose of treating me so cruelly ... an attitude was the most malevolent and suspicious one ... Finally

one demanded to write an open lie that I did stay voluntarily ... During all this time agents were following me and two times one even sent dogs to sniff me ... All this did of course offend me ... After all, during 13 years I remained an invariably faithful citizen of the USSR ... although I could ... naturalize (in England — E.F.) ... I never concealed that I am fully sympathetic with the socialist system in the USSR. The attitude displayed with respect to me is very bad (simply swinishness) (Sic! — E.F.) ... Everything around me looks cloudy now ...

This letter is written by an internally free person and not by a "subject" or slave. Nobody could have dared to write "swinishness" (to whom! to Stalin!). And — he did not have any problem because of this.

One can not escape an impression that he commanded a respect by Stalin who seemed to despise the whole world (perhaps besides Hitler and Churchill), in any case, all intelligents. This respect did show itself very soon.

In the course of one of the recurrent mass purges of Leningrad (the former capital was many times cleaned from "formers," "zinov'evtsy," etc.) one arrested Vladimir Aleksandrovich Fock — at that time he was still a corresponding member of AS but already had a solid international recognition. Next morning (February 12, 1937), at the height of unprecedented terror, Kapitza wrote a new letter to Stalin in which, besides giving a short characterization of Fock as a scientist, listed four reasons explaining why one should not treat scientists in this way in general:

1. This will expand that gap between scientists and the country which unfortunately does exist and which one would so much like to see extinct.
2. The arrest of Fock is an act of rude treatment of a scientist which, like a rude treatment of a machine, spoils his quality. Spoiling Fock's working ability is bringing harm to all the world science.
3. Such treatment of Fock causes our, as well as Western scientists, internal reaction similar to that, for example, at Einstein's expulsion from Germany.
4. We have few scientists like Fock and Soviet science can be proud of him before the world but this becomes difficult when he is put into quod. (Sic! — E.F.) [Kapitza (1990)]

Each of these four reasons amazes by its courage and feeling of independence. Especially remarkable is still the third reason which almost directly likens the Soviet regime to the fascist one and the very style of a free letter including the colloquial "quod." Here Kapitza defends all the scientists.

This was a time when it was no longer allowable to intercede for an arrested person. Such an action made someone a suspicious if not felonious person. There reigned a phrase: "Organy do not err." Nevertheless, strikingly, this letter caused an immediate action. Within three days (this means that Stalin had read the letter immediately) Fock was taken from Leningrad into the office of the main *chekist* of that time Ezhov (imagine his reaction at the question of Fock after he entered the office: "To whom do I have the honor to speak?"). Fock was immediately released from the horrible building in Lubyanka. As told to me later by I.E. Tamm, after going out into the street Fock had come to his place, borrowed money for a railway ticket and left for Leningrad.

It is not difficult to guess that Kapitsa also could not stay passive after the arrest of Landau. Here is his letter to Stalin [Kapitza (1990)]:

April 28, 1939, Moscow

Comrade Stalin!

This morning one arrested a member of scientific staff of the institute, L.D. Landau. In spite of his 29 years he, together with Fock, are the most prominent theoretical physicists of us in the Union. His works on magnetism and quantum theory are often cited both in our and in foreign scientific literature. Only in the last year he published one remarkable paper in which he pointed out to a new source of energy for stellar radiation. This work gives a possible answer to the question: "Why energy of sun and stars is not visibly diminishing over time and has not been exhausted by now." Big future of these ideas by Landau is recognized by Bohr and other leading scientists.

There is no doubt that a loss of Landau will not pass unnoticed either in our institute or for Soviet and world science and will be strongly felt. Of course, knowledge and talent, big as they can be, do not give anybody a right of breaking laws of his country and if Landau is guilty he should answer. Still I am appealing to you to give corresponding instructions for his case to be treated very attentively. It seems to me that one should also take into account the character of Landau which is, simply speaking, bad. He is a trouble-maker and a squabbler, likes to find mistakes by others and after finding them, especially by important elders like our academicians, begins disrespectfully teasing them. He has made quite a few enemies in this way.

Dealing with him in our institute was not easy although he responded to persuasions and was becoming better. I pardoned his escapades because of his exceptional talent. However, with all negative features of Landau's character, it is difficult to imagine that Landau is capable of anything dishonest.

Landau is young and is capable of doing much more in science. Nobody could write all this but another scientist, therefore I am writing to you.

P. Kapitza

This time, however, the result did not seem to be achieved. If one had known what a leaflet was in Landau's file in Lubyanka one would not have been surprised. At the same time, could it be that Stalin was tired of Kapitza's letters? I think that now one can say with confidence that, on the contrary, some very important instruction of Stalin did follow immediately. Otherwise it is impossible to understand such a soft reaction of Luybyanka at such a completely extraordinary event as the ill-fated leaflet and an uncovering of the really anti-Stalin group standing behind it. At that time for almost all other victims of terror, including former highest officials of the party, one invented fantastic crimes, constructed extremely complex scenarios of their alleged actions, invented supposedly existing anti-Stalin groups, parties, organizations. In this case everything was a real fact, more important than inventions to which one was already used to.

Nobody (could be, only so far?) can tell what this instruction precisely was but only Stalin could immediately block the usual practice of "purging with fire" of all the entourage of caught "criminals." Here *nobody* from this entourage did suffer! It is natural to assume that the group itself and the leaflet were provoked by the NKVD. Then, however, the only goal of such a provocation could be a creation of a noticeable court process or another loud but equally bloody large-scale event.

There happened nothing like this. Instead of this interrogators were for many months pulling from Landau a conviction on sabotage on the above-described anecdotal points. In essence instead of decisive repressions they were "dragging out." In addition, at the time where horrible tortures were customary his ones were relatively soft. Gorelik found in the file a piece of paper which he treats as an internal informal note to a higher level, a final report on treating Landau possibly written before his release. Written there: "... was standing for seven hours" (meaning, of course, conveyer interrogation). Only 7 hours in three months before giving out evidence! At those times this was a very soft interrogation. (Let us compare: a German theoretical physicist Hautermans whom we mentioned in the chapter on Heisenberg was arrested half a year before Landau and was in the period of January 11–22, 1938 subjected to a continuous *eleven day long* conveyer interrogation [Frenkel (1997)]). Later: "Threatened, did not beat," etc.

Finally, in conditions when we do not know anything about the back side of the story, if we assume that the leaflet was an element of a broadly thought of provocation of the NKVD then, it could be, when it came to its completion there appeared doubts or the operation was found unsuitable. For example, a further "shooting out" of physicists was found unreasonable. In any case the issue is waiting for the research of historians in the archives. In any case nobody except Stalin who, as stated by Malenkov (see [Kapitza (1994)]), read *all the* letters by Kapitza could not turn the case into such an unusual direction, for example by giving an instruction to end everything quietly.

Kapitza was waiting for a year. However, then he wrote a new letter, this time to a second person in the party and state — Molotov [Kapitza (1990)]:

> Comrade Molotov,
>
> Recently ... I was able to discover a number of new phenomena which would possibly make clear one of the most enigmatic domains of modern physics. However ... I need a help of a theoretician. By us in the Union only Landau had a perfect command over the domain of theory I need, but the trouble is that he is under arrest for already a year.
>
> I was all the time hoping that he will be released, ... I cannot believe that Landau is a state criminal ... It is true that Landau has a very sharp tongue and by misusing it, with all his cleverness, he has made many enemies ... However, with all the badness of his character which I also had to take into account I never noticed any dishonest action from his side.
>
> Of course by telling all this I am interfering into a foreign field of competence, that of the NKVD. Still I think I have to point out the following issues as being abnormal:
>
> 1. Landau is in jail for a year and an investigation is not over, its duration is abnormally long.
> 2. As a director of the place he works at I know nothing on what he is accused of.
> 3. Landau is of puny (sic! — E.F.) health and if he is exterminated for nothing this will be very shameful for us, Soviet people.

Therefore I address you with an appeal:

1. If possible, the NKVD should pay special attention to an acceleration of investigation of Landau's case.
2. If this is impossible then one might use Landau's head for scientific work while he is in Butyrka. One says that this is being done with engineers.

P. Kapitza

This letter not only did not cause anger by high officials (which would have been natural at those times) but, on the contrary, "did work." In a few days Petr Leonidovich got an invitation to Lubyanka to the NKVD deputy of narkom Merkulov "himself" and another well-known butcher, a head of the interrogation division, Kobulov.

As told within narrow circles for already some time, when he entered an immense office on a separated table there lay the volumes of investigation files of Landau and others. In some places one placed layings and Kapitza was politely offered to get familiar with the material to convince himself that Landau was really guilty.

Here, however, Kapitza showed the full scale of his wisdom and character: He categorically refused to read all these files. It is clear why he acted in this way. First, he of course understood that by torture one could beat out of Landau any most absurd confession, for example that he was a Hitler's, or English, or say Bolivian spy, that he prepared a terrorist act against Stalin or wanted to blast the Bolshoy Theater. It was impossible to prove that this was a false confession. Even if this had been absent, if he would have been shown something almost harmless, for example the real pulled out confession in sabotage (discrediting *Diamat* [dialectical materialism] and aspiration of separating UPTI into two institutes) about which Kapitza, of course, did not know he would have been involved in an endless argument of whether this confession meant a crime and a degree of necessary punishment, etc. All this was immediately thrown away due to the firmness of Kapitza. Long persuasions did not help.

Evidently, the question of Landau's release was however predetermined and, certainly, predetermined by Stalin. Everything ended with Landau given out to Kapitza under his responsibility against a receipt [Kapitza (1990)]:

To People's comissar of internal affairs of the USSR comr.

L.P. Beria
April 26, 1939

> I am asking to release from prison an arrested professor of physics, Lev Davidovich Landau, under my personal guarantee.
>
> I guarantee to the NKVD that Landau will not carry out any kind of counterrevolutionary activity against Soviet power in my institute and I will undertake all measures in my capabilities that he will not carry out any counterrevolutionary activity outside the institute as well. In the case I notice from the part of Landau any statements harmful for Soviet power I will immediately inform the NKVD about this.
>
> P. Kapitza

Dau was free and his gratitude to Kapitza remained lifelong. He immediately dived into salutary science. During the remaining two years before the war he did very much. It is interesting that he did indeed explain remarkable experiments of Kapitza with liquid helium about which the latter wrote in his letter to Molotov by creating his remarkable theory of superfluidity of liquid helium. It immediately gave rise to new results, and experiments of Kapitza were expanded in other directions (one could name, e.g. a theoretical discovery of the "second sound" by E.M. Lifshitz and its subsequent confirmation by Kapitza's disciple V.P. Peshkov). Works, new effects were flowing and flowing like a stream (it is for these very works that Landau got a Nobel Prize many years later). Kapitza did not deceive Molotov. It would be interesting to learn whether Stalin and Molotov learned about this. It was clearly of no interest for Merkulov.

I never asked Landau about the details of his arrest and imprisonment. I knew that those released were obliged to keep this in secret. However one could see that he changed, became quiet and more cautious. This was not only a fear for himself but also a feeling of responsibility with respect to Kapitza who had given his guarantee for him. What was inside, I cannot say. At that time we were still not as close as later, I can only recall an episode that elucidates something.

In 1947 when an anti-Semitic company (against "rootless cosmopolites") was in full swing newspapers published, almost every day, articles with "unmaskings" in which one related this "sin" with "low-bowing before abroad" and "concealment of the role of scientists from our country." In October *Literaturnaya gazeta* accused of such concealment V.L. Ginzburg. This threatened to enroll into the company leading to very bad consequences. One composed a protesting letter and physicists-academicians began signing it. I came to Landau to get his signature. He read it, thought for a while and said (I am describing this episode just for citing these memorable words): "I am of course a coward but in this case, perhaps, there is no big danger," and after making some minor corrections, signed the letter. Let me note that another physicist who had also spent some time in imprisonment at the end of the thirties was for a long time convincing me that he was not fearing, prevaricated and did not sign. Before his prison experience Dau, I am sure, would not have called himself a coward. Let me add that the letter with 11 signatures did not cause any reaction from the newspaper.

Still, questions on the inadequately soft reaction of the authorities on the, at that time, almost completely improbable production of a leaflet do remain. If the above-stated assumption on everything having had been a consequence of the direct instruction of Stalin then why all this farce with switching the whole focus on the ridiculous "sabotage"? Gorelik with whom we discussed this question thinks that one simply wanted to make use of "data" accumulated before the arrest (denunciations, reports of "informers" on Landau's conversations, his behavior and activity in UPTI) which were being collected on him like on all noticeable people, even those who were not touched. All this had to be ready to exterminate a person, using these materials when needed. So, one just did not want to leave all this information on Landau without consequences! However, I would treat everything somewhat differently. By focusing on this information gathered already before the leaflet *chekists* did in fact prove to their commanders that they had been vigilant, had not missed Landau and had known he was an "enemy."

Still both explanations seem less convincing to me than an assumption on an immediate direct involvement of Stalin that simultaneously explains many other things: That nothing happened neither to relatives nor to closest disciples and that Landau himself was exposed only to the weakest of then existing tortures (with such a leaflet!), a "mild" sentence to Koretz, etc.

Could Stalin's respect for Kapitza to which we will return below be a sufficient reason for such an instruction? Stalin exterminated many physicists. By that time one had executed such talented and distinguished physicists as L.V. Shubnikov in Kharkov, M.P. Bronstein, V.R. Bursian, V.K. Frederiks in Leningrad, S.P. Shubin in Sverdlovsk, A.A. Vitt from Moscow. Others arrested but later, after much suffering, released, were I.V. Obreimov, Yu.A. Krutkov, A.I. Leipunski, etc.

Nevertheless if one recalls what unlimited resources were spent on development of natural sciences, especially physics, recall the very fact of Kapitza's confinement in the country because his Motherland needed him and the creation of wonderful working conditions for him one should acknowledge that Stalin understood the importance of developing physics as well as other sciences. I have already mentioned a telling fact: During the war Stalin carried out an extremely hard mobilization policy (no exemption was made, e.g. for the only breadwinner in a large family or for old parents like it was done in other countries, literally everybody was mobilized. This went as far as mobilization of several hundreds of thousands of girls, etc.).

However, already on September 15, 1941, three months after the beginning of the war, the state committee for defense presided by Stalin made a decision forbidding mobilization (and in general usage for works not related to the field of expertise) of all university lecturers and members of scientific staff of the institutes including even specialists in humanities — historians of art, philologists, etc. In Hitler's Germany one came to this idea only a year before the end of the war (a decree of Borman returning scientists from the front). In particular only thanks to this were they later able to construct an atomic bomb.

A phrase that Stalin protected physicists would sound scoffing. However he discovered a practical horrible psychological law: Scientists can work very productively even in an atmosphere of total fear, even in imprisonment. The terror with respect to them should only be kept at some minimal level (in other words, killing only few). It seems to me that it is easy to understand the nature of this law: In the atmosphere of horrible terror a full immersion into science is the only possibility for a scientist to keep one's personality. One just needed laboratories and libraries. For them the state did not grudge, it was precisely in this fashion that even in a psychologically heavy atmosphere the provincial physics of pre-revolutionary Russia did reach a world level in Soviet time.

Let us however return to Landau. The cruel lesson that he got, and also fear, made him subdued and later even participate in the work on both atomic (as became clear after half of a century) and hydrogen bombs. Only when Stalin died and Beria was executed he said: "That's it, now I am not afraid and stop working on this." Before that there were however 14 years of fear.[15]

---

[15] In relation to this of extraordinary interest is one episode from the published memoirs of a theoretical physicist Moisei Isaakovich Kaganov, a disciple of Il'ya Mikhailovich Lifshitz, who was in close relations with Landau.

As one can judge from the context when Landau, already after the death of Stalin, mentioned the name of Sakharov, Kaganov asked him 'perhaps a very naive question: "Dau, if you have understood how to make a hydrogen bomb, what would you do?" I remember the answer with the protocol accuracy: "I would not hold out and would calculate everything. If I would get a positive answer I would flush all the papers into the toilet".

The word "now" here means a lot. It shows that through all this years a feeling of Lubyanka, "conveyer interrogation" was living in him and a beam of light directed at his eyes was from time to time blazing up with lesser or higher intensity. In the horrible months before Stalin's death when one prepared a deportation of Jews into Siberian camps that were under construction or already ready this light did once blaze up glaringly brightly. Like previously in Lubyanka, Landau displayed weakness. However, I do not want to describe this in more details. This was not his shame but that of the wretched country and the epoch. Fortunately a sclerosis of Stalin's brain did soon put a full stop to this episode as well.

I have not however told about another psychologically horrible story related to his arrest. The thing is that in the "first draft" of Landau's disciples in Kharkov was another physicist that was not mentioned before — L.M. Pyatigorski. Already in Kharkov Landau began planning his famous course of theoretical physics later accomplished with E.M. Lifshitz. The first "volume" (still a thin book) was issued in Kharkov. The names of the authors on its cover were Landau and Pyatigorski. However, when Landau was arrested, his disciples decided, guided by some very indirect hints, that he was "put into jail" by the only party member among them — Pyatigorski. This was also confidently repeated later. Even I who was at that time very far from Landau knew about it. Pyatigorski stayed in Kharkov and was practically subjected to ostracism (although without telling him this directly — nobody would have believed in his innocence anyway).

There passed years and decades. There died Landau, there died all his closest disciples and coworkers (except for A.I. Akhiezer) — Pomeranchuk, Lifschitz brothers, Migdal, Berestetski, Kompaneets — all of them passed away having been sure that Pyatigorski was a traitor. If I knew about it then many others did know as well.

However, there came the Gorbachev times. A relative of Landau, Maya Bessarab, released a new edition (the fourth) of the book on Landau written by her. She placed there a new text: Now one can tell Landau was arrested because of information by Pyatigorski. Pyatigorski, however, was alive! He sued Bessarab accusing her of slander. The court requested information from the KGB and got an answer that Pyatigorski did not have anything to do with this and obliged Bessarab to publish a refutation which was done. A weak consolation for him. Indeed, for 50 years an innocent man lived with a seal of a traitor who had betrayed his teacher. Convinced in his betrayal there came to the grave his former friends, disciples of Landau,

Landau himself and many, many others. A few years after acquittal he died as well. A small "stroke" of the epoch.

To complete our story, let us however return to the theme of Kapitza versus Stalin.

Already in 1937–1938, using his experience in working with liquefaction of gases at low temperatures, Kapitza made an important technical invention. He created a so-called centrifugal expansion engine which allowed to extract oxygen that was needed in great volumes in metallurgy ("oxygen blowing") from the air many times more cheaply and effectively than by common methods. He had to lead a fierce fight with specialists in this field who defended their traditional method. Of course Kapitza did win. One created a head office of oxygen industry, Glavkislorod, and Kapitza was appointed his head i.e. got a high state position, almost that of a narkom. A large part of his correspondence with members of the government was related to this. One can imagine how much his authority grew in the eyes of Stalin.

When in 1943–1944 there began an unfolding of an atomic epic. Beria became its head. One appointed a special committee under his chairmanship to which one included only two physicists — a scientific head I.V. Kurchatov and, later, P.L. Kapitza. Kapitza, however, was not satisfied with the work of the committee and the complete scientific illiteracy of its chairman and its other members representing the state. As a result on November 25, 1945 there appeared his new letter to Stalin, in those conditions — an absolutely improbable one. Still, it was written. Let us reproduce it here, albeit in excerpts. It deserves it [Kapitza (1990)]:

> Comrade Stalin!
>
> For about four months I am taking part in the sessions and actively participating in the work of the special committee and technical council on the atomic bomb ... I have decided to present you my detailed thoughts on an organization of this work and to ask you once again to exempt me from participating in it ...
>
> A right organization ... is possible only under one condition which is not fulfilled ... it is necessary to have more trust between scientists and state officials. This is an old story by us, remnants of the revolution ... when respect towards science and scientists is not sufficiently emphasized.
>
> Life has shown that I could make someone listen to me only as Kapitza a head of the office at CPK and not as Kapitza a scientist with the world reputation ...

Comrades Beria, Malenkov, Voznesenski behave themselves in the special committee as superpeople. This specially refers to comr. Beria. It is true that he has a conductor's baton in his hands. This is fine. Still, after this, a first violin should be played by a scientist ... A conductor should not only wave a button but understand a score. Beria is very weak in this respect ... I was telling him directly: "You do not understand physics, let us, scientists, decide on these questions." ... In general our dialogues are not very friendly. I offered to teach him physics, to come to our institute. For example, one should not be an artist to understand art...

... I can not be a blind executor because I have already grown up from this position.

My relations with comr. Beria are getting worse and worse and he will doubtlessly be satisfied with my leave ... I have asked from the very beginning not to be involved in this work because I knew right away to which forms it would degenerate here ...

I am counting on your agreement because I know that violence with respect to a scientist does not agree with your point of view (an open flattering aimed at contraposition of Stalin and Beria — indeed, this was Stalin who forcedly did not let Kapitza to return to England — E.F.) ...

Yours, P. Kapitza

P.S. ...

P.P.S. I would like that comr. Beria would read this letter — this is not a denunciation but a helpful critique. I would have told him everything myself but it is difficult to see him.

P.K.

In such a way Kapitza did attack the main butcher of the country and the closest accomplice of Stalin, the only person with whom Stalin was "at familiar you." The cleverest post-postscript (P.P.S.) was of course added because he understood that the letter would be shown to Beria anyway.

This was however too much. Kapitza went too far. Although it is possible to assume that he did not want to take part in the construction of the bomb in principle and chose this way of stepping aside. Indeed, he wrote: "I have asked from the very beginning not to be involved in this work." At the same time I heard from people who knew him well that he pretended at a decisive scientific leadership of the whole project. In any case Kapitza was fired from the position of the head of Glavkislorod and from his own institute.

He moved to his dacha at Nikolina Gora and had been living there "simply as an academician" for nine years before neither Stalin nor Beria were alive. Together with his sons he carried out some experiments in a separate house (later the Academy of Sciences did send him a laboratory assistant with whom he had been already working for many years) and wrote one theoretical paper. Still as a whole this was practically an exile or house arrest.

This is however not all that could be told about the behavior of P.N. Kapitza. His disciple and coworker, the cleverest Elevter Luarsabovich Andronikashvili, told me that soon after Kapitza's move to Nikolina Gora near his dacha there appeared a small military contingent. Elevter told me about his (he said: "... not only mine") assumption that it was placed there following a directive by Stalin to prevent possible unauthorized actions by Beria. I was amazed by this and considered it highly improbable. However, later I read in the memoirs of Anna Alekseevna, the wife of Petr Leonidovich, that they were both afraid of it:

> We were both very careful and our carefulness was probably justified. Indeed, Beria could do away with us in a large number of ways. We noticed some strange things happening to us... Petr Leonidovich did readily admit that one wanted to "eliminate" him. Indeed, in our house we had "government-sent servants" that clearly were watching him. One constantly tried to lure us from the dacha, offered various possibilities of spending time in sanatoriums, various trips. It is possible that only our great carefulness, that we tried to go to Moscow as seldom as possible and never walked alone — that all this allowed us to avoid "an accident." (see [Kapitza (1994), p. 85]).

Now it is especially clear that their apprehension was not without grounds. The same Anna Alekseevna wrote (see [Kapitza (1994), p. 84]):

> Many years later a good acquaintance of ours, army-general Khrulyov (during all the war he was a head of rear front of the whole Red (Soviet) army i.e. occupied a position of immense importance, was an educated person — E.F.) told us about an episode he was a witness of ...
>
> Khrulev was at Stalin's office on some issue. Suddenly Beria enters the office and begins persuading Stalin that Kapitza should be arrested. After being silent for some time Stalin said (I remember this phrase very well): "I'll fire him for you, but you do not touch him."

From the context of the memoirs one can conclude that this conversation took part before Kapitza was fired from all posts and moved to Nikolina Gora.

10.3 Catastrophy

The death of Stalin and the Khrushchev Thaw changed both life in the country and Landau's feelings. Nowadays from declassified archives we know that Landau, like millions of others, remained under the watchful observation of "organy." Reports were published on his overheard not exactly loyal statements (and, what is most sad, in one of these cases it is written that an informer is a person close to Landau), we know now that in his entourage there were many KGB agents. Nevertheless it was no longer as fearful and dangerous as before. P.L. Kapitza returned to his chair of a director of the institute. Scientific work was so far continuing successfully. There appeared a second generation of talented disciples (I.M. Khalatnikov, I.E. Dzyaloshinski, L.P. Gor'kov, L.P. Pitaevski, A.A. Abrikosov) and the latter did already have disciples of their own. The famous Thursday seminar of Landau gathered practically all leading theoretical physicists from Moscow. Years did still not exert a noticeable influence.

At the end of the forties (alas, not by us) there appeared new relativistic covariant renormalizable quantum electrodynamics and Landau whose attitude towards quantum field theory had previously been negative did learn the new technique. At the same time I.E. Tamm, after finishing with thermonuclear fission and the bomb, also studied and used it. He complained to me that at his age (around 60) this was not easy but added that when he had told about it to a younger (at 13 years) Landau he had confessed that it had not been that easy for him either. Nevertheless they did of course master all this and Landau, together with several closest disciples of the first and second generations began doing fundamental works on this. A discovery of the difficulties of electrodynamics ("Moscow zero", E.S. Fradkin from the school of I.E. Tamm from one side and Landau and I.Ya. Pomeranchuk from the other) led him further to attempts of constructing completely new approaches.

All this was combined with a continuing interest to a wide range of problems in quantum physics, especially to superconductivity and low temperature physics. Life continued in the former "before-arrest" style. Dau could again say about some physicist that had consulted with him: "I do not understand why he felt offended by me. I really did not say he was a fool, I only said that his work was stupid..."

And here there happened something terrible. On January 7, 1962, in conditions when the road was icy, Dau went to Dubna by car to his sister (contrary to the advice of an experienced driver, E.M. Lifshitz) and there took place an accident. He sat on the rear seat leaning with his body and head (with his fur hat taken off!) against the door. A lorry hit precisely this door. Dau was literally cracked — traumas of skull, breast, hip area. He was taken to the nearest hospital and was unconscious.

There began a staggering epic of his rescue. Dozens (a hundred?) of his and not his disciples came together to help. Governmental bodies displayed an unheard of activity. A famous neurosurgeon brought from Czechoslovakia wrote in the journal that the traumas were not compatible with life. Another equally significant specialist flown in from Canada left hope. Several times there took place a clinical death, he was taken out from them. Physicists undertook their own measures. The main role in this group was of course played by an extremely businesslike, concise E.M. Lifshitz who was deeply touched by the tragedy. One learned that in one clinic there worked a young untitled doctor, Sergei Nikolaevich Fedorov who "pulled out" people from hopeless situations. One insured that he was invited. He practically moved to Landau's ward, chose himself assistants. His role turned out to be very big.

Physicists organized a precise system of aid. In the hospital there was a special room in which 24 hours a day, changing according to a schedule, on the telephone was somebody who could be relied upon. He had a list of 223 phone numbers — those of specialist doctors, institutions that might be needed, numbers of those who had a car to be sent to fetch a specialist or a rare medicine carried by a regular flight from abroad from an airport, etc. Each had some errand. For example the doctors said that for artificial nutrition one needed a freshly laid egg. One found in Leningradskoe Shosse a women who bred hens and academician Pomeranchuk did himself every day bring such an egg to the hospital. Many came "just in case." Landau, however, remained unconscious.

He (and the whole support system) were moved to Burdenko Neurosurgical Clinic. I lived nearby and often came there. I was twice used when it was necessary to talk, together with E.M. Lifshitz, to doctors, especially if we wanted to persuade them to do something. However, again and again, when we entered the ward in which on a high bed Landau was lying on his back and tried, by words, to cause his reaction, followed the movements of the pupils of his eyes and the eyelashes, everything remained the same. Nevertheless a success was so far just in the fact that he was alive.

Finally there came a day (after three months!) when consciousness began returning to him and then, it seemed, returned. I remember how, sitting in the corridor and waiting when one would bring him back in an armchair from a dentist, I heard his welcoming exclamation coming from some distance. Moreover when later I was sitting by his bed and talking to him he read me, on some occasion, a long verse, admittedly in Danish (his usual childish boasting of knowing foreign languages that I have already mentioned). There came a new and last six a year period of his life. This was however not a real life and in any case not the real Landau.

At last the highest medical authorities solemnly declared that the treatment was over, the patient did recover and one of the bosses even claimed that he recovered up to the level of a provincial professor.

Alas, this was stupid boasting. There was no discussion of scientific work. He was either complaining on pain or was jovial, with not quite an adequate smile, talked profusely, repeated many times that when he would recover he would engage in reforming school textbooks. When I tried to ask him a scientific question on his own old work he evaded the question: "Let the pain in the stomach pass, then we'll talk." Although he once went together with his wife to a Czech resort walking in special orthopedic shoes was not easy for him. Gradually the stream of friends-visitors began running low. It was too painful to see him and there was "no purpose in it." He could repeat many times some old statements.

Once in the summer when he was placed, instead of a sanatorium, into a hospital of the Academy of Sciences and there were almost no visitors I came to him. In the corridor I saw him going, supported from both sides by nurses. Already from a distance he joyfully cried to me: "Zhenya, today I recalled the Dirac equation!" This was as awful as if Shostakovich would joyfully tell me: "Today I have recalled the first bars of the fifth symphony of Beethoven." From this one could however see that he understood that his life was no longer the same.

We went for a walk in the hospital yard (he leaned on a stick and my hand). I asked him on purpose: 'Dau, have you heard the extremely important news, a new neutrino was found, a muon one that is different from the usual one. This is of direct interest to you, would you think about it?' 'Yes,' he said, 'this is really important. Let the pain go away, I'll think about it." Then there was a different conversation which he several times interrupted with the same remark: 'Igor Evgen'evich (Tamm — E.F.) is a very good man. Bu-u-t (significantly and with a cunning smile — E.F.) when Stalin died, he grieved it!' Indeed many (and, I repeat, me myself) thought that people around Stalin were of smaller caliber than he and things would get only worse. Landau, on the contrary, was glad.

On the next day I came again and asked whether he remembered about the second neutrino. No, he forgot everything I had told him. On the next day I asked the same question. This time he said that he remembered it, that it was important but '... let the pain first go away.'

It did not.

On April 1, 1968 Dau died.

Bibliography

Andreev, A.F. (ed.) (1998). *Kapitza, Tamm, Semenov in Memoirs and Letters*. Moscow: Vagrius (In Russian).
Arnold, V.I. (1999). Mathematics and Physics: Parent and Child or Sisters?, *Usp. Fiz. Nauk.*, 169, 1311.
Bethe, H.A. (2000). The German Uranium Project, *Physics Today*, July 2000.
Beyerchen, A. (1977). *Scientists Under Hitler*. New Haven: Yale Univ. Press.
Big Soviet Encyclopedia, 1969. 3rd edn. vol. 24-II, p. 146.
Bohr, A. (1957). Private Communication.
Bonner, E. (1998). Postscriptum. In: *Book on Gorky Exile*, Paris: Ed. de la Presse Libre.
Bothe, W. and Jensen, H. (1944). *Zs. Phys.*, 122, pp. 749–755.
Bullok, A. (1974). *Hitler and Stalin. Life and Power. Comparative Biography*. Smolensk: Rusich (In Russian).
Casimir, H.B. (1983). *Haphazard Reality*. New York: Harper and Row.
Casimir, H. (1998). Landau. In: *Recollections on L.D. Landau*, Moscow: Nauka, pp. 150–160.
Churchill, W. (1955). *Second World War*. Translated from English, vol. 4. Moscow: Voenizdat.
Dalitz, R.G. (1991). PhD thesis of Andrei Sakharov, *Usp. Fiz. Nauk.*, 161, pp. 121–136.
Dirac, P.A. (1937). Reversal operator in quantum mechanics, *Izv. AN SSSR*, Phys.-Math Branch, Nmb. 4/5.
Einstein, A. (1965a). Remarks on cognition theory of Bertrand Russel. In: *Collected Works*. Moscow: Nauka, vol. 4, pp. 248–252 (In Russian).
Einstein, A. (1965b). Physics and reality. In: *Collected Works*. Moscow: Nauka, vol. 4, pp. 200–227 (In Russian).
Einstein, A. (1965c). A Letter to Maurice Solovin. In: *Collected Works*. Moscow: Nauka, vol. 4, pp. 547–575 (In Russian).
Einstein, A. (1967). The nature of reality. In: *Discussion with Rabindranat Tagore. Collection of Scientific Papers*, vol. 4. Moscow: Nauka, pp. 130–132 (In Russian).
Fabelinsky, I.L. (1978). *Usp. Fiz. Nauk.*, 126, p. 124.
Fabelinsky, I.L. (1998). *Usp. Fiz. Nauk.*, 168, p. 1341.
Fabelinsky, I.L. (2000). *Usp. Fiz. Nauk.*, 170, p. 93.
Fabelinsky, I.L. (1982). *To the History of Discovery of Combination Scattering*. Moscow: Znanie (In Russian).

Fainberg, V.Ya. (1995). Passion Towards Science. In: *Recollections on I.E. Tamm.* 3rd edn. Moscow: IzDAT, pp. 24–301.
Feinberg, E.L. (1988). Landau and Others. In: *Recollections on L.D. Landau.* Moscow: Nauka, pp. 253–267.
Feinberg, E.L. (1992). *Two Cultures. Intuition and Logics in Art and Science.* Moscow: Nauka (In Russian) [Expanded German edition (1998): *Zwei Kulturen. Intution und Logik in Kunst und Wissenschaft.* Berlin: Springer].
Feinberg, E.L. (ed.) (1995). *Memoirs on I.E. Tamm.* 3rd edn. Moscow: IzdAt (In Russian).
Frank, C. (ed.) (1993). Introduction. In: *Operation Epsilon: The Farm-Hall Transcripts.* University of California press.
Frayn, M. (1988). *Copenhagen.* London: Methuen.
Frenkel, V.Ya. (1997). *Professor Friedrich Houtermans: Works, Life, Fate.* St. Petersburg: Petersburg Institute for Nuclear Physics of Russian Academy of Sciences.
Frenkel, V.Ya., Moskvichenko, N.Ya. and Savina, G.A. (eds.) (1990). *Physicists on Themselves.* Leningrad: Nauka (In Russian).
Gora, E.K. (1985). One Heisenberg Save, *Sci. News. Lett.*, March 20.
Gorelik, G.E. (1991). My Anti-Soviet activity... One Year From the Life of Landau, *Priroda*, 11, pp. 93–104.
Gorelik, G. (1995). *Meine Antisowietische Taetigkeit... Russische Physiker unter Stalin.* Braunschweig: Vieweg.
Gorelik, G. (2000). *Andrei Sakharov. Science and Freedom*, Moscow-Izhevsk Regular and Chaotic Dynamics (Publishing House): Regular and Chaotic Dynamics (Publishing House) (in Russian).
Groves, L. (1969). *Now One Can Tell About It.* Moscow: Atomizdat (In Russian).
Heisenbeg, W. (1980). *Der Ganze und der Teil* (German edition) [(1989) *Physics and Philosophy. Part and Entirety*, Moscow: Nauka.]
Ioffe, A.F. (1936). *Conditions of My Scientific Work, Izv. An SSSR*, ser. phys. 1/2, pp. 7–33. [Report at the session of AS USSR of March 14–20, 1936].
Irving, D. (1960). *The Virus Wing.* Moscow: Atomizdat (In Russian).
Jungk, P. (1960). *Brighter Than a Thousand Suns.* Moscow: Gosatomizdat (In Russian).
Kapitza, P.L. (1981). Lev Davidovich Landau. In: *Experiment, Theory, Practice*, 3rd edn., Moscow: Nauka.
Kapitza, P.L. (1990). On Science and Power, *Ogonyok*, 137.
Kapitza, P.L. (1994). On Our Life in Cambridge, Moscow and Nikolina Gora. In: *Petr Leonidovich Kapitza. Memoirs. Letters. Documents.* Moscow: Nauka, pp. 64–68.
Keldysh, L.V. (ed.) (1996). He Lived Among Us. In: *Memoirs on A.D. Sakharov*, Moscow: Praktika (In Russian).
Khriplopvich, I.B. (1993). Thorns and Stars of Frietz Houtermans, *Siberian Phys. J.*, 1, pp. 65–79.
Klein, M. (1984). *Mathematics. The Loss of Certainty*, Moscow: Mir (In Russian).
Knickerbocker, H.R. (1932). *The German Crisis*, New York: Farrar and Rinehart.

Kuznetsov, I.V. (ed.) (1952). *Philosophical Questions of Modern Physics*, Moscow: AN USSR Publishing House (In Russian).

Lifshitz, E.M. (1969). Lev Davidovich Landau (1908–1968). in *Collected Works of L.D. Landau*. Moscow: Nauka, pp. 427–447.

Livanova, A. (1983). *Landau*, 4th edn. Moscow: Znanie.

Makhmutkuli (Fragi) (1948). *Selected Verses*. Moscow: Khudozhestvennaya literatura (In Russian).

Markov, M.A. (1986). On one letter of Niels Bohr, *Issues in History of Natural Sciences and Technics* 3, pp. 146–149.

Mott, N.F. and Peierls R.E. (1977). Werner Heisenberg, *Bibliographical Memoirs of the Fellows of Royal Society, London*, 23, pp. 213–251.

Niclaus, K. (1966). *Die Sowietunion und Hitlers Machtgreifung*, Bonn: Rohrscheid.

People's Economy of USSR for 70 years, 1987. Moscow: Finansy i Statistika.

Powers, T. (1993). *Heisenberg's War: The Secret History of the Germanium Atomic Bomb*, New York: Knopf.

Prokhorov, A.M. (ed.) (1979). *Academician L.I. Mandelstam. To the 100-th anniversary*, Moscow: Nauka (In Russian).

Rozhansky, I.D. (1987). Meeting with Niels Bohr, *Issues in History of Natural Sciences and Technics* 1, pp. 120–136.

Rytov, S.M. (ed.) (1948). *L.I. Mandelstam Complete Works*. vol. 1, Moscow: AS USSR (In Russian).

Sakharov, A. (1990). *Gorky, Moscow, Later Everywhere*, New York: Chekhov Publishing.

Shapiro, I.S. (1988). From recollections on L.D. Landau. In: *Recollections on L.D. Landau*, Moscow: Nauka, pp. 283–288.

Singi R., Riess F. (2001). The 1930 Nobel Prize for Physics: A close decision?, *Notes Rec. R. Soc. London*, 55, pp. 267–283.

Snegov, S. (1972). *Unchained Prometeus*. Moscow: Det. Lit.

Snow, C.P. (1977). *Two Cultures*. Moscow: Progress (In Russian).

Tagore, R. (1931). The Nature of Reality. *Modern Review (Calcutta)*, XLIX, pp. 42–43.